Pulses

A. Manickavasagan
Praveena Thirunathan
Editors

Pulses

Processing and Product Development

Springer

Editors
A. Manickavasagan
School of Engineering
University of Guelph
Guelph, ON, Canada

Praveena Thirunathan
School of Engineering
University of Guelph
Guelph, ON, Canada

ISBN 978-3-030-41375-0 ISBN 978-3-030-41376-7 (eBook)
https://doi.org/10.1007/978-3-030-41376-7

This Springer imprint is published by the registered company Springer Nature Switzerland AG
The registered company address is: Gewerbestrasse 11, 6330 Cham, Switzerland

Preface

Health authorities around the world recommend people to maximize their protein intake through vegetable sources (such as pulses) and minimize it from animal sources. Increasing vegetable protein intake is positively associated with the reduction of cardiovascular disease-related mortality and all-cause mortality. Pulse consumption has been shown to improve satiety and metabolism of glucose and lipids, due to their high protein and fibre content, which makes their consumption ideal for preventing and managing obesity.

In recent years, there is an increased demand for pulses and pulse-based products all over the world. Several large-scale collaborative research projects on pulse products have been initiated by government agencies. Similarly, some of the established multinational food companies have started pulse product units recently. Hence, at this point, there is a requirement for a comprehensive book on processing and products of pulses to meet a wide range of audience.

The Food and Agricultural Organization (FAO) of the United Nations has officially listed 16 pulse types. Each pulse type has unique nutritional profiles and sensory attributes. In this book, we have provided 16 comprehensive chapters (each pulse type listed by the FAO as a chapter) on processing and product development. Each chapter explains briefly about the selected pulse, in terms of nutritional composition, structure, and harvesting of the pulse. This is then followed by a discussion on the processing and product development of the pulse. The last section of each chapter provides future trends in processing and product development. We believe this collection will benefit academics, industry professionals, dieticians, and many others in utilizing the full potential of pulses.

We are grateful to all the contributors for writing and submitting their chapters. We also thank the staff of the editorial and production departments of Springer for their support and their efforts to bring this book to publication.

Guelph, ON, Canada

A. Manickavasagan
Praveena Thirunathan

Contents

Contributors

P. Abhirami Department of Food Engineering, Indian Institute of Food Processing Technology, Ministry of Food Processing Industries, Government of India, Thanjavur, Tamil Nadu, India

Allaoua Achouri Agriculture and Agri-Food Canada, Saint-Hyacinthe Research and Development Centre, Saint-Hyacinthe, QC, Canada

Shafia Ashraf Division of Food Science and Technology, Sher-e-Kashmir University of Agricultural Science & Technology, Jammu, India

K. A. Athmaselvi Indian Institue of Food Processing Technology, Thanjavur, Tamil Nadu, India

Rejaul Hoque Bepary DRDO-Defence Food Research Laboratory, Mysore, India

D. Betancur-Ancona Facultad de Ingeniería química — Campus de Ciencias Exactas e Ingenierías, Universidad Autónoma de Yucatán, Mérida, Yucatán, México

Sanika Bhokarikar Department of Food Process Engineering, School of Bioengineering, SRMIST, Kattankulathur, Tamil Nadu, India

A. Castellanos-Ruelas Facultad de Ingeniería química — Campus de Ciencias Exactas e Ingenierías, Universidad Autónoma de Yucatán, Mérida, Yucatán, México

Sindhu School of Engineering, University of Guelph, Guelph, ON, Canada

L. Chel-Guerrero Facultad de Ingeniería química — Campus de Ciencias Exactas e Ingenierías, Universidad Autónoma de Yucatán, Mérida, Yucatán, México

V. Chelladurai Department of Agricultural Engineering, Bannari Amman Institute of Technology, Sathyamangalam, India

C. Erkinbaev Department of Biosystems Engineering, University of Manitoba, Winnipeg, MB, Canada

Ashoka Gamage Department of Chemical and Process Engineering, Faculty of Engineering, University of Peradeniya, Peradeniya, Sri Lanka

Isha Gupta Division of Food Science and Technology, Sher-e-Kashmir University of Agricultural Science & Technology, Jammu, India

G. Jeevarathinam Department of Food Technology, Bannari Amman Institute of Technology, Sathyamangalam, India

Lamia L'Hocine Agriculture and Agri-Food Canada, Saint-Hyacinthe Research and Development Centre, Saint-Hyacinthe, QC, Canada

Gayani W. Loku Hetti Arachchige Agriculture and Agri-Food Canada, Saskatoon Research and Development Centre, Saskatoon, SK, Canada

Terrence Madhujith Department of Food Science and Technology, Faculty of Agriculture, University of Peradeniya, Peradeniya, Sri Lanka

A. Manickavasagan School of Engineering, University of Guelph, Guelph, ON, Canada

Delphine Martineau-Côté Agriculture and Agri-Food Canada, Saint-Hyacinthe Research and Development Centre, Saint-Hyacinthe, QC, Canada

Shabir Ahmad Mir Department of Food Science and Technology, Government College for Women, Srinagar, India

Rangarajan Jagan Mohan Department of Nutrition and Dietetics, Jamal Mohamed College (Autonomous), Trichy, Tamil Nadu, India

Dept of Food Product Development, Indian Institute of Food Processing Technology, Governmnet of India, Thanjavur, Tamil Nadu, India

Martin Mondor St-Hyacinthe Research and Development Centre, Agriculture and Agri-Food Canada, St-Hyacinthe, QC, Canada

Khalid Muzaffar Department of Food Science and Technology, Government College for Boys, Baramulla, India

Na Zhang College of Food Engineering, Harbin University of Commerce, Harbin, China

G. Mohan Naik Department of Food Engineering, Indian Institute of Food Processing Technology, Ministry of Food Processing Industries, Government of India, Thanjavur, Tamil Nadu, India

M. Ngadi Mcgill University, Montreal, QC, Canada

G. Peraza-Mercado Instituto de Ciencias Biomédicas (Departamento de Ciencias Químico-Biológicas), Universidad Autónoma de Ciudad Juárez, Ciudad Juárez, Chihuahua, México

T. S. Rathna Priya School of Engineering, University of Guelph, Guelph, ON, Canada

Dhritiman Saha School of Engineering, University of Guelph, Guelph, ON, Canada

M. Sandoval-Peraza Universidad del Valle de México, Escuela de Ciencias de la Salud, Mérida, Yucatán, Mexico

Facultad de Ingeniería química — Campus de Ciencias Exactas e Ingenierías, Universidad Autónoma de Yucatán, Mérida, Yucatán, México

Arumugam Sangeetha Department of Nutrition and Dietetics, Jamal Mohamed College (Autonomous), Tiruchirappalli, Tamil Nadu, India

T. Senthilkumar Mcgill University, Montreal, QC, Canada

Subajiny Sivakanthan Department of Agricultural Chemistry, Faculty of Agriculture, University of Jaffna, Jaffna, Sri Lanka

Sajad Ahmad Sofi Division of Food Science and Technology, Sher-e-Kashmir University of Agricultural Science and Technology, Jammu, India

Aryasree Sukumar Department of Food Process Engineering, School of Bioengineering, SRMIST, Kattankulathur, Tamil Nadu, India

N. Venkatachalapathy Department of Food Engineering, Indian Institute of Food Processing Technology, Ministry of Food Processing Industries, Government of India, Thanjavur, Tamil Nadu, India

Rizliya Visvanathan Department of Nutrition, Dietetics and Food, Monash University, Notting Hill, VIC, Australia

Dadasaheb D. Wadikar DRDO-Defence Food Research Laboratory, Mysore, India

Janitha P. D. Wanasundara Agriculture and Agri-Food Canada, Saskatoon Research and Development Centre, Saskatoon, SK, Canada

About the Editors

A. Manickavasagan, PhD, P.Eng. (Canada), obtained his PhD in biosystems engineering from the University of Manitoba, Canada. At present, he is an Associate Professor at the University of Guelph, Canada. He has published more than 60 scientific papers in peer-reviewed journals and edited 7 books. He has supervised more than 10 postgraduate students (PhD and MSc). At present, he is supervising/co-supervising 11 PhD students.

He is the founder and president of the Whole Grains Research Foundation in India. Through this foundation, he is promoting whole grains research in Asia. He has organized 8 conferences/symposiums/workshops in the field of whole grains.

Praveena Thirunathan, M.ASc, obtained her master's degree in biological engineering from the University of Guelph. She has over 4 years of research and industry experience, having worked at both Nestle and Cargill, and has also published 2 papers. She was awarded the Arrell Food Institute Food for Thought Research Assistantship to work on an agricultural challenge together with an industry partner.

Adzuki Bean

Sindhu and A. Manickavasagan

Introduction

The adzuki bean (*Vigna angularis*) is a legume crop, mostly bushy and upright, typically 1–2 ft. in height. The adzuki bean belongs to the Fabaceae family; it is commonly known as adzuki bean, red bean, and red mung bean. The wild forms of adzuki bean are supposed to be originated in Japan for over 6000 years ago, whereas the cultivated varieties were developed 4000 years ago. It is widely grown in the Yangtze River valley in China (FAO 2007). It occupies second rank after soybean among dry bean crop in Japan. These beans are also stated to be used as a soil improvement crop and animal feed crop (FAO 2007).

The bean has mainly straw-coloured pods, but sometimes black- or brown-coloured pods have also been reported. It is cylindrical in shape, 6–12 cm in length, and 5 mm in diameter. Each pod contains approximately 5–12 seeds and constitutes half of the total weight of the pods. The seed has a variation in the colour from maroon, straw, brown, and black. The skin of the bean is tough, whereas the cotyledon is normally smooth. The seed coat of the bean is red in colour and contains polyphenols, flavonol glycosides, and catechins (Ariga and Asao 1981; Hori et al. 2009; Yoshida et al. 2005).

At present, various varieties of adzuki bean are cultivated in more than 30 countries which include Korea, New Zealand, India, Taiwan, Thailand, and the Philippines. The variation in the features is dependent on the cultivar type, time of growing, time of harvesting, and climatic conditions (Gohara et al. 2016).

In Japan, there are approximately 60 types of adzuki bean grown and traded; the seed size ranges from 5 to 8 mm in length and 3–5 mm in width (Chilukuri and Swanson 1991). The seeds can be allocated into two types—small size (regular) is

Sindhu · A. Manickavasagan (✉)
School of Engineering, University of Guelph, Guelph, ON, Canada
e-mail: mannamal@uoguelph.ca

A. Manickavasagan, P. Thirunathan (eds.), *Pulses*,
https://doi.org/10.1007/978-3-030-41376-7_1

greater than 4.2 to less than 4.8 mm in length, known as Erimo type, and large size is greater than 4.8 mm in length, named as Dainagon.

Overview of Production, Export, and Import

Adzuki bean is grown in a place with annual precipitation in between 530 and 1730 mm and temperature range of 7.8–27.8 °C. The preferable soil pH is 5.0–7.5 (Duke 1981). The major adzuki bean producers are Japan, China, Taiwan, and South Korea. According to Sacks (1977), adzuki bean is the sixth largest grown pulses in Japan. The bean is cultivated in an area of 670,000, 120,000, 30,000, and 20,000 ha, in China, Japan, Korean peninsula, and Taiwan, respectively (Rubatzky and Yamaguchi 1997). In the United States, the bean is grown in a very limited quantity; the sole purpose is for the production of seeds and to be exported to Japan.

As such, there is not much production database available for adzuki bean. Japan imports the bean from China, Taiwan, the United States, Thailand, and Canada; it produces 100,000 tons/year and further consumes 140,000 tons/year. In the oriental market, the seed and seed flour are a commodity of high economic value (PlantUse English Contributors 2015).

The major area of adzuki bean production in Japan is Hokkaido, and the yield varies according to the duration of growing season and weather conditions. In China, the major areas are Wuqing County, Hebei Province, Jilin Province, Tai Lai County of Heilongjiang Province, north of the Huaihe River, and near Qinling. The areas in Taiwan include Pingtung and Kaohsiung provinces. In South Korea, it is the fourth vital legumes grown in terms of area and production. In the United States, it is supposed to be introduced in the year 1854 by the Perry Expedition. The United States uses it as a fodder crop (Hoshikawa 1985; Sacks 1977).

Chemical and Nutritional Qualities

The adzuki bean is a vital source of proteins, minerals, carbohydrates, fibres, and vitamins (Tjahjadi et al. 1988). It is rich in polyphenols, catechins, and chlorogenic acid (Yoshida et al. 1996, Ariga and Hamano 1990; Ariga 1988). Adzuki bean has few antinutritional factors like phytates, α-galactosides, and trypsin inhibitors whose quantity relies on the different cultivars of the crop.

The chemical composition of adzuki bean greatly varies in accordance with the variety of the crop. The nutrient contents of adzuki bean (Tables 1 and 2), i.e. the content of carbohydrates, proteins, lipids, vitamins, and minerals, are very much related to other legumes. The protein content varies with the seed size (Kato et al. 2000). The key storage proteins of adzuki bean are glycoproteins (Sakakibara et al. 1979). It has been reported that the amylose/amylopectin ratio of adzuki bean is 34.9/65.1 and 27.9/72.1, respectively (Biliaderis et al. 1980; Su and Chang 1995).

Table 1 Adzuki bean nutrition profile: (*Vigna angularis*) per 100 g (Source: USDA National Nutrient data base 2019)

Principle	Nutrient value
Energy	329 Kcal
Carbohydrates	62.90 g
Protein	19.87 g
Total fat	0.53 g
Cholesterol	0 mg
Dietary fibre	12.7 g
Vitamins	
Folates	622 μg
Niacin	2.630 mg
Pantothenic acid	1.471 mg
Pyridoxine	0.351 mg
Riboflavin	0.220 mg
Thiamin	0.455 mg
Vitamin A	17 IU
Vitamin C	0 mg
Electrolytes	
Sodium	5 mg
Potassium	1254 mg
Minerals	
Calcium	66 mg
Copper	1.094 μg
Iron	4.98 mg
Magnesium	127 mg
Manganese	1.730 mg
Phosphorus	381 mg
Selenium	3.1 μg
Zinc	5.04 mg

The starch obtained from adzuki bean has properties similar to corn starch in terms of colour, swelling power, solubility, amylose content, and gel strength. The shape of starch granules was found to be smooth, round, oval to kidney, or irregular (Reddy et al. 2017). The viscosity of its hot paste was more than corn starch and far more than wheat starch. The adzuki bean starch is very much different from the potato starch in most of its properties other than colour and amylose content. Thus from the study, it was concluded that unmodified adzuki bean starch can be used as thickeners and water-binding agents in the place of corn and wheat starch (Tjahjadi and Breene 1984).

Like other pulses, cysteine and methionine are limiting amino acids in adzuki bean. It has 25% saturated fats and 75% unsaturated fatty acids. It has a minimum of 0.4% and a maximum of 2.1% fat content which is comparatively very less than the fat content of groundnut (40.1% fat) (Su and Chang 1995). While in the case of total lipid content of adzuki bean, it is less than 2% and is alike to other foods such

Table 2 Amino acid profile of adzuki bean (*Vigna angularis* var. Tjahjadi et al. 1988)

Amino acid	g/16 g Nitrogen
Cystine/2	2.02
Aspartic acid	11.33
Threonine	3.74
Serine	4.53
Glutamic acid	17.70
Proline	5.51
Glycine	3.74
Alanine	4.10
Valine	5.63
Methionine	1.78
Methionine and cystine	2.79
Isoleucine	5.02
Leucine	8.70
Tyrosine	3.31
Phenylalanine	6.31
Phenylalanine and tyrosine	9.62
Histidine	3.55
Lysine	8.45
Ammonia	1.71
Arginine	7.78
Tryptophan	0

as rice, rye, bean, and chickpea, with the benefit that the fat of adzuki bean has PUFA n-6/n-3 having great advantage in the human nutrition (Yoshida et al. 2010). The adzuki bean releases a distinct sweet beany aroma upon cooking. The characteristic odour of bean is due to the following chemical compounds—hydrocarbons, alcohols, acids, phenols, and many more. The principal odour compound is maltol that gives a prominent sweet caramel-like flavour (Tokitomo and Kobayashi 1988). Adzuki bean is rich source of polyphenols amounting to 370 mg per 100 g of dry bean.

Medicinal Values of Adzuki Bean

In China, the adzuki bean is used for the treatment of diseases like kidney trouble, constipation, abscesses, certain tumours, threatened miscarriage, retained placenta, and non-secretion of milk. The leaves of the bean are suitable for lowering fever, whereas the sprouted bean is used to prevent abortion that is caused by injury. The compounds present in adzuki bean lowers blood pressure, cholesterol, and triglyceride levels, which finally lead to a healthy heart. It has several other health benefits that include improved bone health to prevent the spread of cancer cells. This bean is used at a large scale in traditional China as a diuretic, an antidote, and as a medication for dropsy and beriberi (Chiang Su New Medical College 1977). The adzuki bean has powerful antioxidant, anti-inflammatory, anti-diabetic, and anti-hypercholesteraemic properties (Luo et al. 2016; Shigenori et al. 2008; Tao et al. 2011; Tomohiro et al. 2009). The investigators have also found that the 40% ethanol part of the hot water extract from the bean stops the proliferation of stomach cancer cells in humans (Itoh et al. 2002).

Post-harvest Processing

The post-harvest handling and storage is a vital phase for long-term usage of the bean with all the properties intact. The unit operations involved in the post-harvest processing of adzuki bean are represented in Fig. 1. The various processes that are being carried out after the harvesting stage of bean to maintain its quality during long-term application are summarized in this section.

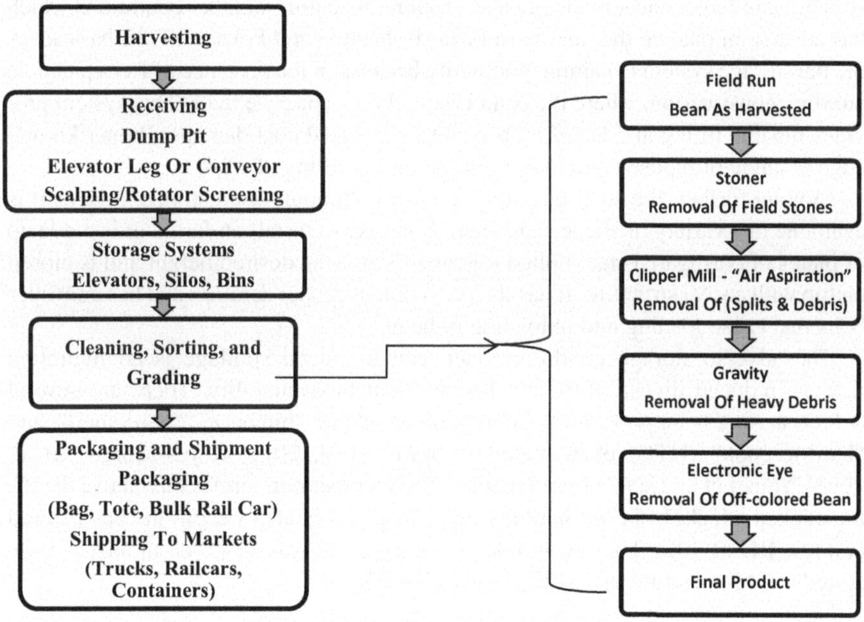

Fig. 1 Unit operations in raw adzuki bean handling and cleaning (Source—Kelly et al. 2013)

Storage

After harvesting, the bean is loaded on the trucks and then conveyed to the storage receipt elevator. It is then passed to the aspirator to remove any kind of impurities; this initial step removes foreign materials which can have spoilage in the bean during long-term storage. Most often the bean is stored in steel silos where the quality parameters are strictly monitored to retain its quality, as the weather fluctuation can have negative impact on the bean. There is a provision of continuous aeration in the silos to maintain a proper temperature balance and prevent the development of any kind of off-flavour or odour.

The adzuki bean is then transferred to the large storage silos for medium- or long-term storage before further cleaning is carried out. There is a central elevator which lifts the bean from the pits and deposits it on conveyor to be transferred into the large silos. Inside each large silo, there is bean ladder which helps in the proper flow of bean inside with sliding and circular motion and thus reducing the damage caused to the seed coats. The temperature below 18% is best for the bean before storage in the silos. The bean is then separated on the basis of size and colour before being taken for the shipment (Sacklin 1985). The bean is normally stored in silos which can be made up of wood, steel, or concrete. Among the entire concrete one is having high strength and larger capacity. It can also sustain fluctuating weather conditions which has adverse impact on the quality of bean (Roberston and Frazier 1978). Nowadays the flat storage system is gaining popularity because of the presence of free span pole building construction, where the bean is stored on a concrete floor. This system provides flexible filling and keeps an account on the seed coat damage. Proper knowledge of angle of repose is necessary for the proper filling of silos.

Moving further, the steel bins are the most commonly used storage area as it is available in a variety of shapes and sizes. It is easy to install on farms. It is made up of plates of corrugated steel bolted together to give the desired height and is closed on top with a steel structure. It has the provision of proper aeration and has conveyor belts that cause loading and unloading of bean.

The adverse storage conditions can lead to fungal spoilage, seed hydration defects, reduced digestibility, and less nutrient bioavailability. There are several defects arising from poor storage condition which are “bin burn”, “hard-shell”, and “hard-to-cook” (HTC) phenomena (Pirhayati et al. 2011; Reyes-Moreno et al. 1993). Yousif et al. (2002) found that at a high temperature preferably above 30 °C, the cooked adzuki bean has hard texture. The effect is also seen in the bean stored at a low RH of 40%. The favourable cooking quality was observed in adzuki bean stored at a temperature of 10–20 °C and 65% RH.

Handling

During conveying, the bean is transferred to the elevator, and then a known quantity of it is taken for quality analysis and to make sure the absence of any foreign particles. The most prominent effect of loading and unloading is on the seed coat of bean. The most frequent damage in bean occurs during conveyor transfer due to mechanical dropping and shattering. The poor design of bucket lifts causes damage to the seed by shearing action. The bean dropping has a detrimental consequence on the seed coat damage (Shahbazi et al. 2012). It is suggested to have less drop distance to minimize damage. The presence of bean ladders greatly diminished the seed coat damage by decreasing the free fall of the bean. A conveyor belt is comparatively better than auger conveyor in comparison to the damage caused to the bean.

Receiving and Cleaning

The bean is fed into the local elevator; it is then weighed and then transferred into a handling pit which is available at the floor of the receiving area. Now the next cleaning process is carried out; the first operation here is the application of high-velocity air to separate out any foreign material from the bean which is lighter in weight compared to bean. Moving further, the next step is removal of heavier materials which involves the application of a gravity table to remove stones and mud balls. It separates out heavier particles by vibratory motion, and on the basis of density, the bean and stones are separated out. The stone gets collected at the edge of the table. The separation on the basis of size is carried out on a sieve which divides the whole lot into oversized and undersized materials. This process of screening provides uniformity in a lot of beans and has a good economic value in the market. The next stage is the separation on the basis of colour which involves electric eyes that have a series of photoelectric cells. The bean is passed across lanes near the photoelectric cells, where a blast of air the rejects the discoloured bean. This process of colour sorting causes uniformity in the bean leading to a commodity of high economic value.

Drying

This is the process of reducing the moisture content from the bean by supplying heated air through forced convection (Afonso Júnior and Corrêa 1999). This step leads to enhanced storage life by providing physical, chemical, and microbial stability. But higher rate of moisture removal can have an adverse effect on seed quality (Almeida et al. 2009). Resende et al. (2012) reported that the drying time reduces by increasing the air temperature; the air temperature of 60 °C and 70 °C adversely affects the physiological and technological seed quality. Hence the favourable temperature for adzuki bean drying without causing any damage to the properties must not exceed 50 °C.

The process of drying also alters the physical properties of adzuki bean like an increase in sphericity and surface-volume ratio and decrease in volume, unitary volumetric contraction, surface area, and projected area (Mendes et al. 2016). The freeze-dried adzuki bean hydrolysate powder contained higher antioxidant activities as compared to ungerminated adzuki bean; also the DPPH radical scavenging activity estimated through electron spin resonance spectrometry of the freeze-dried adzuki bean was higher as compared to the colorimetric method (Sangsukiam and Duangmal 2017).

Packaging

The adzuki bean is carefully examined prior to weighing and filling into blue poly-lined cartons. The cartons are checked for the best before and production date, weight, and presence of metals, properly palletised, stretch wrapped, and then stored at a temperature of −18 °C till the time it is dispatched. Apart from this the package is also checked for the physical defects, case sealing, case coding, and print quality (Foodnet Limited 2018). The bean if not stored properly can lead to spoilage and loss in organoleptic characteristics.

Effect of Storage on Quality Factors

1. Effect on Cooking—The adzuki bean stored at high-temperature hydrate unequally during cooking which leads to irregular bean softening and finally results in the phenomenon of hard-to-cook (HTC) bean. These beans have reduced nutritional and organoleptic quality. The inappropriate storage of beans decreases the water imbibition capacity, which leads to decrease in the yield of Ann (Yousif et al. 2007).
2. Starch Quality—The fresh adzuki bean with a moisture content of 11% achieves 50% gelatinization, whereas adzuki bean stored for 1 year having a moisture content of 8% achieves 18% gelatinization. The unfavourable storage of Bloodwood and Erimo varieties at high temperature and reduced relative humidity leads to a rise in the starch gelatinization onset and peak temperature; it indicates the resistance of starch within the granules for gelatinization. Thus it causes loss of adzuki bean cooking quality (Yousif et al. 2007).
3. Texture—There was double-fold increase in the cooked adzuki bean hardness after storing at 30 °C and elevated relative humidity of 75% in comparison to the beans which were stored at 5 °C with increased relative humidity, 5 °C with reduced relative humidity of 45%, and 30 °C with reduced relative humidity. It was observed that hardness was because water was not able to enter into the cotyledon tissue of the bean and was in the free spaces of the bean that is the seed coat and the cotyledon fissure. Hence, finally it was observed that the bean uptakes water; however the cotyledons were not able to hydrate accurately (Ozawa 1978).

4. Protein Quality—The storage of bean at elevated temperature of 30 °C led to an increase in the presence of higher molecular weight proteins as estimated by size exclusion HPLC (Yousif et al. 2003). The change in the qualitative and quantitative electrophoretic protein pattern was observed in the bean stored at a high temperature of 30 °C and increased relative humidity of 85%. This change was reported to be because of the denaturation of the protein during storage period (Hussain et al. 1989). Further there was a decrease in the onset temperature and gelatinization peak temperature from the protein extracted from the bean which was stored at 40% relative humidity in contrast to those kept at a relative humidity of 65%. This shows the occurrence of destabilization in the protein at high relative humidity (Yousif et al. 2003).

Novel Processing Techniques

1. High Hydrostatic Pressure—The high hydrostatic pressure (HHP) treatment is a non-thermal method which is given distinct importance in the food industry since it reduces the cooking time without having any adverse effect on the nutritional value of the food (Bello et al. 2014). The application of HHP in adzuki bean led to a rise in the effective diffusion coefficient and decrease in the water absorption time. The value of effective diffusion coefficient of HHP-treated adzuki bean (6.7×10^{-11} m^2 s^{-1}) was found to be similar to that of various untreated grains that comprise soybean ($1.8–4.2 \times 10^{-10}$ m^2 s^{-1}), green bean ($0.8–3.6 \times 10^{-10}$ m^2 s^{-1}), brown rice (3.9×10^{-11} m^2 s^{-1}), barley (1.0×10^{-11} m^2 s^{-1}), wheat (4.0×10^{-11} m^2 s^{-1}), and amaranth (2.6×10^{-12} m^2 s^{-1}). Hence it can be said that the HHP has a significant impact on the structural modifications in the adzuki bean. This information will surely lead to a surge in the demand of adzuki bean for cooking purposes (Ueno et al. 2015).
2. Ozonation—The application of ozone in grains leads to degradation of mycotoxin and inactivation of microorganisms and helps in insect disinfestation. On the other hand, the oxidizing property of ozone causes a change in the constituent of the grain and thus has an impact on both product characteristics and process behaviour. In the case of adzuki beans, there was no significant effect of ozonation found in the hydration kinetics, phenolic compounds, and antioxidant capacity. Therefore, ozonation can be an important approach for the control of grain contaminants without affecting its quality (Santos Alexandre et al. 2018).

Product Development

In East Asia, adzuki bean is consumed in numerous ways for more than 2000 years owing to its maroon seed colour and mild flavour; it was conventionally served on the eve of celebrations such as weddings, birthdays, or New Year parties (McClary et al. 1989). It is normally made into a component of sweet soups like *zenzai* and *sarashi ame*, which is when mixed with rice called as *azuki-mochi* and *sekihan*, when mixed with sprouts called as *moyashi*. In China and Japan, the adzuki seeds are used as traditional foods like amanatto and wagashi (Gohara et al. 2016). The adzuki bean is very common ingredients of confectioneries in Eastern Asia, particularly in Japan. In this section, there is a description of some of the food and non-food applications of the adzuki bean.

Meat Extender and Fat Replacers

Owing to its high content of protein and fibre, adzuki bean has a great possibility to be used in meat product as a meat extender and fat replacer. The use of different proportions of adzuki bean flour in the place of cornflour and fat led to the production of a meatball with reduced fat and good acceptability. The adzuki bean flour has the capability to bind water, as the meatball with 100% (w/w) of the flour showed maximum cooking yield and more moisture content as compared to the meatballs with no content of flour. This replacement of cornflour and fat in meatballs with adzuki bean flour led to a reduction in the calorie count, keeping protein and carbohydrate quantity similar to the meatball without the flour. The organoleptic test for the meatballs with 25% (w/w) and 50% (w/w) of adzuki bean flour did not show any prominent variation from the meatball without flour, but they received highest overall acceptability among the panel members. Further the meatballs with 50% (w/w) replacement of cornflour with adzuki bean flour were reported to have enhanced physicochemical properties and better organoleptic properties as compared to the meatballs without adzuki bean flour (Aslinah et al. 2018).

Bakery and Confectionery Products

1. Ann—The paste made from adzuki bean is known as Ann, it is normally sweet in taste (Kato et al. 2000). The seed coat of the bean is responsible for the unique colour of the bean paste. There are two forms of adzuki bean paste, one is made by removing the seed coat called "Koshi-Ann", and the other one is made with the seed coat called "Tsubu-Ann." The preparation process for Koshi-Ann involves boiling of bean in freshwater, then draining the extra water further reboiling it. After that the seed coat is removed and approximately 40% of sugar is added to it. Then it is mixed properly, packaged, and stored at chilling or freezing condition. The Tsubu-Ann is made in a similar way like Koshi-Ann but with seed coat intact. The boiling step involved here causes tenderization of the seed coat and cotyledon. This process further is responsible for the distinct colour, flavour, and texture of the product that is again linked to the variety, age, and storage conditions of the adzuki bean (Yousif et al. 2007). In Japan, "Ann" is

available as peanut butter; it also comes in smooth and chunky forms with textures ranging like baked bean. In Japan, the manufacturer calls Ann as Asian chocolate because it has same role as chocolate in the western confectionery. The common use of Ann is as a filling for the snack buns, cakes, steamed dumplings, and doughnuts. The Ann, when sandwiched in between the two silver-dollar pancakes, is called as the dorayaki snack cake. The Ann is served as the main ingredient along with the sticky rice in the conventional Japanese wagashi sweet during the tea ceremony. The Ann has also found a new application in frozen desserts, where it is used as a topping for ice cream also as a flavouring agent (Ag Marketing Resource Center 2019). The bakery product that is commonly made with wheat or rice flour that has Ann filled within it is called manju. The Ann-filled glutinous rice cake is called as daifuku. The bun with Ann filling is called Anpan. Finally the baked pastry with Ann filling is called as monaka.

2. Anko—Anko is a traditional Japanese product made up of soaked and boiled adzuki bean paste; further johakuto a Japanese sugar or regular granulated sugar is added to it. It is used as an ingredient in Japanese confectionery products (Borchgrevink 2013).

Extruded Products

The use of adzuki bean flour has a beneficial effect on the commercial and nutritional quality of the Chinese steamed bread. The incorporation of extruded adzuki bean flour reduces the development time, stability time, and the farinograph quality number in the bread, whereas the water absorption and softening degree increase in the bread. The pasting properties of the product decreased with addition of the extruded adzuki bean flour. Also, this flour incorporation reduces the lightness and strengthens the hardness of the bread. The α-glucosidase inhibitory activity of the protein was enhanced in the bread by the addition of the flour. This inhibitory effect increased to 39.88% which is five times more as compared to the bread made from wheat flour (Chen et al. 2019).

Beverages

1. Adzuki-Flavoured Pepsi Drink—In 2009, Pepsi Japan released an adzuki-flavoured Pepsi drink where they blended the aroma of adzuki into the original Pepsi flavour. The product was the second seasonal drink in the year 2009, followed by the shiso-flavoured cola released in the summer. The blending of the adzuki bean flavour in the carbonated drink has a refreshing and unique tang. This product is mainly for the people in the age group of 20–30 (Japan Today 2009).

2. Probiotic Yogurt—A probiotic-based yogurt with high nutritional value and low production cost is also available; it includes skimmed milk and adzuki bean (Qing et al. 2009).
3. Granuliform Adzuki Bean Set Yogurt—In China, there is a beverage from the bean known as granuliform adzuki bean set yogurt. The optimized parameters for the preparation of yogurt are reported to be 10% of adzuki bean, 8% of milk powder, 8% of sugar, 4% of inoculums size, 4 h of fermentation time, and 42 °C of fermentation temperature (Li et al. 2011b).

Germinated Product

The sprouted adzuki bean is known as a vegetable. In China, the sprouted adzuki bean is called as Yuweiya if seeding for around 3 days and is called as Yuweimiao if seeding for around 7 days. The sprouted adzuki bean has a soft and somewhat crispy texture with an attractive fragrance (Li et al. 2011a).

Future Trends in Processing and Product Development

The problem of deep-rooted malnutrition can be cured, and overall human health can be maximized by taking into consideration the dietary quality of adzuki bean. In order to maintain the equilibrium with increasing yields and household techniques, there is a need to diversify the food production both at countrywide and domestic level (Singh and Raghuvanshi 2012). Unawareness of people is to the extent that they are not using few agricultural foods as main food despite their high nutritional quality. Adzuki bean is also one of such things that have many nutritious and medical properties but underestimated (Yao et al. 2012). Dried adzuki bean has the potent to supply few health-improving nutrients related to plant-based diets. The adzuki bean is a vital source of protein and is rich in various numbers of micronutrients, comprising potassium, magnesium, folate, iron, and zinc. Vegetarians have important role of dried bean in their diets because it contributes to some of the health benefits.

The prevailing consumer outlook on the adoption of healthy products is evident. Hence, consumers are drifting towards traditional and healthy eating habits. Adzuki bean is considered to be nutritional powerhouse as it is rich in fibre, protein, and folic acid. Adzuki bean can be used in various forms such as extract and paste. The lipid fraction of the bean proved to have anti-atherogenic, anti-thrombogenic, and hypocholesterolemic effects, whereas the ratios PUFA: SFA and n-6:n-3 was considered suitable for the proper activity of the systems in the human body. Thus there is a need to popularize this underutilized bean and more research should be undertaken to explore its full potential.

References

Afonso Júnior, P. C., & Corrêa, P. C. (1999). Comparação de modelos matemáticos para descrição da cinética de secagem em camada fina de sementes de feijão. *Revista Brasileira de Engenharia Agrícola e Ambiental, 3*(3), 349–353.

Ag Marketing Resource Center. (2019). Retrieved October 31, 2019, from https://www.agmrc.org/commodities-products/vegetables/azuki-beans

Almeida, D. P., et al. (2009). Cinética de secagem do feijão adzuki (Vigna angularis). *Global Science and Technology, 2*(1), 72–83.

Ariga, T. (1988). Antioxidative properties of procyanidins B-1 and B-3 from azuki beans in aqueous system. *Agricultural and Biological Chemistry, 52*, 2722–2817.

Ariga, T., & Asao, Y. (1981). Isolation, identification and organoleptic astringency of dimeric proanthocyanidins occurring in Azuki beans. *Agricultural and Biological Chemistry, 45*(12), 2709–2712.

Ariga, T., & Hamano, M. (1990). Radical scavenging action and its mode in procyanidins B-1 and B-3 from azuki beans to peroxyl radicals. *Agricultural and Biological Chemistry, 54*(10), 2499–2504.

Aslinah, L. N. F., Yusoff, M. M., & Ismail-Fitry, M. R. (2018). Simultaneous use of adzuki beans (Vigna angularis) flour as meat extender and fat replacer in reduced-fat beef meatballs (bebola daging). *Journal of Food Science and Technology, 55*(8), 3241–3248.

Bello, E. F. T., Martínez, G. G., Ceberio, B. F. K., Rodrigo, D., & López, A. M. (2014). High pressure treatment in foods. *Food, 3*(3), 476–490.

Biliaderis, C. D., Maurice, T. J., & Vose, J. R. (1980). Starch gelatinization phenomena studied by differential scanning calorimetry. *Journal of Food Science, 45*(6), 1669–1673.

Borchgrevink, C. P. (2013). *Culinary perspective of dry beans and pulses. Dry beans and pulses: Production, processing, and nutrition*. Chichester: Wiley-Blackwell.

Chen, Y., Yang, X., Guo, H., Li, J., & Ren, G. (2019). Effect of extruded adzuki bean flour on the quality and α-glucosidase inhibitory activity of Chinese steamed bread. *Food Science & Nutrition, 7*(10), 3244–3252.

Chiang Su New Medical College. (1977). *"Dictionary of Chinese Crude Drugs" (in Chinese)* (p. 1090). Shanghai: Shanghai Scientific Technologic Publisher.

Chilukuri, A., & Swanson, B. G. (1991). Microstructure of adzuki beans (Vigna angularis cv. Express). *Food Structure, 10*(2), 3.

Duke, J. A. (1981). Vigna angularis (Willd.) Ohwi & Ohashi. In *Handbook of legumes of world economic importance* (pp. 288–293). New York: Plenum Press.

Food and Agricultural Organization of the United Nations. (2007). Help to build a world without hunger. Retrieved October 31, 2019, from http://ecocrop.fao.org/ecocrop/srv/en/cropView?id=2147

Foodnet Limited. (2018). Retrieved from https://www.foodnet.ltd.uk/

Gohara, A. K., Souza, A. H. P. D., Gomes, S. T. M., Souza, N. E. D., Visentainer, J. V., & Matsushita, M. (2016). Nutritional and bioactive compounds of adzuki beans cultivars using chemometric approach. *Ciência e Agrotecnologia, 40*(1), 104–113.

Hori, Y., Murakoso, T., Fukumura, M., Toriizuka, K., & Ida, Y. (2009). Constituents and antioxidative activity of a hot-water extract of adzuki (Vigna angularis) beans. *Journal of Japanese Society of Nutrition and Food Science, 62*(1), 3–11.

Hoshikawa, K. (1985). Azuki beans (in Japanese). In *Edible crops* (pp. 460–471). Tokyo: Yokendo Publisher.

Hussain, A., Watts, B. M., & Bushuk, W. (1989). Hard-to-cook phenomenon in beans: Changes in protein electrophoretic patterns during storage. *Journal of Food Science, 54*(5), 1367–1368.

Itoh, T., Itoh, Y., Mizutani, M., Fujishiro, K., Furuichi, Y., Komiya, T., et al. (2002). A hot-water extract of adzuki (*Vigna angularis*) induces apotosis in cultured human stomach cancer cells. *Nippon Shokuhin Kagaku Kogaku Kaishi, 49*, 339–344; in Japanese.

Japan Today. (2009). Retrieved from https://japantoday.com/category/features/new-products/azuki-flavored-pepsi
Kato, J., Yousif, A. M., Deeth, H. C., Suzuki, M. M., Caffin, N. A., & Meguro, T. (2000). Differences in the cooking quality between two adzuki varieties harvested in Australia and stored at different temperatures. *Journal of Cookery Science of Japan, 33*(2), 257–266.
Kelly, J. D., Cichy, K. A., Siddiq, M., & Uebersax, M. A. (2013). *Dry bean breeding and production technologies. Dry beans and pulses: Production, processing, and nutrition* (pp. 23–54). Chichester: Wiley-Blackwell.
Li, L., Liu, B., & Zheng, X. (2011a). Bioactive ingredients in adzuki bean sprouts. *Journal of Medicinal Plants Research, 5*(24), 5894–5898.
Li, N., Li, Y., & Zhu, Y. (2011b). Development of granuliform adzuki bean set yogurt. *China Dairy Industry, 39*(3), 62–64.
Luo, J., Cai, W., Wu, T., & Xu, B. (2016). Phytochemical distribution in hull and cotyledon of adzuki bean (Vigna angularis L.) and mung bean (Vigna radiate L.) and their contribution to antioxidant, anti-inflammatory and anti-diabetic activities. *Food Chemistry, 201*, 350–360. https://doi.org/10.1016/j.foodchem.2016.01.101.
McClary, D. C., Raney, T. L., & Lumpkin, T. A. (1989). Japanese food marketing channels: A case study of azuki beans and azuki products. Washington State Univ., IMPACT Center Rpt. 29. Pullman.
Mendes, U. C., Resende, O., Donadon, J. R., Almeida, D. P., Da Rocha, A. C., & Oliveira, D. E. C. (2016). Effect of drying on the physical properties of adzuki bean. Semina: Ciências Agrárias, 37(6), 3871–3880.
Ozawa, Y. (1978). On the viability of stored adzuki beans (Phaseolus angularis) and its hardness after cooking (in Japanese, English summary). *Report of the National Food Research Institute, 33*, 37–40.
Pirhayati, M., Soltanizadeh, N., & Kadivar, M. (2011). Chemical and microstructural evaluation of 'hard-to-cook' phenomenon in legumes (Pinto Bean and Small-type Lentil). *International Journal of Food Science and Technology, 46*, 1884–1890.
PlantUse English Contributors. (2015). "Vigna angularis (PROTA)," PlantUse English. Retrieved October 31, 2019, from https://uses.plantnetproject.org/e/index.php?title=Vigna_angularis_(PROTA)&oldid=198622
Qing, K., Zhiqiang, D., Wei, Z., Yan, W. (2009). Study on adzuki bean yogurt beverage. *China Brewing, 2*.
Reddy, C. K., Luan, F., & Xu, B. (2017). Morphology, crystallinity, pasting, thermal and quality characteristics of starches from adzuki bean (Vigna angularis L.) and edible kudzu (Pueraria thomsonii Benth). *International Journal of Biological Macromolecules, 105*, 354–362.
Resende, O., Almeida, D. P., Costa, L. M., Mendes, U. C., & Sales, J. D. F. (2012). Adzuki beans (Vigna angularis) seed quality under several drying conditions. *Food Science and Technology, 32*(1), 151–155.
Reyes-Moreno, C., Paredes-López, O., & Gonzalez, E. (1993). Hard-to-cook phenomenon in common beans—A review. *Critical Reviews in Food Science & Nutrition, 33*(3), 227–286.
Roberston, L. S., & Frazier, R. D. (1978). Dry bean production – Principles and practices. Michigan State University Extension Bulletin E-1251. East Lansing. 225 p.
Rubatzky, V. E., & Yamaguchi, M. (1997). *World vegetables: Principles, production, and nutritive values* (2nd ed.). New York: Chapman & Hall.
Sacklin, J. (1985). Quality control for edible dry beans. In Technical Conference on Dry Bean Research (1985, San Francisco, CA, USA). Proceedings.
Sacks, F. M. (1977). A literature review of Phaseolus angularis—the adzuki bean. *Economic Botany, 31*, 9–15.
Sakakibara, M., Aoki, T., & Noguchi, H. (1979). Isolation and characterization of 7S protein-I of Phaseolus angularis (Adzuki bean). *Agricultural and Biological Chemistry, 43*(9), 1951–1957.
Sangsukiam, T., & Duangmal, K. (2017). A comparative study of physico-chemical properties and antioxidant activity of freeze-dried mung bean (Vigna radiata) and adzuki bean (Vigna

angularis) sprout hydrolysate powders. *International Journal of Food Science & Technology, 52*(9), 1971–1982.

Santos Alexandre, A. P., Vela-Paredes, R. S., Santos, A. S., Costa, N. S., Canniatti-Brazaca, S. G., Calori-Domingues, M. A., & Augusto, P. E. D. (2018). Ozone treatment to reduce deoxynivalenol (DON) and zearalenone (ZEN) contamination in wheat bran and its impact on nutritional quality. *Food Additives & Contaminants: Part A, 35*(6), 1189–1199.

Shahbazi, F., Analooei, M., & Saffar, A. (2012). Mechanical damage to pinto bean seeds as affected by moisture content, impact velocity and seed orientation. *International Journal of Food Engineering, 7*(6), 8.

Shigenori, N., Yusuke, S., Chihiro, S., Jun, K., Hiroshi, K., & Kazunori, H. (2008). Suppression of serum cholesterol levels in mice by adzuki bean polyphenols. *Food Science and Technology Research, 14*, 217–220.

Singh, P., & Raghuvanshi, R. S. (2012). Finger millet for food and nutritional security. *African Journal of Food Science, 6*(4), 77–84.

Su, H. L., & Chang, K. C. (1995). Physicochemical and sensory properties of dehydrated bean paste products as related to bean varieties. *Journal of Food Science, 60*(4), 794–814.

Tao, Y. U., Ahn, H. M., Shen, T., Yoon, K. J., Jang, H. J., & Lee, Y. J. (2011). Anti-inflammatory activity of ethanol extract derived from Phaseolus angularis beans. *Journal of Ethnopharmacology, 137*, 1197–1206.

Tjahjadi, C., & Breene, W. M. (1984). Isolation and characterization of adzuki bean (Vigna angularis cv Takara) starch. *Journal of Food Science, 49*(2), 558–562.

Tjahjadi, C., Lin, S., & Breene, W. M. (1988). Isolation and characterization of adzuki bean (Vigna angularis cv Takara) proteins. *Journal of Food Science, 53*(5), 1438–1443.

Tokitomo, Y. (1988). Odor of cooked Japanese adzuki beans. *Japanese Journal of Agricultural Chemistry, 62*, 17–22.

Tomohiro, I., Misato, K., Fumihiko, H., & Yukio, F. (2009). Hypoglycemic effect of hot-water extracts of adzuki (Vigna angularis) in spontaneously diabetic KK-Ay mice. *Nutrition Journal, 25*, 134–141.

US Department of Agriculture, Agricultural Research Service (2019). Food Data Central. https://fdc.nal.usda.gov/fdc-app.html#/food-details/173727/nutrients

Ueno, S., Shigematsu, T., Karo, M., Hayashi, M., & Fujii, T. (2015). Effects of high hydrostatic pressure on water absorption of adzuki beans. *Food, 4*(2), 148–158.

Yao, Y., Cheng, X., Wang, S., Wang, L., & Ren, G. (2012). Influence of altitudinal variation on the antioxidant and antidiabetic potential of azuki bean (Vigna angularis). *International Journal of Food Sciences and Nutrition, 63*(1), 117–124.

Yoshida, H., Tomiyama, Y., Yoshida, N., Shibata, K., & Mizushina, Y. (2010). Regiospecific profiles of fatty acids in triacylglycerols and phospholipids from adzuki beans (Vigna angularis). *Nutrients, 2*(1), 49–59.

Yoshida, K., Kondo, T., Ito, M., & Kondo, T. (2005). Analysis of polyphenols in water extract of red adzuki bean, Vigna angularis. *ITE Letters on Batteries New Technologies and Medicine, 6*(3), C1.

Yoshida, K., Sato, Y., Okuno, R., Kameda, K., Isobe, M., & Kondo, T. (1996). Structural analysis and measurement of anthocyanins from colored seed coats of Vigna, Phaseolus, and Glycine legumes. *Bioscience, Biotechnology, and Biochemistry, 60*(4), 589–593.

Yousif, A. M., Batey, I. L., Larroque, O. R., Curtin, B., Bekes, F., & Deeth, H. C. (2003). Effect of storage of adzuki bean (Vigna angularis) on starch and protein properties. *LWT-Food Science and Technology, 36*(6), 601–607.

Yousif, A. M., Deeth, H. C., Caffin, N. A., & Lisle, A. T. (2002). Effect of storage time and conditions on the hardness and cooking quality of adzuki (Vigna angularis). *LWT-Food Science and Technology, 35*(4), 338–343.

Yousif, A. M., Kato, J., & Deeth, H. C. (2007). Effect of storage on the biochemical structure and processing quality of adzuki bean (Vigna angularis). *Food Reviews International, 23*(1), 1–33.

Bambara Groundnut

Dhritiman Saha and A. Manickavasagan

Introduction

Bambara groundnut (*Vigna subterranea*) is the third most important pulses produced in Africa after groundnut and cowpea (Gulzar and Minnaar 2016; Bamshaiye et al. 2011). It is believed to have originated in West Africa but is also found in several parts of South America, Asia, and Oceania (Baudoin and Mergeai 2001). The production of bambara groundnut in Africa was about 0.33 million tons during 2014–2015. Nigeria leads in production followed by Burkina Faso, Ghana, Mali, Cameroon, and Ivory Coast (Hillocks et al. 2012). It is difficult to obtain reliable production statistics for bambara groundnut, because the crop is mainly grown for domestic consumption and local markets. It is a small herb having lateral stems with height of 0.30–0.35 m and is a self-pollinating and annual crop (Amadou et al. 2001; NRC 2006; Bamshaiye et al. 2011). Two major botanical varieties of the crop are mainly available, *V. subterranea* var. subterranean and *V. subterranea* var. spontanea, whereby the former includes the cultivated varieties and the latter includes the wild varieties (Swanevelder 1998; Gulzar and Minnaar 2016). Bambara groundnut is generally classified as a bean, but, like peanut, the bambara seeds are scraped from the ground (Bamshaiye et al. 2011). Bambara groundnut is highly tolerant to drought conditions and can survive under unfavorable climatic and soil conditions (Baryeh 2001; Azam-Ali et al. 2001; Amadou et al. 2001; Bamshaiye et al. 2011) as well as being resistant to diseases and pests (NRC 2006). Besides, the crops require no fertilizer and hence are also regarded as low-cost crop, and root nodules present help in fixing atmospheric nitrogen, resulting in the fulfillment of the nitrogen requirements of the crop (NRC 2006). Bambara groundnut is generally used for crop rotation due to its nitrogen-fixing ability (NRC 2006).

D. Saha · A. Manickavasagan (✉)
School of Engineering, University of Guelph, Guelph, ON, Canada
e-mail: mannamal@uoguelph.ca

A. Manickavasagan, P. Thirunathan (eds.), *Pulses*,
https://doi.org/10.1007/978-3-030-41376-7_2

Source: ayurtimes.com

(a) (b)

Fig. 1 (**a**) bambara groundnut pods, (**b**) shelled bambara groundnut

Bambara groundnut contains considerable amount of proteins, carbohydrates, fats, and minerals (Poulter 1981). It is composed of carbohydrates (53–69%), protein (17–25%), oil (6.5–8.5%), and mineral salts (2.5–3.5%) (Baryeh 2001; NRC 2006; Steve Ijarotimi and Ruth Esho 2009; Bamshaiye et al. 2011; Hillocks et al. 2012). Bambara groundnut has more protein content than groundnut, pigeon pea, and cowpea (Brough and Azam-Ali 1992). It also contains the highest soluble fiber (4–12%) among all other beans (NRC 2006; Bamshaiye et al. 2011). In addition, it is quite rich in phosphorous (39.6 mg/100 g), magnesium (20.9 mg/100 g), potassium (50.7 mg/100 g), and calcium (77.7 mg/100 g) (Olaleye et al. 2013; Yao et al. 2015) (Fig. 1).

Nutritive Value and Antinutritive Factors

The high protein content in bambara groundnut can be effectively used in several food formulations in improving their nutritional profile. Like most other legumes, bambara groundnut protein lacks sulfur-containing amino acids (cysteine and methionine) and tryptophan in terms of essential amino acid requirements for children. However, it has been reported that bambara groundnut contains high methionine content (up to 2 g/100 g of protein) than any other beans (Aremu et al. 2006; NRC 2006; Bamshaiye et al. 2011). The other essential amino acids in bambara groundnut proteins either meet or exceed the optimum level set by the Food and Agricultural Organization (FAO) for a good-quality protein for growing children (Adebowale et al. 2011).

Very few data are available on the impact of processing on the protein digestibility of bambara groundnut. Roasting and cooking operations were found to decrease the protein digestibility, which might be due to crosslinking of protein and the presence of granule-bound starch and polyphenols (Yagoub and Abdalla 2007). However, germination resulted in increased protein digestibility due to decrease in antinutritional factors and protein hydrolysis by proteolytic enzymes present in germinating seedlings (Yagoub and Abdalla 2007). It has been reported that low levels of antinutritional factors (trypsin inhibitor 6.75–19.08 units/mg) and other compounds like phytic acid (46 mg/100 g), oxalate (3 mg/100 g), and phytin phosphorus (13 mg/100 g) are present in bambara groundnut (Poulter 1981; Steve Ijarotimi and Ruth Esho 2009; Bamshaiye et al. 2011). The presence of tannins (0.36–0.94 g/100 g) further adds up to the list of antinutritional factors present in bambara groundnut (Poulter 1981; Bamshaiye et al. 2011; Adegunwa et al. 2014) in bambara groundnut. Different pre-treatments like soaking, blanching, boiling, roasting, cooking, autoclaving, and addition of salts gave promising results in reduction of the antinutritional factors present in bambara groundnuts (Yagoub and Abdalla 2007; Afoakwa et al. 2007; Adegunwa et al. 2014). Besides, the toxic compounds like phytohalmagglutinin, cyanogenic glucoside or alkaloid are absent in bambara groundnuts (Bamshaiye et al. 2011).

Post-harvest Processing

Shelling or Dehulling

The shelling or dehulling methods involve the use of indigenous technologies. Drying of pods is mainly carried out under sun, and, hence, overdrying of pods leads to poor shelling output. Most of the village communities use mortar and pestle to break the dry pods. In some communities, the pods are filled inside jute or hessian bags and pounded with sticks on flat surfaces, while in others they are heaped on the ground and trodden with feet for shelling the pods. Other methods include the use of stone to crack the pods on a flat surface.

Winnowing

Winnowing of the hull from the seed-hull mixture is carried out after the shelling operation to obtain clean seeds. This involves the dropping of seed-hull mixture of the shelled material from a certain height, against the wind. The lighter particles or hull is blown off by the wind, while the relatively heavier seeds get collected in a container placed on the ground below the height.

Roasting or Cooking

Dehulled seeds are generally roasted at 140 °C for 20 min using a roaster. Roasted and cooled seeds are milled to flour and stored till further use.

Fermentation

Cleaned whole bambara groundnut seeds are generally soaked in water in the ratio of 1:3 and allowed to ferment at room temperature for 3 days. After fermentation, the water is decanted, and seeds are washed in water and dehulled by pressing between the fingers. Cotyledons are washed in several rinses of water, followed by drying at 65 °C for 8 h using a convection oven. Seeds are then cooled, milled to flour, packed in polyethylene bags, and stored till further use (Olagunju et al. 2018).

Milling

Dehulled seeds are milled using a coarse grinder, and the material obtained after grinding is then sieved to obtain the flour of bambara groundnut. The flour is then packed in polyethylene bags and stored till further use.

Storage

Bambara groundnut is generally stored in both shelled and unshelled conditions. The unshelled pods are generally stored in pots when the quantity of bambara groundnut is small. Large quantities of bambara groundnut are stored in granaries. Sometimes, the product is also treated with chemicals and then stored in bags, drums, and some other available containers to control the infestation of pests and insects (Fig. 2).

Product Development

Bambara groundnut has an optimal combination of good-quality protein and other well-balanced nutrients (carbohydrate, fats, and mineral contents) which can effectively be used in several food formulations. The seeds (immature) of bambara groundnut are consumed fresh, boiled, or grilled and sometimes eaten in combination with groundnut or maize. The matured seeds are roasted, boiled, or milled into

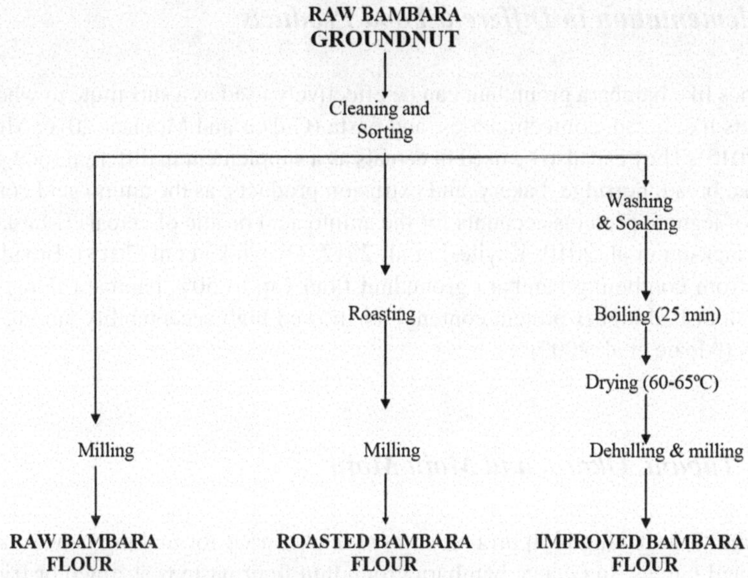

Fig. 2 Post-harvest processing of bambara groundnut

flour. Thereafter the seeds or milled flour are eaten alone or in combination with maize (Swanevelder 1998; NRC 2006; Charles 2010). Bambara groundnuts are also roasted and used as a salted snack in some parts of Africa. The boiled seeds are also canned (Baudoin and Mergeai 2001) and have good market value in Africa (NRC 2006; Gulzar and Minnaar 2016). Zimbabwe and Ghana have reported successful commercial canning of bambara groundnuts in gravy (Mkandawire 2007; Gulzar and Minnaar 2016). Some of the products from bambara groundnut are given below.

Nasima

The flour of dehulled bambara groundnut seeds is used to make high viscous porridge called "nasima" (Gulzar and Minnaar 2016; Mkandawire 2007).

Chipele

The bambara groundnut seeds are first roasted followed by boiling. The boiled seeds are then crushed and eaten as a delight (Mkandawire 2007; Gulzar and Minnaar 2016).

Supplementation in Different Food Products

Legumes like bambara groundnut can be effectively used as a substitute to wheat in products like bread, confectionaries, and pasta (Gulzar and Minnaar 2016; Multari et al. 2015). They can also be used in cereals as a supplement in different food products like bread, porridge, bakery, and extrusion products, as the amino acid composition of legume proteins accounts for the amino acid profile of cereals (Maruatona 2008; Jackson et al. 2010; Kayitesi et al. 2012; Oyeyinka et al. 2018). Bread prepared from combining bambara groundnut flour (up to 50% bambara flour) with wheat flour had higher protein content and showed high acceptability among consumers (Alozie et al. 2009).

Akla, Tubani, Okara, and Moin Moin

Bambara flour having high protein content is also used for making pancakes and soups and purees. In Ghana, bambara groundnut flour paste is steamed or fried to make traditional dish like "akla" (also called koose) and "tubani," respectively (Nti 2009). It is also used in the form of a bean fritter called "okara" and as a savory pudding called "moin moin." The doughy paste of bambara flour is steamed or boiled to form "okpa" (Barimalaa et al. 2005; NRC 2006; Gulzar and Minnaar 2016).

Bambara Groundnut Milk

Bambara groundnut milk is highly nutritious with high consumer acceptability and good taste (NRC 2006). Bambara groundnuts are soaked overnight and dehulled and resoaked for 24 h. Raw milk is obtained by homogenizing soaked beans with hot water (1:2 w/v) followed by straining through muslin. The obtained milk has shown higher acceptability than pigeon pea, soybean, and cowpea milk, and its lighter color is much appreciated (Brough et al. 1993). It also contains relatively lower amount of trypsin inhibitor when compared with other legume milks (Poulter and Caygill 1980) and hence has a better scope of use in different food applications. Besides, bambara groundnut milk possess very good emulsion stability, and curds which are prepared from soymilk can easily be prepared from bambara groundnut milk (Poulter and Caygill 1980). Bambara groundnut milk fermented using lactic acid bacteria has shown high acceptability among consumers (Murevanhema 2012; Gulzar and Minnaar 2016).

Extruded Products

The extruded meat analogs or texturized vegetable proteins made from bambara groundnut flour have shown high consumer acceptability. It could be an effective source of cheap protein to substitute some of the meat proteins in the diet and can replace minced meat in several recipes. The extruded products contain less fat and sodium and provide important quantities of dietary fiber (Gulzar and Minnaar 2016; Charles 2010). Besides, extruded snack from blends of bambara groundnut flour, cassava starch, and corn bran flour have also found acceptability among consumers.

Trends in Processing and Product Development

The seeds of bambara groundnut are very hard, and their cooking time is longer (Tweneboah 2000; Gulzar and Minnaar 2016), thereby limiting its processing at the commercial level. Besides the extended cooking time, antinutritional factors and dehulling are also the major challenges of bambara groundnut for applications in food (Hillocks et al. 2012). The existing pre-treatments method need to be further improved for enhancing the nutritional quality of bambara groundnut. It is also important to explore efficient, cost-effective, and environmentally friendly pre-treatment methods for processing of bambara groundnut. Further, the development of proper post-harvest machinery like bambara groundnut sheller, solar dryer for drying of pods, and proper storage facilities like use of silo bin will relieve the processors of the tedium of the present methods, reduce seed damage, and check the attack of mold, aflatoxin, and other microorganisms.

Bambara groundnut lacks sulfur-containing amino acids (Evans and Bandemer 1967; Brough and Azam-Ali 1992). Thus, combining bambara groundnut with a staple food, such as maize, which contains higher levels of cysteine and methionine, is a recommended nutritional strategy (Akpapunam and Darbe 1994).

Plant-based protein sources are generally considered as a lower-quality protein source which limits their application in different food products or formulations. However, there is a growing trend of using isolated plant proteins over animal proteins for environmental, economic, and religious reasons (Wong et al. 2013). However, the partial or full substitution of animal proteins by these proteins depends on the functionality of the proteins in different food applications. Though it has been reported that bambara groundnut proteins have inferior functional properties (foaming, emulsification, and gelling) than egg white and soy proteins (Brough et al. 1993), succinylation and acetylation have shown encouraging results in improving the foaming, emulsifying, and solubility properties of bambara proteins (Lawal et al. 2007). However, further research is required to substantiate the functional applicability of the bambara proteins using model food systems.

Bambara groundnuts are also potentially capable of substituting soybean in animal (fish, rabbit, dairy cattle, etc.) feed; however extensive studies are needed to formulate the recommended level that can be tolerated by these farm animals without affecting their performance (Multari et al. 2015). Since this crop contains the least amount of antinutritional factors, pre-treated bambara can also be introduced in poultry and pig diets.

References

Adebowale, Y. A., Schwarzenbolz, U., & Henle, T. (2011). Protein isolates from bambara groundnut (Voandzeia subterranean L.): Chemical characterization and functional properties. *International Journal of Food Properties, 14*, 758–775.

Adegunwa, M., Adebowale, A., Bakare, H., & Kalejaiye, K. (2014). Effects of treatments on the antinutritional factors and functional properties of bambara groundnut (*Voandzeia subterranea*) flour. *Journal of Food Processing and Preservation, 38*, 1875–1881.

Afoakwa, E. O., Budu, A. S., & Merson, A. B. (2007). Response surface methodology for studying the effect of processing conditions on some nutritional and textural properties of bambara groundnuts (Voandzei subterranea) during canning. *International Journal of Food Sciences and Nutrition, 58*, 270–281.

Akpapunam, M., & Darbe, J. (1994). Chemical composition and functional properties of blends of maize and bambara groundnut flours for cookie production. *Plant Foods for Human Nutrition, 46*(2), 147–155.

Alozie, Y. E., Iyam, M. A., Lawal, O., Udofia, U., & Ani, I. F. (2009). Utilization of bambara groundnut flour blends in bread production. *Journal of Food Technology, 7*, 111–114.

Amadou, H., Bebeli, P., & Kaltsikes, P. (2001). Genetic diversity in bambara groundnut (Vigna subterranea L.) germplasm revealed by RAPD markers. *Genome, 44*, 995–999.

Aremu, M. O., Olaofe, O., & Akintayo, T. E. (2006). A comparative study on the chemical and amino acid composition of some Nigerian under-utilized legume flours. *Pakistan Journal of Nutrition, 5*, 34–38.

Azam-Ali, S., Sesay, A., Karikari, S., Massawe, F., Aguilar-Manjarrez, J., Bannayan, M., & Hampson, K. (2001). Assessing the potential of an underutilized crop-A case study using bambara groundnut. *Experimental Agriculture, 37*, 433–472.

Bamshaiye, O., Adegbola, J., & Bamshaiye, E. (2011). Bambara groundnut: An under-utilized nut in Africa. *Advances in agricultural biotechnology, 1*, 60–72.

Barimalaa, I. S., Agoha, G., Oboh, C. A., & Kiin-Kabari, D. B. (2005). Studies on bambara groundnut flour performance in Okpa preparation. *Journal of the Science of Food and Agriculture, 85*, 413–417.

Baryeh, E. A. (2001). Physical properties of bambara groundnuts. *Journal of Food Engineering, 47*, 321–326.

Baudoin, J., & Mergeai, G. (2001). Grain legumes. In R. Raemaeker (Ed.), *Crop production in tropical Africa* (pp. 313–317). Brussels: Directorate Generale for International Cooperation.

Brough, S., Azam-Ali, S., & Taylor, A. (1993). The potential of bambara groundnut (Vigna subterranea) in vegetable milk production and basic protein functionality systems. *Food Chemistry, 47*, 277–283.

Brough, S. H., & Azam-Ali, S. N. (1992). The effect of soil moisture on the proximate composition of bambara groundnut (Vigna subterranea (L) Verdc). *Journal of the Science of Food and Agriculture, 60*(2), 197–203.

Charles, O. O. (2010). *Sensory properties of extruded meat analogue from bambara groundnut flour at high moisture contents. Department of Food Science and Technology, College of Food Science and Human Ecology*. Abeokuta: University of Agriculture.

Evans, R. J., & Bandemer, S. L. (1967). Nutritive value of legume seed proteins. *Journal of Agricultural and Food Chemistry, 15*(3), 439–443.

Gulzar, M., & Minnaar, A. (2016). Underutilized protein resources from African legumes. In S. R. Nadathur, J. P. D. Wanasundara, & L. Scanlin (Eds.), *Sustainable protein sources* (pp. 197–205). London: Academic Press. https://doi.org/10.1016/B978-0-12-802778-3.00012-3.

Hillocks, R., Bennett, C., & Mponda, O. (2012). Bambara nut: A review of utilisation, market potential and crop improvement. *African Crop Science Journal, 20*, 1–16.

Jackson, J. C., Duodu, K. G., Holse, M., Lima de Faria, M. D., Jordaan, D., Chingwaru, W., et al. (2010). The morama bean (Tylosema esculentum): A potential crop for Southern Africa. *Advances in Food and Nutrition Research, 61*, 187–246.

Kayitesi, E., de Kock, H. L., Minnaar, A., & Duodu, K. G. (2012). Nutritional quality and anti-oxidant activity of marama-sorghum composite flours and porridges. *Food Chemistry, 131*, 837–842.

Lawal, O., Adebowale, K., & Adebowale, Y. (2007). Functional properties of native and chemically modified protein concentrates from bambarra groundnut. *Food Research International, 40*, 1003–1011.

Maruatona, G. N. (2008). *Physico-chemical, nutritional and functional properties of defatted marama bean flour* (MSc dissertation, University of Pretoria, Pretoria).

Mkandawire, C. H. (2007). Review of bambara groundnut (Vigna subterranea (L.) Verdc.) production in Sub-Sahara Africa. *Agricultural Journal, 2*, 464–470.

Multari, S., Stewart, D., & Russell, W. R. (2015). Potential of fava bean as future protein supply to partially replace meat intake in the human diet. *Comprehensive Reviews in Food Science and Food Safety, 14*, 511–522.

Murevanhema, Y. Y. (2012). *Evaluation of bambara groundnuts (Vigna subterrenea (L.) Verdc.) milk fermented with lactic acid bacteria as a probiotic beverage* (MSc dissertation, Cape Peninsula University of Technology).

NRC (National Research Council). (2006). *Lost crops of Africa. Volume II: Vegetables*. Washington, DC: The National Academies Press.

Nti, C. A. (2009). Effects of bambara groundnut (Vigna subterranea) variety and processing on the quality and consumer appeal for its products. *International Journal of Food Science & Technology, 44*, 2234–2242.

Olagunju, O., Mchunu, N., Durand, N., Alter, P., Montet, D., & Ijabadeniyi, O. (2018). Effect of milling, fermentation or roasting on water activity, fungal growth, and aflatoxin contamination of Bambara groundnut (Vigna subterranea (L.) Verdc). *LWT - Food Science and Technology, 98*, 533–539.

Olaleye, A. A., Adeyeye, E. I., & Adesina, A. J. (2013). Chemical composition of bambara groundnut (V. subterranea L. Verdc) seed parts. *Bangladesh Journal of Scientific and Industrial Research, 48*(3), 167–178.

Oyeyinka, S. A., et al. (2018). Value added snacks produced from Bambara groundnut (Vigna subterranea) paste or flour. *LWT - Food Science and Technology, 88*, 126–131.

Poulter, N. H. (1981). Properties of some protein fractions from bambara groundnut (*Voandzeia subterranean). Journal of the Science of Food and Agriculture, 32*, 44–50.

Poulter, N. H., & Caygill, J. C. (1980). Vegetable milk processing and rehydration characteristics of bambara groundnut [Voandzeia subterranea (L.) thouars]. *Journal of the Science of Food and Agriculture, 31*, 1158–1163.

Steve Ijarotimi, O., & Ruth Esho, T. (2009). Comparison of nutritional composition and anti-nutrient status of fermented, germinated and roasted Bambara groundnut seeds (*Vigna subterranea*). *British Food Journal, 111*, 376–386.

Swanevelder, C. J. (1998). *Bambara, food for Africa: Vigna subterranea (bambara groundnut)*. South Africa: National department of Agriculture, ARC-Grain Crops Institute.

Tweneboah, C. (2000). *Modern agriculture in the tropics, food crops*. Ghana: Co-Wood Publishers.

Wong, A., Pitts, K., Jayasena, V., & Johnson, S. (2013). Isolation and foaming functionality of acid-soluble protein from lupin (Lupinus angustifolius) kernels. *Journal of the Science of Food and Agriculture, 93*, 3755–3762.

Yagoub, A., & Abdalla, A. A. (2007). Effect of domestic processing methods on chemical composition, in vitro digestibility of protein and starch and functional properties of bambara groundnut (*Voandzeia subterranea*) seed. *Research Journal of Agriculture and Biological Sciences, 3*, 24–34.

Yao, D. N., Kouassi, K. N., Erba, D., Scazzina, F., Pellegrini, N., & Casiraghi, M. C. (2015). Nutritive evaluation of the bambara groundnut Ci12 landrace [*Vigna subterranea* (L.) Verdc. (Fabaceae)] produced in Coˆte d'Ivoire. *International Journal of Molecular Sciences, 16*, 21428–21441.

Broad Bean (Faba Bean)

Lamia L'Hocine, Delphine Martineau-Côté, Allaoua Achouri, Janitha P. D. Wanasundara, and Gayani W. Loku Hetti Arachchige

Introduction

Broad bean or faba bean (*Vicia faba*) is one of the oldest domesticated pulses. This grain legume from the Fabaceae family is native to North Africa and Southwest Asia, where it has been grown and used as human food and stock feed for millennia (Cubero 1974). From there, faba bean spread around the globe and is nowadays cultivated in almost all regions of the world. The 2017 world production of faba beans was 4.84 million tonnes from 2.46 million ha, China being the biggest producer accounting for approximately 37% of the total production (1.8 million tonnes), followed by Ethiopia (0.9 million tonnes), Australia (0.4 million tonnes) and the United Kingdom (0.3 million tonnes) (FAO STAT 2017). In Canada and the United States, the crop was introduced late in the 1970s (McDonald 1974; Ells et al. 1978) and accounts for a very small percentage of the global production. Australia, France and the United Kingdom dominate the export of dry faba bean grain, while the import market for human consumption is relatively small (300,000 tonnes/year) and is dominated by the Middle East Mediterranean region, particularly Egypt (Gnanasambandam et al. 2012).

Differences in worldwide specific adaptation factors and selection of plant agronomic traits have resulted in an impressive genetic diversity within faba bean species (Göl et al. 2017). Four groups, on the basis of seed size, have been defined:

L. L'Hocine (✉) · D. Martineau-Côté · A. Achouri
Agriculture and Agri-Food Canada, Saint-Hyacinthe Research and Development Centre, Saint-Hyacinthe, QC, Canada
e-mail: lamia.lhocine@canada.ca; delphine.martineau-cote@canada.ca; allaoua.achouri@canada.ca

J. P. D. Wanasundara · G. W. Loku Hetti Arachchige
Agriculture and Agri-Food Canada, Saskatoon Research and Development Centre, Saskatoon, SK, Canada
e-mail: janitha.wanasundara@canada.ca; gayani.lokuhettiarachchige@canada.ca

A. Manickavasagan, P. Thirunathan (eds.), *Pulses*,
https://doi.org/10.1007/978-3-030-41376-7_3

major, equine, minor and paucijuga (Cubero 1984). The largest seeds emerged in South Mediterranean countries and China and are known as broad beans. Medium-seeded types (equine) are grown throughout the Middle East, North Africa and in Australia, while small seeds, known as field or horse bean, are found in Ethiopia (Duc 1997; Alghamdi et al. 2012). The term faba bean (*Vicia faba* L.) has been accepted to refer the whole species and is used throughout this chapter.

The whole seed is constituted of the seed coat, two cotyledons and the embryo, representing 15%, 84% and 1% of whole seed mass, respectively (Çalışkantürk Karataş et al. 2017). Depending on the cultivar, the weight of 100 seeds may vary from 21 g to 196 g and the size from 10 × 13 mm to 20 × 30 mm (Santos et al. 2018). The colour of the seeds also varies between white, green and brown (Santos et al. 2018). The cotyledon is the main reserve of protein (25–32%) and carbohydrate (40–50%) (Hood-Niefer et al. 2012; Mattila et al. 2018), while the firm and thick seed coat is mainly composed of fibre and contains most of the phenolic compounds such as tannins and minerals (Çalışkantürk Karataş et al. 2017).

Faba bean attracts a growing attention due to nutritional, agronomic and economical advantages. This legume distinguishes itself by a high protein content (Raikos et al. 2014) and a balanced amino acids profile, except for a low level of methionine and cysteine (Raikos et al. 2014). It is however particularly rich in lysine (Hood-Niefer et al. 2012). Furthermore, it is a rich source of other beneficial nutrients including dietary fibres (Çalışkantürk Karataş et al. 2017), ash (Khan et al. 2015), phenolic compounds (Khalil and Mansour 1995; Turco et al. 2016), 3,4-dihydroxyphenylalanine (L-DOPA) (Etemadi et al. 2018) and γ-aminobutyric acid (GABA) (Coda et al. 2017). Faba bean is also less prone to develop off-flavours than soybeans and peas, due to its lower endogenic lipoxygenase activity (Rackis et al. 1979; Chang and McCurdy 1985) and a low-fat content (Mattila et al. 2018). Faba bean also has light cotyledons, which makes it suitable to incorporate in several food matrices to increase protein content and nutritional value.

Besides their nutritional significance, faba bean appears to offer definite agronomic advantages in comparison to other legumes. It is one of the pulses with the ability to grow over a wide range of climatic and soil conditions, including boreal climate (Lizarazo et al. 2015). It has also the highest capacity to fix atmospheric nitrogen by a root nodule symbiosis with *Rhizobium* bacteria (Senberga et al. 2017), thus diminishing substantially the amount of artificial fertilizers needed. It is therefore a suitable crop to introduce in land rotation (Aschi et al. 2017) to increase soil quality (Dubova et al. 2018).

The crop use in food and feed applications has been, however, restricted due to the presence of antinutrients, such as enzyme inhibitors, phytates, condensed tannins, oligosaccharides and the pyrimidine glycosides vicine and convicine (Jamalian 1999). Vicine and convicine are the most relevant faba bean antinutrients; they are precursors of the aglycones divicine and isouramil, the main factors of favism, a genetic condition which may lead to acute haemolysis in humans who are deficient in glucose-6-phosphate dehydrogenase (G6PD) (McMillan et al. 2001; Cappellini and Fiorelli 2008; Schuurman et al. 2009). Glucose-6-phosphate deficiency affects around 330 million people worldwide (Nkhoma et al. 2009), predominantly men (Hagag et al. 2018). Vicine and convicine are also responsible for a reduced growth rate and feed efficiency in monogastric animals (Olaboro et al. 1981; Grosjean et al. 2000; Vilariño et al. 2009; Rizzello et al. 2016).

The reduction of faba bean antinutrients through the development of low antinutrient cultivars (Singh et al. 2019) or through processing techniques such as cooking, fermentation, soaking, germination or use of degrading enzymes has been the task for many studies (Crépon et al. 2010; Cardador-Martínez et al. 2012; Pulkkinen et al. 2015; Rizzello et al. 2016).

This chapter describes current postharvest processing technologies and their incidence on faba bean nutritional, physicochemical and sensorial properties and product shelf life. It also overviews faba bean utilization and discusses future trends and prospects that can open up new application scenarios and promote fast adaptation and stable positioning of faba bean in the plant-based food market as one of the most valuable protein resources for both human food and animal feed industries.

Postharvest Processing

Faba beans can be harvested immature at "milky stage" and consumed fresh as a vegetable. This use is, however, limited due to a short shelf life in this state (Collado et al. 2019). Faba beans are therefore mainly harvested at maturity when their nutritional quality is maximized. The seeds are then dried for storage. At this development stage, the seeds are rich in protein (31.2%), carbohydrates (63.3%) and ash (3.4%) and contain a low lipid concentration (2.1%) (Mattila et al. 2018).

Dried faba beans can be processed and consumed in many ways as presented in Fig. 1. The seeds can be consumed whole after soaking and cooking and/or in ready-to-eat formats such as frozen or canned (Hefni et al. 2015). Dried seeds can as well be milled to obtain a flour that can be easily added in many food products. Gluten-free products enrichment, for instance, is a promising application of faba bean flour. The flour can also be fractionated into protein-, starch- and fibre-rich fractions to play specific roles in food matrices. The processes applied to faba bean can be favourable to improve its nutritional, physicochemical and sensorial properties, as well as its shelf life. Nutritional value can be maximized by increasing health beneficial component level, such as γ-aminobutyric acid (GABA), 3.4-dihydroxyphenylalanine (L-DOPA) and folate, and/or by reducing antinutrient content like phytic acid, trypsin and α-amylase inhibitors, lectins, vicine and convicine, oxalate and tannins. These antinutrients can interfere with normal digestion process, decrease absorption of essential nutrients and/or induce toxicity (Shi et al. 2018; Cardador-Martínez et al. 2012; Hejdysz et al. 2016; Weihua et al. 2015).

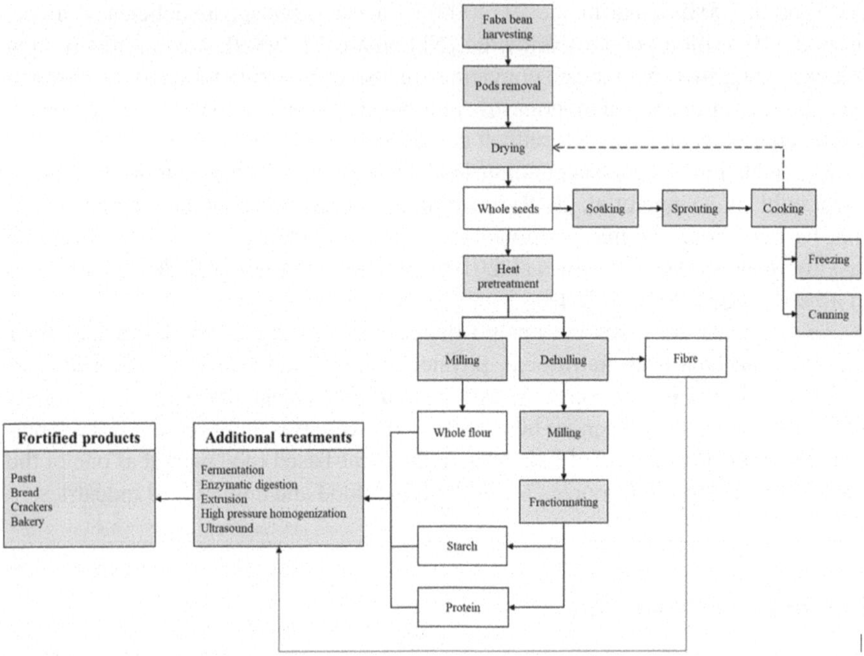

Fig. 1 Faba bean processing overview

Drying

The first treatment applied to the seeds following harvesting is drying, which is either done naturally in fields or artificially. Drying is a crucial step because its efficiency affects product quality and conservation, and artificial drying is preferable to assure moisture content uniformity among seeds (Amiri Chayjan and Shadidi 2014). The drying process requires a significant amount of energy and remains a costly step that needs to be optimized. An effective technology to perform drying is to use a fluidized bed, where the temperature and the velocity of the hot airflow are optimized to achieve rapid drying and to reduce energy consumption. It was shown that faba bean drying in a fluidized bed follows an Aghbashlo kinetic and that moisture diffusivity is maximal using an airflow of 3 m/s at 70 °C (semi-fluidized bed conditions) (Amiri Chayjan and Shadidi 2014). However, the lowest specific energy consumption (energy required to remove 1 kg of moisture) was obtained with an airflow velocity of 0.7 m/s (fix-bed conditions) at 70 °C. The drying kinetic simulation can contribute to find the optimal operational point of the drying process to reduce the cost of this step (Amiri Chayjan and Shadidi 2014). Faba bean seed moisture has to be monitored closely to reduce spoiling risks due to microorganism growth and endogenous enzyme activation (Amiri Chayjan and Shadidi 2014). Also, moisture content and temperature throughout storage affect seed rehydration, cooking quality (Nasar-Abbas et al. 2008) and seed colour (Nasar-Abbas et al. 2009). Long-term storage at high temperature and high-moisture content was shown to enhance seed browning and thus reduce faba bean economical value (Nasar-Abbas et al. 2009). Incomplete seed rehydration leads to texture deterioration following cooking (seed firmness increasing) (Abdel-Aal et al. 2018). The ideal moisture content of faba beans during storage is below 14% (Mehrabi Kooshki et al. 2018), and temperature should be below 25 °C (Nasar-Abbas et al. 2009).

Soaking

Soaking is another essential step that allows seed rehydration prior to cooking. This step enables a higher cooking quality, a lower cooking time and the reduction of some antinutrient concentration via endogenous enzyme activation and leaching (Abdel-Aal et al. 2018). Appropriate rehydration during soaking reduces seed hardness and improves the texture of cooked seeds. Rehydration velocity and capacity are affected by both soaking media temperature and pH. The effect of temperature variation on reaction constant of the soaking process can be modelled by the Arrhenius equation, which means that an exponential relation exists between temperature and reaction constant (Haladjian et al. 2003). Optimization of soaking conditions can reduce soaking time and maximize rehydration rate to assure a high-quality cooking. A combination of an alkaline pH (9.0) and a high temperature (65 °C) was reported as being the ideal condition (Haladjian et al. 2003). Furthermore, a 4 h soaking with a water-to-seed ratio of 5:1 reduces by 5% haemagglutinin activity and by 17.4–38.1% oxalate content. However, no significant effect was observed on phytic acid level (Shi et al. 2018). Soaking must be combined with other processing steps to further decrease antinutrient level.

Sprouting

Sprouting is a natural germination process by which seeds put out shoots to initiate plant growth. It is widely used in pulse and cereal processing to enhance nutritional value and organoleptic properties (Xie et al. 2014; Erba et al. 2019). Endogenous enzyme activation is responsible for the biochemical changes that occur in the seeds during germination (Nkhata et al. 2018). Sprouting takes place following soaking, in the dark and in a high-moisture atmosphere (Setia et al. 2019; Xie et al. 2014). Fast sprouting (24–72 h) is interesting from an industrial point of view to reduce spoiling risks and diminish processing time (Setia et al. 2019). Sprouting does not modify significantly the content of major nutrients (proteins, lipids, starch, ash and fibres) but changes intermolecular structures between those major constituents and as a result increases their respective digestibility. Before soaking, starch granules are trapped in a tight matrix composed of protein and fibre. Soaking and sprouting causes the relaxation of this structure, due to seed rehydration and hydrolytic enzyme activation. For example, endogenous α-amylase activity increased from 0.4 to 1.1 U/g after 72 h sprouting (Setia et al. 2019). Starch granules become more accessible to digestion, which improved substantially starch digestibility. Resistant starch content decreases from 25.7% to 9.3%, while slow-digestible starch increases from 4.7% to 9.9% and fast-digestible starch from 12.4% to 23.7%. Protein digestibility also augmented following sprouting but in a less marked manner (from 78.0% to 80.4%) (Setia et al. 2019).

Sprouting does not change significantly ash level but has an interesting effect on mineral bioavailability. Faba beans are rich in minerals; however, their absorption is limited due to high antinutrient content (Rubio et al. 1994; Luo and Xie 2014). A 24 h soaking followed by 48 h sprouting was shown to increase iron, calcium and copper bioavailability from 6.34% to 31.45%, 32.83% to 38.75% and 35.37% to 39.83%, respectively. However, no significant impact on manganese bioavailability (33.17% to 33.83%) and a decrease in zinc bioavailability were observed (15.20% to 2.74%) (Xie et al. 2014). This effect is possibly caused by endogenous phytase activation during sprouting, which results in an important reduction of phytic acid. Phytic acid is composed of several negatively charged phosphate groups, which have a high affinity for cations (Lopez et al. 2002). Ions have to be in their free form to be absorbed by intestinal cells; their bioavailability are therefore compromised by phytic acid binding (Lopez et al. 2002). Alonso et al. (2000) have shown that 72 h sprouting reduces phytic acid concentration by 60% (Alonso et al. 2000). Sprouting can also enhance folate content in faba bean seeds. It was shown that sprouting prior to cooking permits to reach a folate content of 194 μg/100 g in canned faba bean, which represents a 52% increment compared to standard processed canned faba bean (Hefni et al. 2015). Overall, sprouting is an interesting step to improve nutritional value of faba beans. Sprouting is therefore a simple and efficient process to enhance the nutritional value of faba beans.

Cooking

Cooking is a major step in faba bean processing, allowing to reach a desirable palatability and to increase protein and starch digestibility (Anderson et al. 1994). Typically, presoaked faba beans are placed in hot or boiling water at atmospheric pressure until the seeds reach the targeted tenderness (Hefni et al. 2015; Lafarga et al. 2019). Cooking also increases faba bean nutritional value by significantly reducing heat-labile antinutrients. Cooking for 1 h at 95 °C of presoaked whole and dehulled faba bean seeds decreases their haemagglutinin activity by 98% (Shi et al. 2018), thereby eliminating toxicity risks related to lectins contained in faba bean. Vicine and convicine are also reduced on average by 22.53% and 18.86%, respectively, which is, however, not sufficient to exclude risks of developing favism. Cooking also lowers partially phytic acid content (between 19.09% and 38.3%) and oxalate content (between 31.85% and 45.81%) (Shi et al. 2018), while trypsin inhibitors activity decreases by more than 40% following various heat treatments and most efficiently under pressure cooking (autoclave) (Yu-Wei and Wei-Hua 2013). Reduction of these antinutrients results in an increase of protein digestibility by 1.7–7.3% depending on the heat treatment used. Combination of various treatments increases further protein digestibility. Yu-Wei and Wei-Hua (2013) reported that dehulling combined to soaking and pressure cooking increased faba bean protein digestibility between 11.2% and 17.4% of different faba bean cultivars (Yu-Wei and Wei-Hua 2013). In addition, heat treatment induces starch gelatinization, which increases starch digestibility by making it more accessible to amylolytic enzymes. Resistant starch proportion decreases from 25.7% to 3.0% after cooking and fast-digestible starch increase from 4.7% to 37.2% (Setia et al. 2019).

Cooking quality plays also an important role in sensorial properties of the product. In fact, seed consistency after cooking is an important attribute that affects consumer acceptability. The impact of cooking temperature on seed hardness has been studied. A high-temperature cooking (115 °C for 20 min) increases hull hardness but has no effect on the whole seed texture (Revilla and Vivar-Quintana 2008). It was also shown that additives, such as EDTA and NaCl, added to cooking water increased the shelf life of canned products, but had no impact on seed hardness (Revilla and Vivar-Quintana 2008). In addition, it was demonstrated that there are significant correlations between seed hardness after cooking and nutrient content (proteins [$r = 0.72$], starch [−0.86] and ash [−0.80]), rehydration capacity ($r = -0.71$) and the final viscosity of flour ($r = -0.70$), of acid ($r = -0.91$) and of alkaline extract ($r = -0.70$). Hardness was also higher in the high-tannin faba bean cultivars as compared to the tannin-free varieties (snowdrop and snowbird) (Abdel-Aal et al. 2018). Flour viscosity changes during heat treatment are representative of starch structural change during heat treatment, which has a direct impact on seed firmness following cooking. Measurement of the final flour viscosity and rehydration capacity after soaking are therefore simple measures that can be used in the industry to anticipate seed quality after cooking (Abdel-Aal et al. 2018).

Roasting

Faba bean seeds can also be roasted to modify their nutritional properties. Roasting is a dry thermal process performed in an oven at a temperature over 100 °C (Cardador-Martínez et al. 2012; Siah et al. 2014). It is used as a heat pretreatment prior to dry milling (Siah et al. 2014). Roasting was shown to diminish faba bean phenolic compound content and modify their specific profile (Siah et al. 2014). It also reduces vicine and convicine content but in a less effective manner than traditional cooking. However, both heat treatments eliminate L-DOPA, a heat-labile molecule, from the seeds (Cardador-Martínez et al. 2012). L-DOPA is an amino acid formed by biotransformation of tyrosine. It is a dopamine precursor which has the capacity to cross the blood-brain barrier. It has health beneficial effects and is used to treat Parkinson's disease symptoms (Waller and Sampson 2018). L-DOPA removal is hence a negative collateral effect of heat processing.

Microwave Treatment

Microwave treatment distinguishes itself from other thermal treatments by taking advantage of electromagnetic waves. As roasting, it is used as a thermal pretreatment prior to dry milling (Jiang et al. 2016). The main advantage of the microwave over roasting is the low time required, which decrease processing time (Jiang et al. 2016; Hernández-Infante et al. 1998). However, it's drawback is the lack of cooking uniformity (Hernández-Infante et al. 1998). It was shown that faba bean microwave cooking (1.5 min with a power of 950 W) is sufficient to inactivate endogenous lipoxygenases and peroxidases (Jiang et al. 2016). Lipoxygenase inhibition is desirable because it is responsible for unsaturated fatty acid degradation into unpleasant aroma compounds. Peroxidase is used as an indicator, because it is one of the most thermostable enzymes found in faba beans. On the other hand, extensive microwave heat treatment was also shown to increase lipid autoxidation, resulting in a rancid off-flavour development. It also reduces faba bean protein digestibility as a result of excessive protein denaturation during this process (Jiang et al. 2016). Moreover, microwave cooking also induces the formation of an insoluble complex between protein and starch that was observed under microscopy and which amount correlated precisely with the decrease of protein solubility (Jiang et al. 2016). The decrease in protein solubility can affect protein extraction yield and alter some protein functionality properties as foaming and gelling (Jiang et al. 2016). Nevertheless, microwave cooking reduces seed hardness, which makes milling easier and decreases mean flour particle size. A better quality flour was therefore obtained following microwave cooking as compared to raw dried seeds. Microwave process optimization is therefore necessary to obtain the desired effect (Jiang et al. 2016).

Dry Milling

To facilitate faba bean integration in various food products, dried seeds can be milled. The whole flour can be either used as is or further fractionated into protein-, starch- and fibre-rich fractions. Faba bean flour and enriched fractions can also undergo many processes to improve their nutritional, physicochemical and organoleptic properties and also their shelf life. The seeds can either be directly milled or pretreated as described above (soaking, sprouting, cooking and roasting) prior to milling to diminish the seed hardness and improve the milling quality and sensorial properties of the flour (Jiang et al. 2016). The seeds can also be dehulled before milling. The hull and the cotyledons are separated to obtain flour enriched in starch and protein (cotyledons) and a fraction enriched in fibre (hull). Based on reported studies, dehulling removes most (between 59.2% and 92.3%) of the phenolic compounds, including tannins (Yu-Wei and Wei-Hua 2013; Alonso et al. 2000). Tannin reduction is desirable, because they were shown to bind protein and form insoluble complexes reducing protein digestibility and bioavailability (Ortiz et al. 1993). Tannins are also responsible for an undesirable bitter taste (Collado et al. 2019) and are associated with faba bean browning phenomenon during conservation (Nasar-Abbas et al. 2009). Moreover, since whole and dehulled faba bean flour exhibited different aroma profiles in both low- and high-tannin faba bean varieties, dehulling was also suggested as an effective way to improve aroma profile of faba bean flour (Akkad et al. 2019). Dehulling can be done by crushing coarsely the seed with a roller mill and separating the hull from the cotyledon by sieving and air densification. This process allows a high recovery of 89% for the hull and 96% for the cotyledon at an industrial scale (the roller mill capacity was 500 kg/h, 250 kg/h for the sieving process and 200 kg/h for air densification) (Meijer et al. 1994). However, dehulling does not reduce phytic acid, oxalate, lectin, vicine and convicine, trypsin and α-amylase inhibitors. These antinutrients are mainly found in the cotyledons and would require other treatments.

Fractionation of Faba Bean Flour

Faba bean flour components can be separated physically by either wet or dry fractionation processes generating fractions rich in starch, protein or fibre (Colonna et al. 1980; Schutyser et al. 2015). This process generates therefore many food ingredients that can be used in several food matrices to enhance nutritional value or to contribute to the texture, stability and/or techno-functionality of food products (Vioque et al. 2012; Felix et al. 2018). The type and conditions of the fractionation process used have an important impact on the recovered fractions' quality and functionality. The wet process generally implies a first aqueous or alkaline extraction step to solubilize proteins, which are then separated from starch and insoluble material by a centrifugation or a sedimentation step. Proteins are concentrated from the supernatant by isoelectric precipitation via pH adjustment, and the starch is washed to eliminate residual fibre and proteins (Vioque et al. 2012; Li et al. 2019).

Studies were conducted to optimize wet process conditions to extract faba bean proteins. The results demonstrated that a pH increase during the extraction step enhances protein solubilization (76% at pH 7.0 and 85% at pH 10.0), that adding 0.3 M NaCl reduces protein solubility and that temperature variation between 10 and 20 °C has no effect (McCurdy and Knipfel 1990). Even though alkaline pH increase protein yield, the author suggested using pH 7.0 to reduce protein denaturation and functionality alteration. Harsher conditions can, however, be used to maximize protein recovery. The pH used for protein precipitation was also changed from 4.0 to 5.3. Protein recovery was higher at lower pH (4.0), but protein purity was reduced due to nonprotein compounds co-precipitation. The optimal pH to maximize both recovery and purity was 4.7. A purity of 92.5% and a pure protein recovery of 48% were obtained with the optimized process (McCurdy and Knipfel 1990). Other studies have used similar conditions to extract faba bean proteins. A purity of 94% and a pure protein yield of 77% were obtained using an extraction at pH 9.5 and an isoelectric precipitation at pH 4.5 (Singhal et al. 2016), and a purity of 92% was obtained with an extraction at pH 10.5 and a precipitation at pH 4.0 (Vioque et al. 2012).

The technology used for protein drying also has an incidence on protein functionality (Cepeda et al. 1998). It was shown that spray drying gives smaller and more uniform faba bean protein powder particle size as compared to freeze-drying. Higher protein solubility at neutral pH was also obtained for spray-dried fraction in comparison to freeze-drying, although it did not exceed 50% with the two processes and was further decreased at high protein concentration. Protein extraction at alkaline conditions may cause protein denaturation, which is reflected by a solubility loss (Cepeda et al. 1998). Yang et al. (2018) have shown that the solubility of an alkaline (pH 11) which precipitated faba bean protein isolate was increased by a high-pressure homogenization (15,000 psi) treatment which induced dissociation of faba bean protein aggregates. This process could increase protein solubility from 35% to 99% at neutral pH (Yang et al. 2018). Therefore, high-pressure homogenization is an additional step that could compensate for some wet process inconveniences. Moreover, this treatment increases foaming stability and capacity but has a negative impact on emulsifying properties. Another possibility is the use of high-intensity ultrasound treatment. This process was shown to result in a reduction of the faba bean protein isolate particle size with an increase in protein solubility and foaming properties. It was also shown that ultrasound treatment induced protein structural changes as reflected by an increase in beta sheet amount that resulted in a 3.6% reduction of in vitro protein digestibility (Martínez-Velasco et al. 2018).

During fractionation, lipids and phenolic-enriched fractions can also be obtained by hexane and 75% acetone extraction, respectively (Vioque et al. 2012). Lipid removal reduces the risk of off-flavour development during storing, and polyphenol elimination reduces browning risks due to quinolone formation (Vioque et al. 2012). Nonetheless, high quantity utilization of such solvents in a plant remains questionable. Food industry intends to move away from solvent extraction to adopt a more eco-friendly and safer process and also to limit risks of toxic solvent residuals in the end products (Cheng et al. 2019). Many eco-innovative alternative technologies are possible, such as supercritical fluid extraction and enzyme-assisted aqueous extraction (Cheng et al. 2019; Kumar et al. 2017; Martin et al. 2015). However, these processes have not been applied to faba bean up to date.

Faba bean component fractionation can also be conducted by dry processing. Air classification is used to separate flavour particles regarding their size. Protein particles are smaller than starch granules and are mostly recovered in the light effluent, whereas starch is recovered in the heavy one. Dry fractionation of faba bean flour resulted in 46.7% of the initial flour mass being recovered in the protein-rich fraction, 52.4% recovered in the starch-rich fraction and 0.9% of the initial flour lost during processing (Coda et al. 2015). Only this process cannot lead to ingredient purity as high as those obtained by wet processing. It also results in higher levels of starch damage (Li et al. 2019). Coda et al. (2015) reported a 67% protein recovery in the protein-rich fraction, with a protein purity of 51.49% and a 65.82% purity of the starch-rich fraction on air classification of faba bean. Other studies on faba bean, pea and lentil have also reported similar purity of 56.4% for protein-rich fraction (Felix et al. 2018) and 65.3% for starch-rich fraction (Li et al. 2019). However, protein concentrate obtained by the dry fractionation process exhibited a high solubility of over 80% at pH 6.5, reaching 100% at pH 9.0 (Felix et al. 2018), which is notably higher than wet processing isolate. During dry fractionation, protein-rich fraction is also enriched in lipids, fibre and ash (Coda et al. 2015). Lipid enrichment can be problematic, being important precursors of off-flavour development. The same tendency is observed for antinutrients (vicine, convicine, phytic acid, condensed tannins and trypsin inhibitors). Further treatment, prior or following isolate preparation, would therefore be desirable to increase fractions' quality and organoleptic properties.

Extrusion

Extrusion is a process extensively used in the industry that consists of fast cooking at high temperature under high shear rate and a high-moisture environment (Tiwari and Jha 2017). This process is efficient to reduce spoilage risks and to remove heat-labile antinutrients such as trypsin and chymotrypsin inhibitors (−98.9 and −52.8%, respectively), lectins (−99.6%) and α-amylase inhibitors (−100%) (Alonso et al. 2000). Extrusion can also destabilize complex molecular structures and therefore affect the physicochemical properties of the flour and the various fractions (Rahman et al. 2015). This treatment increases also starch and protein digestibility in a more substantial way as compared to soaking and sprouting (Alonso et al. 2000). In broilers, protein digestibility is enhanced from 80.1–89.3% to 87.9–91.6% following extrusion and starch digestibility from 70.7–81.2% to 95.1–99.0% depending on the faba bean variety (Hejdysz et al. 2016).

The effect of other thermal treatments under varying moisture conditions has also been investigated. The effect of annealing (65 °C and 60% moisture) and heat-moisture treatment (120 °C and 30% moisture) on faba bean protein digestibility was compared. Annealing caused an increase of in vitro protein digestibility from 76% to 82%, and heat-moisture treatment decreases digestibility to 73% (Estefania et al. 2018). The low moisture content and high temperature may enhance insoluble complex formation between protein and starch, which may explain digestibility reduction when moisture is restrained.

Fermentation

Fermentation is another cost-effective process that can be used to enhance nutritional properties of faba bean flour and fractions. A 48 h fermentation at 30 °C with *Lactobacillus plantarum* with an initial density of 10^7 cfu/g and a ratio sample:water of 50:50 decreases faba bean vicine and convicine content by over 91% and trypsin inhibitors by over 40% in protein-rich faba bean fraction obtained by air classification (Coda et al. 2015). This has for effect to increase significantly in vitro protein digestibility. Fermentation also increases γ-aminobutyric acid (GABA) content by more than 50 times compared to raw fractions and twice as compared to fractions treated in the same conditions without inoculum. This result suggests that hydration and warm conditions activate endogenous glutamic acid decarboxylase which transforms glutamic acid into GABA. GABA is a neurotransmitter amino acid which has many health benefits, including blood pressure reduction in patients with hypertension. The consumption of a reasonable serving (10 g) of faba bean fermented flour is enough to obtain the health beneficial effects associated to GABA (Coda et al. 2015). Chandra-Hioe et al. (2016) have demonstrated that 20 h faba bean flour fermentation at 30 °C with *Lactobacillus delbrueckii* and *Streptococcus thermophilus* with a ratio water:sample of 30:10 was able to slightly decrease trypsin inhibitor activity and increase to some extent protein digestibility, but those effects were not significant (Chandra-Hioe et al. 2016). These results suggest that the bacteria strain used for fermentation and fermentation conditions and duration are important factors to optimize in order to reach targeted nutritional improvements.

Specific enzymes can also be used to target more precisely some antinutrients. Digestion with phytases and tannases (100 U/kg of flour) prior to flour consumption permits to increase free phosphorous (+118%) and decrease hydrolysable (−48%) and condensed tannins (−67%) in faba bean flour. This treatment increases phosphorous absorption in rats as compared to raw flour, from −56% to 81%. This represents important advantages from both environmental and nutritional perspectives. However, this treatment is not sufficient by itself to improve protein digestibility (Weihua et al. 2015).

Enzymatic Modifications of Faba Bean for Food Product Fortification

Food products enrichment with faba bean flour is advantageous due to its high nutritional quality. Faba bean flour was incorporated in various products, including crackers, pasta, bread and other bakery products (Tazrart et al. 2016; Rosa-Sibakov et al. 2016; Millar et al. 2017; Coda et al. 2017; Xu et al. 2017; Wang et al. 2018). However, some challenges can be encountered due to the lack of gluten. Gluten network remains unique and is responsible for a desirable texture in many cereal-containing products. Some processing steps were used to modify faba bean flour structure to mimic the gluten network to a certain extent. For instance, *Weissella confusa* was added to sourdough fermentation of faba bean flour to form dextran in situ (Wang et al. 2018). Dextran is an exopolysaccharide with colloid properties. Its presence increases flour thickness and viscosity (Xu et al. 2017) as well as bread specific volume (Wang et al. 2018). This transformation allows the addition of faba bean flour up to 43% in bread formulation without altering bread texture and specific volume, as compared to wheat bread (Wang et al. 2018). An enzymatic process with a transglutaminase is also possible. Transglutaminase forms new covalent bonds between lysine and glutamine side chains, which changes protein structural properties. This structural change allows to entrap further starch granules into the protein matrices. This treatment was attempted to create a gluten-free pasta containing 30% faba bean flour in its formulation. Transglutaminase treatment was effective to control pasta adhesiveness but was not sufficient to improve pasta hardness and chewiness (Rosa-Sibakov et al. 2016).

Product Development

Faba bean (*Vicia faba*) is consumed as a staple as well as a subsidiary food in the global food cultures. As one of the oldest crops in the world, the use of faba bean (broad bean, horse bean, field bean) for food and food ingredient development dates back to over 5000 years in China. Cultivation of faba bean has been reported in Egypt 3000 years ago and by the Hebrews in biblical times and later by Greeks and Romans.

Traditional Uses

Faba bean use could be in a home cooking setting or at an industrial level. Traditionally, the dry whole seed with or without seed coat; split seeds; whole ground seed; germinated seed; soaked and boiled seed; soaked, boiled and mashed seed; fried seed and fractions of starch and protein fraction are used. In China, immature faba bean pods (large seeded) are a vegetable, but in the Mediterranean region, only a small amount is consumed fresh.

Fried faba bean seeds are a popular snack in China; similarly, roasted seeds are popular in India (Duke 1981). Other common forms are stewed beans, whole bean soups, boiled germinated bean soups and boiled and fried (Jambunathan et al. 1994). Boiled, hulled and mashed faba bean flavoured with oil, sugar or fruit peels is used as stuffing for dumplings and steamed bread in China. Soaked seeds are boiled, mashed and fried with spices such as cumin, chilies and herbs which make a popular breakfast food (*ful medames*) and also consumed any time of the day in Egypt. *Falafel* is deep-fried faba bean cotyledon paste (sometimes mixed with chick pea) with some vegetables and spices and is a popular food in the Mediterranean. Canned whole faba beans in salt water are a commercial practice in North America and are used in various dishes, including salads and Mexican-style chilli-like preparations.

Fermented or bio-converted faba bean is used for different types of sauces and pastes. Faba bean pastes with added flavourings such as chilli, sesame, chicken, ham and beef are popular in the Sichuan and Anhui provinces of China (Lang et al. 1993). Basic faba bean paste preparation involves soaking and short-term germination, dehulling, steaming, inoculation with fungal/bacterial starter culture, shaping into cakes and drying. Sun drying is common in the traditional preparations. Duration of fermentation could be as short as 3 months or as long as 8 years and brings colour changes and new flavour development. Addition of salt and chilli during fermentation process brings unique colour, taste and flavour notes to the paste. The bean paste is dark and coarse, and the flavour characteristics are described as pungent with notes of burnt caramel and fried onion, with slightly sour, bitter and heat hidden with the saltiness, e.g. Doubanjiang of Sichuan province of China. Bacterial communities present in Doubanjiang (Chinese red pepper paste) have been identified as dominated by lactic acid bacteria (LAB) and *Enterobacteriaceae*family, and core microbiome (>1% average relative abundance) is mainly assigned to *Proteobacteria*, *Cyanobacteria* and *Firmicutes* phyla. A total of 29 volatile compounds mainly alcohols, esters, aldehydes, ketones and phenols consisted the flavour volatile profile of this paste showing strong relationship between volatile compounds and bacterial taxa. For example, *Pseudomonas* is highly associated with 3-methylbutanal, 2-methylbutanal, benzeneacetaldehyde and 2-acetylpyrrole ($P < 0{\cdot}001$) and significantly correlated with 5-methylfurfural and 5-methyl-2-phenyl-2-hexenal ($P < 0{\cdot}01$) (Li et al. 2016).

Separation (physical modification) of faba bean to generate starch- and protein-rich fractions has been a traditional practice in China and used in making different foods. Soaked seeds are ground with water to obtain a slurry with fine particles. Addition of excess water and separation of liquid, allowing starch to sediment and then removing excess water by filtering and drying of the washed sediment, generate faba bean starch. Faba bean starch alone or mixed with others can be further processed to make bean vermicelli, noodles, sheet jelly noodles, etc. Sesame and sugar are added to faba bean starch to make refreshments such as sweet bean pudding and sweet sesame bean pudding. Water left from starch sedimentation contains protein and can be recovered by lowering pH. This protein curd/paste can be dried and used as substrate for fermentation with suitable fungi and makes sauces (Lang et al. 1993). Pretreatment of seeds with heat, especially microwave heating for 1.5 min at 950 W, inactivates endogenous peroxidase and lipoxygenase and reduces beany flavour notes and maintains milling properties and protein solubility (Jiang et al. 2016).

Beyond Traditional Uses

Faba bean is a protein-rich as well as a gluten-free option for product development and formulations. Of the pulses available for the food product developers, faba bean and faba bean ingredients provide unique opportunities to develop products that can claim "good source of protein". Under Code of Federal Regulations Title 21 (FDA 2018), safe food use of faba bean is approved as broad beans.

Enhanced use of faba bean in food beyond using whole seed or ground whole seed flour can be achieved by converting faba bean into components and new ingredients by means of biological (e.g. microbiological, enzymatic, sprouting), chemical (e.g. organic or mineral acids, various salts, polysaccharides) or physical (milling and air classification, pH-based aqueous separation) processes or their combinations. Nutritionally, these processes help to increase the level of macro- and micronutrients and decrease components that interfere nutritional value of faba bean by means of concentrating and biochemical transformation. Simultaneously, the biochemical and physical alterations that occur in macromolecules change overall functionalities of the resulting ingredient, therefore expanding application potential.

Utilization of Faba Bean Fractions

Faba bean ingredients/fractions generated through wet or dry fractionation can be used in different applications or further converted to new ingredients using chemical and/or biological conversion processes.

Starch-rich fraction is primarily for noodle-type product processing which has been traditionally practiced in China, nowadays a viable alternative to mung bean starch. Research at VTT Technical Research Centre of Finland (https://www.vttresearch.com/) has shown that a combination of mechanical separation and bioconversion (fermentation) results in improving nutritional and functional, primary flavour of faba bean and allows expanded use as a food ingredient. Fermentation of dough (50:50, w:v) with *Lactobacillus plantarum* VTT E-133328 (initial cell density ~107 cfu/g of dough) at 30 °C for 24 h was able to reduce vicine and convicine by 91%, 40% reduction of condensed tannin and trypsin inhibitor. It is reported that this conversion of faba bean can be included in bread (100% faba bean flour) and pasta (70% faba bean flour) (Sözer 2014).

Utilization of Spouted/Germinated Seeds

Allowing soaked seeds (usually 12 h) for short termination (~48 or 72 h) has been reported to reduce the levels of components affecting nutritional value of Faba bean (Coda et al. 2015). Trypsin inhibitor activity and level of tannins can be reduced nearly 90% upon 12 h soaking and 48 h germination (Sharma and Sehgal 1992). Germination reduces stachyose level of faba bean than heat processing (Khalil and Mansour 1995). Reduction of the levels of vicine and convicine (maximum 19%) and oligosaccharides (raffinose by 100%, stachyose by 60%, verbascose by 80%) increased water- and oil-holding capacities, and decreased peak viscosity in flour slurries upon 72 h germination (Wei 2019) shows that germinated faba bean seeds could provide biochemically transformed base material for ingredient development.

Use in Pasta

Faba bean flour (25–35%) can be a good alternative for supplementation or blending with other gluten-free flours to prepare gluten-free pasta products. One of the main technical challenges is the textural changes of weakened pasta structure that comes with faba bean flour (Petitot et al. 2010b). Although high-temperature drying strengthens protein network and improves pasta quality, starch digestibility can be affected. Faba bean ingredients introduce more globulin proteins low in S-amino acids that have less cross-linking potential than wheat proteins and result in increased cooking loss (e.g. from 4.71 mg/100 g in control pasta to 7.28 mg/100 g in 50% substituted pasta). Allowing faba bean proteins in the flour to cross-link with transglutaminase-assisted reactions (20 nkat/g flour dry matter, 10 min) has helped to improve hardness, chewiness, cohesiveness and resilience of pasta, while adhesiveness was reduced to the level of 100% semolina pasta (Petitot et al. 2010b).

Faba bean inclusion improves nutritional value of pasta compared to traditional durum wheat. Pasta formulated with whole seed flour, protein enriched fraction or lactic acid bacteria fermented seed flour, had 1.8 times more protein than regular semolina pasta and resistant starch contents of 1.5, 1.8 and 1.1 g/100 g dry pasta (durum wheat pasta had 0.5 g/100 g dry semolina pasta) (Rosa-Sibakov et al. 2016). Incorporation of faba bean flour (10%, 30% and 50%) lowered starch hydrolysis resulting in lower pasta glycaemic index (GI) for enriched pasta, 91.9%, 83.4% and 71.3%, respectively, compared to traditional pasta (95.9) and white bread (100) level (Tazrart et al. 2016). The protein content increase can be ~ 2.25% for every 10% of broad bean flour added and improves protein level of the final product (21% against 13.7% in 50% enriched pasta and the control, respectively). Dietary fibre, resistant starch (from 1.4% in the control to 2.5% in 50% pasta) and ash and mineral (calcium, iron and zinc) levels can also be enhanced. Faba bean flour added pasta had increased redness than regular ones and showed reduced yellowness. Presence of proanthocyanidins (Baginsky et al. 2013) and lack of carotenes (Tazrart et al. 2016) in faba flour, respectively, may be related to these changes in colour parameters. Use of corn-faba bean flour blends (70:30) and extrusion cooking (28% moisture, 100 °C, Brabender 10 DN single-screw extruder with 3:1 compression ratio) provide a pasta-like product with highly acceptable nutritional and functional quality (Giménez et al. 2013). Shorter cooking times for spaghetti prepared with faba beans in three substitution levels as 10%, 20% and 30% have been reported (Giménez et al. 2012) than that for similar product prepared with 35% broad bean flour (Petitot et al. 2010b); however these cooking times were shorter than 100% wheat spaghetti (Petitot et al. 2010b). The increased stickiness of faba bean flour-enriched pasta may be related to the high dietary fibre level of broad bean flour which causes the formation of discontinuities and cracks inside the pasta leading to a weakened structure (Petitot et al. 2010a, b). Faba bean flour can be used up to 35% incorporation level in traditional durum wheat pasta with improved nutritional quality (Rosa-Sibakov et al. 2016). A nutritional evaluation of mixed faba bean-wheat pasta diet in growing rats demonstrated an improvement in faecal protein digestibility and protein utilization, muscle weight, blood nutritional markers and rat growth performance in comparison with rats fed with a gluten-wheat pasta diet. The enrichment of wheat-based pasta with faba bean increased the content of lysine, threonine and branched amino acids by 97, 23 and 10%, respectively, resulting in a more adequate essential amino acid profile for growth processes (Laleg et al. 2019).

Use in Bakery Products

In wheat bread formulation, supplementing wheat flour with faba bean flour at 30% level caused a slight decrease in the loaf volume and increase in hardness compared to 100% wheat flour bread (Coda et al. 2017). Fermented faba bean flour (40:60, flour/water) with *Pediococcus pentosaceus* I02 at 20 °C for 24 h (sourdough fermentation) in bread formulation at 30% level gave a final product with comparable crumb porosity as 100% wheat flour bread and with improved nutritional quality. Faba bean sourdough prepared with *Weissella confusa* VTT E-143403 (E3403) fermentation incorporates microbial dextran to the dough. This natural hydrocolloid produced in situ was able to improve viscoelastic properties of the composite faba bean (43%)-wheat flour bread with increase in specific volume (~21%) and decrease in crumb hardness compared to 100% wheat flour bread (Wang et al. 2018). Replacement of wheat flour (10%, 15% and 20%) with faba bean starches resulted in decreased loaf volume, inferior crumb grains and pale crust in breads but without any objectionable flavour or odour (Morad et al. 1980). Use of starch of germinated faba bean seeds gave further reduction in the loaf volume. Incorporation of faba bean protein products in wheat bread formulations (Canadian hard red wheat) showed that protein product obtained through micellar mass preparation complemented gluten functionality better than protein concentrates or isolates obtained by isoelectric precipitation (Youssef and Bushuk 1986).

Use in Dairy Alternatives

Faba bean has been used in developing non-dairy milks, eggless mayonnaise, dips, cheese, yoghurts, kefir and ice cream (Gugger et al. 2016). The process of obtaining suitable faba bean base material for using in these products involves hydrating starch-rich seed, removing excess water and heating of flour slurry with α-amylase under controlled pH allowing degradation of seed structure and separation of insoluble fibre. Fermentation of this liquefied flour with microbes (*Lactobacillus bulgaricus*, *L. bifidus*, *L. lactis*, *L. casei*, *L. acidophilus*, *L. cremoris*, *L. lactis*, *Streptococcus thermophilus* and *Bifidobacterium bifidus*) brings biochemical changes of the fermenting substrate including changes to the native bean flavours. Addition of oils, viscosity modifiers, vitamins, oils, gel formers, stabilizers, sequestrants and desired flavours can generate a cultured dairy alternative such as yoghurt. It is noted that citric acid in the form of citrus juice 0.1–5 weight percent of total weight of the product helps to mask residual flavours while bringing acceptable flavour notes.

Use in Protein Curds

Wet fractionation of faba bean starch generates protein-rich soluble fraction. Heat treatment of diluted protein solution (3% protein) to reach 85 °C and then adding $CaSO_4$ slurries (~ 80 °C) generates a protein curd upon cooling (Cai et al. 2002). Comparison of yield and texture properties of the final product with other legumes showed the performance of faba bean curd was similar to soy in terms of yield and the values for hardness, springiness and cohesiveness.

Other Applications

Products such as protein bars, protein-enriched bread, gluten-free coatings, pasta and bread containing faba bean are available for the consumer. Products such as hummus-type preparations, split, fried or roasted form, are also popular. Faba bean flours made from cotyledons only (38% protein) and from protein enriched fractions (67% protein), used as an extender in low-fat emulsion-type meat gels (e.g. low-fat pork bologna) at an inclusion level of less than 3% does not compromise the texture properties and sensory acceptability of the final product (Wei 2019).

Future Trends in Processing and Product Development

The importance of pulses is now well established in the agricultural system, and the transition to a more plant-based diet is also widely recognized as being environmentally more sustainable. As a result of its high nutritional value, diverse uses and ability to grow under a wide range of climatic and soil conditions, faba bean is considered as one of the best performing crops under global warming and climate change scenario for increasing both sustainability and world supply of plant protein and international trade (Warsame et al. 2018; Akkad et al. 2019). Faba bean is used as a staple food in many parts of the developing world (Africa, parts of Asia and Latin America); its current consumption and utilization in Western industrialized countries remain however limited. But the increasing utilization of pulse-based food products to satisfy consumers' concerns on health, environmental and ethical food production practices will pull more pulses and diverse pulse-based products in the food industry to satisfy the market push appropriately. This creates an excellent opportunity for faba bean, which is increasingly regarded as a promising alternative protein source to animals and soybean.

Through molecular breeding strategies (Gnanasambandam et al. 2012; Multari et al. 2015), tannin-free and vicine- and convicine-reduced cultivars are now available (Torres et al. 2010). Genetic improvement approach, however, is not an easy task as faba bean possesses one of the largest genomes among cultivated legume plants with only limited gene-based genetic maps being available (Ellwood et al. 2008; Alghamdi et al. 2012; Warsame et al. 2018). The identification of new markers in faba bean would therefore be an essential step to increase the efficiency of faba bean breeding programmes and enable the establishment of marker-assisted selection for economically important traits, such as the increase of protein content and quality or the reduction of antinutrients (Torres et al. 2010; Alghamdi et al. 2012). Identification of the genes related to the vicine and convicine, and understanding their biosynthesis and transport pathways will lead to generation of germ plasm completely devoid of or with lower levels of these compounds than the existing varieties (Khazaei et al. 2019; Singh et al. 2019). Besides, the application of processing technologies is regarded as an interesting avenue in achieving faba bean products with optimum nutritional value and consumer acceptability (Multari et al. 2015; Rizzello et al. 2016). Fermentation, for example, was successful in reducing the content of antinutrients in faba bean while at the same time improving its nutritional attributes (Coda et al. 2015, 2017; Chandra-Hioe et al. 2016; Rizzello et al. 2016). Continuing efforts in the fractionation of faba bean components (proteins, fibres and starches) to produce various functional food ingredients would be effective in increasing the utilization of pulses in various food products. Application to faba bean of advanced eco-innovative technologies for this purpose is also proposed as versatile, rapid, safe and effective to conventional processing for improving protein extractability and their techno-functional and nutritional properties. Examples are electrostatic separation, aqueous two-phase system extraction and enzyme-, microwave-, ultrasound- and high-pressure-assisted extraction, among others (Pojić et al. 2018). Moreover, these technologies are rated clean and ecologically friendly, thereby enabling clean label status (Tiwari 2015). Electrostatic separation, for example, is a new dry fractionation process which involves charging the particles and separating them in an electric field (Hemery et al. 2011). This process allowed the fractionation of legume flour to obtain protein- and carbohydrate-enriched fractions and resulted in a 15% increase of protein fraction enrichment in comparison to classical air classification process (Wang et al. 2016; Tabtabaei et al. 2017). An improvement of wet fractionation process was also achieved by solubilizing proteins in saline solutions, which were then recovered by salt removal using ultrafiltration and diafiltration. Such process maintained high protein functionality by preserving their native state (Hadnadjev et al. 2017). Novel pressurized solvents and ultrasound- and microwave-assisted extractions, using water as the solvent, are more efficient and economical than conventional technologies for the extraction of various phytochemicals including proteins (Karki et al. 2009; Pojić et al. 2018). Microwave- and ultrasound-assisted alkaline extraction yielded 77% and 136% more protein from peanut when compared to conventional alkaline extraction with a purity of 86% (Ochoa-Rivas et al. 2017). Enzyme-assisted protein extraction despite its complexity and high cost is also proposed as a milder extraction method with lower environmental impact in comparison to acid- and alkaline-assisted extraction (Jung et al. 2006; Sari et al. 2013; Pojić et al. 2018). It also results in a protein ingredient of greater quality (Liu et al. 2016; Ochoa-Rivas et al. 2017; Pojić et al. 2018).

In conclusion, faba bean, as a high protein pulse with favourable processing and agronomic characteristics, is one of the most versatile pulses with great potential to help respond to global needs for nutritious, sustainable, renewable and affordable proteins. To fully realize its potential, faba bean would have to meet both producers' expectations and consumers' demand for nutritious, healthy, safe, tasty and ready-to-eat food products. This implies continuing research and development efforts in genetics and varietal improvement and agronomy and management of cropping systems, as well as novel processing technologies to address the agronomic, nutritional and sensory constraints of faba bean products.

References

Abdel-Aal, E.-S. M., Ragaee, S., Rabalski, I., Warkentin, T., & Vandenberg, A. (2018). Nutrient content and viscosity of Saskatchewan-grown pulses in relation to their cooking quality. *Canadian Journal of Plant Science, 99*(1), 67–77. https://doi.org/10.1139/cjps-2018-0140.

Akkad, R., Kharraz, E., Han, J., House, J. D., & Curtis, J. M. (2019). Characterisation of the volatile flavour compounds in low and high tannin faba beans (Vicia faba var. minor) grown in Alberta, Canada. *Food Research International, 120*, 285–294. https://doi.org/10.1016/j.foodres.2019.02.044.

Alghamdi, S. S., Migdadi, H. M., Ammar, M. H., Paull, J. G., & Siddique, K. H. M. (2012). Faba bean genomics: Current status and future prospects. *Euphytica, 186*(3), 609–624. https://doi.org/10.1007/s10681-012-0658-4.

Alonso, R., Aguirre, A., & Marzo, F. (2000). Effects of extrusion and traditional processing methods on antinutrients and in vitro digestibility of protein and starch in faba and kidney beans. *Food Chemistry, 68*(2), 159–165. https://doi.org/10.1016/S0308-8146(99)00169-7.

Amiri Chayjan, R., & Shadidi, B. (2014). Modeling high-moisture Faba bean drying in fixed and semi-fluidized bed conditions. *Journal of Food Processing and Preservation, 38*(1), 200–211. https://doi.org/10.1111/j.1745-4549.2012.00766.x.

Anderson, J. C., Idowu, A. O., Singh, U., & Singh, B. (1994). Physicochemical characteristics of flours of faba bean as influenced by processing methods. *Plant Foods for Human Nutrition, 45*(4), 371–379. https://doi.org/10.1007/BF01088087.

Aschi, A., Aubert, M., Riah-Anglet, W., Nélieu, S., Dubois, C., Akpa-Vinceslas, M., & Trinsoutrot-Gattin, I. (2017). Introduction of Faba bean in crop rotation: Impacts on soil chemical and biological characteristics. *Applied Soil Ecology, 120*, 219–228. https://doi.org/10.1016/j.apsoil.2017.08.003.

Baginsky, C., Peña-Neira, T., Cáceres, A., Hernández, T., Estrella, I., Morales, H., & Pertuzé, R. (2013). Phenolic compound composition in immature seeds of fava bean (Vicia faba L.) varieties cultivated in Chile. *Journal of Food Composition and Analysis, 31*(1), 1–6. https://doi.org/10.1016/j.jfca.2013.02.003.

Cai, R., McCurdy, A., & Baik, B.-K. (2002). Textural property of 6 legume curds in relation to their protein constituents. *Journal of Food Science, 67*(5), 1725–1730. https://doi.org/10.1111/j.1365-2621.2002.tb08713.x.

Çalışkantürk Karataş, S., Günay, D., & Sayar, S. (2017). In vitro evaluation of whole faba bean and its seed coat as a potential source of functional food components. *Food Chemistry, 230*, 182–188. https://doi.org/10.1016/j.foodchem.2017.03.037.

Cappellini, M. D., & Fiorelli, G. (2008). Glucose-6-phosphate dehydrogenase deficiency. *The Lancet, 371*(9606), 64–74. https://doi.org/10.1016/S0140-6736(08)60073-2.

Cardador-Martínez, A., Maya-Ocaña, K., Ortiz-Moreno, A., Herrera-Cabrera, B. E., Dávila-Ortiz, G., Múzquiz, M., Martín-Pedrosa, M., Burbano, C., Cuadrado, C., & Jiménez-Martínez, C. (2012). Effect of roasting and boiling on the content of Vicine, Convicine and L-3,4-dihydroxyphenylalanine in Vicia faba L. *Journal of Food Quality, 35*(6), 419–428. https://doi.org/10.1111/jfq.12006.

Cepeda, E., Villarán, M. C., & Aranguiz, N. (1998). Functional properties of faba bean (Vicia faba) protein flour dried by spray drying and freeze drying. *Journal of Food Engineering, 36*(3), 303–310. https://doi.org/10.1016/S0260-8774(98)00061-2.

Chandra-Hioe, M. V., Wong, C. H. M., & Arcot, J. (2016). The potential use of fermented chickpea and faba bean flour as food ingredients. *Plant Foods for Human Nutrition, 71*(1), 90–95. https://doi.org/10.1007/s11130-016-0532-y.

Chang, P. R. Q., & McCurdy, A. R. (1985). Lipoxygenase activity in fourteen legumes. *Canadian Institute of Food Science and Technology Journal, 18*(1), 94–96. https://doi.org/10.1016/S0315-5463(85)71727-0.

Cheng, M.-H., Rosentrater, K. A., Sekhon, J., Wang, T., Jung, S., & Johnson, L. A. (2019). Economic feasibility of soybean oil production by enzyme-assisted aqueous extraction processing. *Food and Bioprocess Technology, 12*(3), 539–550. https://doi.org/10.1007/s11947-018-2228-9.

Coda, R., Melama, L., Rizzello, C. G., Curiel, J. A., Sibakov, J., Holopainen, U., Pulkkinen, M., & Sozer, N. (2015). Effect of air classification and fermentation by Lactobacillus plantarum VTT E-133328 on faba bean (Vicia faba L.) flour nutritional properties. *International Journal of Food Microbiology, 193*, 34–42. https://doi.org/10.1016/j.ijfoodmicro.2014.10.012.

Coda, R., Varis, J., Verni, M., Rizzello, C. G., & Katina, K. (2017). Improvement of the protein quality of wheat bread through faba bean sourdough addition. *LWT-Food Science and Technology, 82*, 296–302. https://doi.org/10.1016/j.lwt.2017.04.062.

Collado, E., Venzke Klug, T., Martínez-Hernández, G. B., Artés-Hernández, F., Martínez-Sánchez, A., Aguayo, E., Artés, F., Fernández, J. A., & Gómez, P. A. (2019). Nutritional and quality changes of minimally processed faba (Vicia faba L.) beans during storage: Effects of domestic microwaving. *Postharvest Biology and Technology, 151*, 10–18. https://doi.org/10.1016/j.postharvbio.2019.01.008.

Colonna, P., Gallant, D., & Mercier, C. (1980). Pisum sativum and Vicia faba carbohydrates: Studies of fractions obtained after dry and wet protein extraction processes. *Journal of Food Science, 45*(6), 1629–1639. https://doi.org/10.1111/j.1365-2621.1980.tb07578.x.

Crépon, K., Marget, P., Peyronnet, C., Carrouée, B., Arese, P., & Duc, G. (2010). Nutritional value of faba bean (Vicia faba L.) seeds for feed and food. *Field Crops Research, 115*(3), 329–339. https://doi.org/10.1016/j.fcr.2009.09.016.

Cubero, J. I. (1974). On the evolution of Vicia faba L. *Theoretical and Applied Genetics, 45*(2), 47–51. https://doi.org/10.1007/BF00283475.

Cubero, J. I. (1984). Taxonomy, distribution and evolution of the faba bean and its mild relatives. In J. R. Witcombe & W. Erskine (Eds.), *Genetic resources and their exploitation — Chickpeas, faba beans and lentils* (pp. 131–144). Dordrecht: Springer Netherlands. https://doi.org/10.1007/978-94-009-6131-9_12.

Dubova, L., Alsiņa, I., Ruža, A., & Šenberga, A. (2018). Impact of faba bean (Vicia faba L.) cultivation on soil microbiological activity. *Agronomy Research, 16*(5), 2016–2025. https://doi.org/10.15159/AR.18.195.

Duc, G. (1997). Faba bean (Vicia faba L.). *Field Crops Research, 53*(1), 99–109. https://doi.org/10.1016/S0378-4290(97)00025-7.

Duke, J. A. (1981). *Handbook of legumes of world economic importance*. New York: Plenum Press.

Ells, J. E., Hansen, J. C., & Bare, D. (1978). Faba beans (Vicia faba) -A promising new crop for the San Luis Valley. Colorado State University, Agric. Exp. Sta, Fort Collins (Colo.).

Ellwood, S. R., Phan, H. T. T., Jordan, M., Hane, J., Torres, A. M., Avila, C. M., Cruz-Izquierdo, S., & Oliver, R. P. (2008). Construction of a comparative genetic map in faba bean (Vicia faba L.); conservation of genome structure with Lens culinaris. *BMC Genomics, 9*, 380–380. https://doi.org/10.1186/1471-2164-9-380.

Erba, D., Angelino, D., Marti, A., Manini, F., Faoro, F., Morreale, F., Pellegrini, N., & Casiraghi, M. C. (2019). Effect of sprouting on nutritional quality of pulses. *International Journal of Food Sciences and Nutrition, 70*(1), 30–40. https://doi.org/10.1080/09637486.2018.1478393.

Estefania, C. M. C., Ivan, V. T. J., Maria, C. T. L., Rosa-Millan, J. D. L., & Othona, S. S. S. (2018). Physicochemical characteristics, ATR-FTIR molecular interactions and in vitro starch and protein digestion of thermally-treated whole pulse flours. *Food Research International, 105*, 371–383. https://doi.org/10.1016/j.foodres.2017.11.029.

Etemadi, F., Hashemi, M., Randhir, R., ZandVakili, O., & Ebadi, A. (2018). Accumulation of l-DOPA in various organs of faba bean and influence of drought, nitrogen stress, and processing methods on l-DOPA yield. *The Crop Journal, 6*(4), 426–434. https://doi.org/10.1016/j.cj.2017.12.001.

FAO. (2017). Food and agriculture data. Retrieved May 19, 2019, from http://www.fao.org/faostat/en/#home

FDA. (2018). Retrieved May 19, 2019, from https://www.accessdata.fda.gov/scripts/cdrh/cfdocs/cfCFR/CFRSearch.cfm?CFRPart=112&showFR=1&subpartNode=21:2.0.1.1.12.1

Felix, M., Lopez-Osorio, A., Romero, A., & Guerrero, A. (2018). Faba bean protein flour obtained by densification: A sustainable method to develop protein concentrates with food applications. *LWT, 93*, 563–569. https://doi.org/10.1016/j.lwt.2018.03.078.

Giménez, M. A., Drago, S. R., De Greef, D., Gonzalez, R. J., Lobo, M. O., & Samman, N. C. (2012). Rheological, functional and nutritional properties of wheat/broad bean (Vicia faba) flour blends for pasta formulation. *Food Chemistry, 134*(1), 200–206. https://doi.org/10.1016/j.foodchem.2012.02.093.

Giménez, M. A., González, R. J., Wagner, J., Torres, R., Lobo, M. O., & Samman, N. C. (2013). Effect of extrusion conditions on physicochemical and sensorial properties of corn-broad beans (Vicia faba) spaghetti type pasta. *Food Chemistry, 136*(2), 538–545. https://doi.org/10.1016/j.foodchem.2012.08.068.

Gnanasambandam, A., Paull, J., Torres, A., Kaur, S., Leonforte, T., Li, H., Zong, X., Yang, T., & Materne, M. (2012). Impact of molecular technologies on faba bean (Vicia faba L.) breeding strategies. *Agronomy, 2*(3), 132–166.

Göl, Ş., Doğanlar, S., & Frary, A. (2017). Relationship between geographical origin, seed size and genetic diversity in faba bean (Vicia faba L.) as revealed by SSR markers. *Molecular Genetics and Genomics, 292*(5), 991–999. https://doi.org/10.1007/s00438-017-1326-0.

Grosjean, F., Bourdillon, A., Rudeaux, F., Bastianelli, D., Peyronnet, C., Duc, G., & Lacassagne, L. (2000). Feeding value for poultry of isogenic fababeans (Vicia faba L) involving zero-tannin and zero-vicine genes. *Sciences et Techniques Avicoles*, (32), 17–23.

Gugger, E. T., Galuska, P., & Tremaine, A. (2016). Legume-based dairy substitute and consumable products. US Patent 2016/0309732 A1 27 Oct 2016.

Hadnadjev, M., Dapčević Hadnađev, T., Pojić, M., Šarić, B., Mišan, A., Jovanov, P., & Sakač, M. (2017). Progress in vegetable proteins isolation techniques: A review. *Food and Feed Research, 44*, 11–21. https://doi.org/10.5937/FFR1701011H.

Hagag, A. A., Badraia, I. M., Elfarargy, M. S., Elmageed, M. M. A., & Abo-Ali, E. A. (2018). Study of glucose-6-phosphate dehydrogenase deficiency: 5 years retrospective Egyptian study. *Endocrine, Metabolic and Immune Disorders - Drug Targets, 18*(2), 155–162. https://doi.org/10.2174/1871530317666171003160350.

Haladjian, N., Fayad, R., Toufeili, I., Shadarevian, S., Sidahmed, M., Baydoun, E., & Karwe, M. (2003). pH, temperature and hydratation kinetics of faba beans (Vicia Faba L.). *Journal of Food Processing & Preservation, 27*(1), 9–20. https://doi.org/10.1111/j.1745-4549.2003.tb00497.x.

Hefni, M. E., Shalaby, M. T., & Witthöft, C. M. (2015). Folate content in faba beans (Vicia faba L.)-effects of cultivar, maturity stage, industrial processing, and bioprocessing. *Food Science & Nutrition, 3*(1), 65–73. https://doi.org/10.1002/fsn3.192.

Hejdysz, M., Kaczmarek, S. A., & Rutkowski, A. (2016). Extrusion cooking improves the metabolisable energy of faba beans and the amino acid digestibility in broilers. *Animal Feed Science and Technology, 212*, 100–111. https://doi.org/10.1016/j.anifeedsci.2015.12.008.

Hemery, Y., Holopainen, U., Lampi, A.-M., Lehtinen, P., Nurmi, T., Piironen, V., Edelmann, M., & Rouau, X. (2011). Potential of dry fractionation of wheat bran for the development of food ingredients, part II: Electrostatic separation of particles. *Journal of Cereal Science, 53*(1), 9–18. https://doi.org/10.1016/j.jcs.2010.06.014.

Hernández-Infante, M., Sousa, V., Montalvo, I., & Tena, E. (1998). Impact of microwave heating on hemagglutinins, trypsin inhibitors and protein quality of selected legume seeds. *Plant Foods for Human Nutrition, 52*(3), 199–208. https://doi.org/10.1023/A:1008033610737.

Hood-Niefer, S. D., Warkentin, T. D., Chibbar, R. N., Vandenberg, A., & Tyler, R. T. (2012). Effect of genotype and environment on the concentrations of starch and protein in, and the physicochemical properties of starch from, field pea and faba bean. *Journal of the Science of Food and Agriculture, 92*(1), 141–150. https://doi.org/10.1002/jsfa.4552.

Jamalian, J. (1999). Removal of favism-inducing factors vicine and convicine and the associated effects on the protein content and digestibility of faba beans (Vicia faba L)†. *Journal of the Science of Food and Agriculture, 79*(13), 1909–1914. https://doi.org/10.1002/(SICI)1097-0010(199910)79:13<1909::AID-JSFA454>3.0.CO;2-H.

Jambunathan, R., Blain, H. L., Dhindsa, K. S., Hussein, L. A., Kogure, K., Li-Juan, L., & Youssef, M. M. (1994). Diversifying use of cool season food legumes through processing. In F. J. Muehlbauer & W. J. Kaiser (Eds.), *Expanding the production and use of cool season food legumes: A global perspective of persistent constraints and of opportunities and strategies for further increasing the productivity and use of pea, lentil, faba bean, chickpea and grasspea in different farming systems* (pp. 98–112). Dordrecht: Springer Netherlands. https://doi.org/10.1007/978-94-011-0798-3_4

Jiang, Z. Q., Pulkkinen, M., Wang, Y. J., Lampi, A. M., Stoddard, F. L., Salovaara, H., Piironen, V., & Sontag-Strohm, T. (2016). Faba bean flavour and technological property improvement by thermal pre-treatments. *LWT - Food Science & Technology, 68*, 295–305. https://doi.org/10.1016/j.lwt.2015.12.015.

Jung, S., Lamsal, B. P., Stepien, V., Johnson, L. A., & Murphy, P. A. (2006). Functionality of soy protein produced by enzyme-assisted extraction. *Journal of the American Oil Chemists' Society, 83*(1), 71–78. https://doi.org/10.1007/s11746-006-1178-y.

Karki, B., Lamsal, B. P., Grewell, D., Pometto, A. L., III, van Leeuwen, J., Khanal, S. K., & Jung, S. (2009). Functional properties of soy protein isolates produced from ultrasonicated defatted soy flakes. *Journal of the American Oil Chemists' Society, 86*(10), 1021–1028. https://doi.org/10.1007/s11746-009-1433-0.

Khalil, A. H., & Mansour, E. H. (1995). The effect of cooking, autoclaving and germination on the nutritional quality of faba beans. *Food Chemistry, 54*(2), 177–182. https://doi.org/10.1016/0308-8146(95)00024-D.

Khan, M. A., Ammar, M. H., Migdadi, H. M., El-Harty, E. H., Osman, M. A., Farooq, M., & Alghamdi, S. S. (2015). Comparative nutritional profiles of various faba bean and chickpea genotypes. *International Journal of Agriculture and Biology, 17*(3), 449–457. https://doi.org/10.17957/IJAB/17.3.14.990.

Khazaei, H., Purves, R. W., Hughes, J., Link, W., O'Sullivan, D. M., Schulman, A. H., Björnsdotter, E., Geu-Flores, F., Nadzieja, M., Andersen, S. U., Stougaard, J., Vandenberg, A., & Stoddard, F. L. (2019). Eliminating vicine and convicine, the main anti-nutritional factors restricting faba bean usage. *Trends in Food Science and Technology, 91*, 549–556. https://doi.org/10.1016/j.tifs.2019.07.051.

Kumar, S. P. J., Prasad, S. R., Banerjee, R., Agarwal, D. K., Kulkarni, K. S., & Ramesh, K. V. (2017). Green solvents and technologies for oil extraction from oilseeds. *Chemistry Central Journal, 11*, 9–9. https://doi.org/10.1186/s13065-017-0238-8.

Lafarga, T., Villaró, S., Bobo, G., Simó, J., & Aguiló-Aguayo, I. (2019). Bioaccessibility and antioxidant activity of phenolic compounds in cooked pulses. *International Journal of Food Science & Technology, 54*(5), 1816–1823. https://doi.org/10.1111/ijfs.14082.

Laleg, K., Salles, J., Berry, A., Giraudet, C., Patrac, V., Guillet, C., Denis, P., Tessier, F. J., Guilbaud, A., Howsam, M., Boirie, Y., Micard, V., & Walrand, S. (2019). Nutritional evaluation of mixed wheat-faba bean pasta in growing rats: Impact of protein source and drying temperature on protein digestibility and retention. *The British Journal of Nutrition, 121*(5), 496–507. https://doi.org/10.1017/S0007114518003586.

Lang, L. J., Yu, Z. H., Zheng, Z. J., Xu, M. S., Ying, H. Q. (1993). *Faba bean in China: State-of-the-art review*. International Center for Agricultural Research in the Dry Areas (ICARDA), Aleppo.

Li, L., Yuan, T. Z., Setia, R., Raja, R. B., Zhang, B., & Ai, Y. (2019). Characteristics of pea, lentil and faba bean starches isolated from air-classified flours in comparison with commercial starches. *Food Chemistry, 276*, 599–607. https://doi.org/10.1016/j.foodchem.2018.10.064.

Li, Z., Dong, L., Huang, Q., & Wang, X. (2016). Bacterial communities and volatile compounds in Doubanjiang, a Chinese traditional red pepper paste. *Journal of Applied Microbiology, 120*(6), 1585–1594. https://doi.org/10.1111/jam.13130.

Liu, J.-J., Gasmalla, M. A. A., Li, P., & Yang, R. (2016). Enzyme-assisted extraction processing from oilseeds: Principle, processing and application. *Innovative Food Science and Emerging Technologies, 35*, 184–193. https://doi.org/10.1016/j.ifset.2016.05.002.

Lizarazo, C. I., Lampi, A., Jingwei, L., Sontag-Strohm, T., Piironen, V., & Stoddard, F. L. (2015). Nutritive quality and protein production from grain legumes in a boreal climate. *Journal of the Science of Food and Agriculture, 95*(10), 2053–2064. https://doi.org/10.1002/jsfa.6920.

Lopez, H. W., Leenhardt, F., Coudray, C., & Remesy, C. (2002). Minerals and phytic acid interactions: Is it a real problem for human nutrition? *International Journal of Food Science & Technology, 37*(7), 727–739. https://doi.org/10.1046/j.1365-2621.2002.00618.x.

Luo, Y., & Xie, W. (2014). Effect of soaking and sprouting on iron and zinc availability in green and white faba bean (Vicia faba L.). *Journal of Food Science and Technology, 51*(12), 3970–3976. https://doi.org/10.1007/s13197-012-0921-7.

Martin, L., Skinner, C., & Marriott, R. J. (2015). Supercritical extraction of oil seed rape: Energetic evaluation of process scale. *The Journal of Supercritical Fluids, 105*, 55–59. https://doi.org/10.1016/j.supflu.2015.04.017.

Martínez-Velasco, A., Lobato-Calleros, C., Hernández-Rodríguez, B. E., Román-Guerrero, A., Alvarez-Ramirez, J., & Vernon-Carter, E. J. (2018). High intensity ultrasound treatment of faba bean (Vicia faba L.) protein: Effect on surface properties, foaming ability and structural changes. *Ultrasonics Sonochemistry, 44*, 97–105. https://doi.org/10.1016/j.ultsonch.2018.02.007.

Mattila, P., Mäkinen, S., Eurola, M., Jalava, T., Pihlava, J. M., Hellström, J., & Pihlanto, A. (2018). Nutritional value of commercial protein-rich plant products. *Plant Foods for Human Nutrition, 73*(2), 108–115. https://doi.org/10.1007/s11130-018-0660-7.

McCurdy, S. M., & Knipfel, J. E. (1990). Investigation of faba bean protein recovery and application to pilot scale processing. *Journal of Food Science, 55*(4), 1093–1094. https://doi.org/10.1111/j.1365-2621.1990.tb01606.x.

McDonald, B. E. (1974). In: First National Faba bean Conference, Winnipeg, Manitoba (Canada), 21–22 February 1974.

McMillan, D. C., Bolchoz, L. J. C., & Jollow, D. J. (2001). Favism: Effect of divicine on rat erythrocyte sulfhydryl status, hexose monophosphate shunt activity, morphology, and membrane skeletal proteins. *Toxicological Sciences, 62*(2), 353–359. https://doi.org/10.1093/toxsci/62.2.353.

Mehrabi Kooshki, M., Moradi, A., Balouchi, H., & Amiri Fahliani, R. (2018). Evaluation of the germination performance and biochemical indices of faba bean (Vicia faba L.) seeds stored at different temperatures and moisture contents. *Journal of Plant Process and Function, 7*(25), 17–28.

Meijer, M. M. T., Ogink, J. J. M., & van Gelder, W. M. J. (1994). Technological-scale dehulling process to improve the nutritional value of faba beans. *Animal Feed Science and Technology, 46*(1), 1–10. https://doi.org/10.1016/0377-8401(94)90060-4.

Millar, K. A., Barry-Ryan, C., Burke, R., Hussey, K., McCarthy, S., & Gallagher, E. (2017). Effect of pulse flours on the physiochemical characteristics and sensory acceptance of baked crackers. *International Journal of Food Science and Technology, 52*(5), 1155–1163. https://doi.org/10.1111/ijfs.13388.

Morad, M. M., Leung, H. K., Hsu, D. L., & Finney, P. L. (1980). Effect of germination on physicochemical and bread-baking properties of yellow pea, lentil, and faba bean flours and starches. *Cereal Chemistry, 6*, 390–396.

Multari, S., Stewart, D., & Russell, W. R. (2015). Potential of fava bean as future protein supply to partially replace meat intake in the human diet. *Comprehensive Reviews in Food Science and Food Safety, 14*(5), 511–522. https://doi.org/10.1111/1541-4337.12146.

Nasar-Abbas, S. M., Plummer, J. A., Siddique, K. H. M., White, P., Harris, D., & Dods, K. (2008). Cooking quality of faba bean after storage at high temperature and the role of lignins and other phenolics in bean hardening. *LWT - Food Science & Technology, 41*(7), 1260–1267. https://doi.org/10.1016/j.lwt.2007.07.017.

Nasar-Abbas, S. M., Siddique, K. H. M., Plummer, J. A., White, P. F., Harris, D., Dods, K., & D'Antuono, M. (2009). Faba bean (Vicia faba L.) seeds darken rapidly and phenolic content falls when stored at higher temperature, moisture and light intensity. *LWT - Food Science & Technology, 42*(10), 1703–1711. https://doi.org/10.1016/j.lwt.2009.05.013.

Nkhata, S. G., Ayua, E., Kamau, E. H., & Shingiro, J.-B. (2018). Fermentation and germination improve nutritional value of cereals and legumes through activation of endogenous enzymes. *Food Science & Nutrition, 6*(8), 2446–2458. https://doi.org/10.1002/fsn3.846.

Nkhoma, E. T., Poole, C., Vannappagari, V., Hall, S. A., & Beutler, E. (2009). The global prevalence of glucose-6-phosphate dehydrogenase deficiency: A systematic review and meta-analysis. *Blood Cells, Molecules & Diseases, 42*(3), 267–278. https://doi.org/10.1016/j.bcmd.2008.12.005.

Ochoa-Rivas, A., Nava-Valdez, Y., Serna-Saldívar, S. O., & Chuck-Hernández, C. (2017). Microwave and ultrasound to enhance protein extraction from Peanut flour under alkaline conditions: Effects in yield and functional properties of protein isolates. *Food and Bioprocess Technology, 10*(3), 543–555. https://doi.org/10.1007/s11947-016-1838-3.

Olaboro, G., Marquardt, R. R., & Campbell, L. D. (1981). Isolation of the egg weight depressing factor in faba beans (Vicia faba L. var. minor). *Journal of the Science of Food and Agriculture, 32*(11), 1074–1080. https://doi.org/10.1002/jsfa.2740321106.

Ortiz, L. T., Centeno, C., & Treviño, J. (1993). Tannins in faba bean seeds: Effects on the digestion of protein and amino acids in growing chicks. *Animal Feed Science and Technology, 41*(4), 271–278. https://doi.org/10.1016/0377-8401(93)90002-2.

Petitot, M., Barron, C., Morel, M. H., & Micard, V. (2010a). Impact of legume flour addition on pasta structure: Consequences on its in vitro starch digestibility. *Food Biophysics, 5*(4), 284–299. https://doi.org/10.1007/s11483-010-9170-3.

Petitot, M., Boyer, L., Minier, C., & Micard, V. (2010b). Fortification of pasta with split pea and faba bean flours: Pasta processing and quality evaluation. *Food Research International, 43*(2), 634–641. https://doi.org/10.1016/j.foodres.2009.07.020.

Pojić, M., Mišan, A., & Tiwari, B. (2018). Eco-innovative technologies for extraction of proteins for human consumption from renewable protein sources of plant origin. *Trends in Food Science and Technology, 75*, 93–104. https://doi.org/10.1016/j.tifs.2018.03.010.

Pulkkinen, M., Gautam, M., Lampi, A.-M., Ollilainen, V., Stoddard, F., Sontag-Strohm, T., Salovaara, H., & Piironen, V. (2015). Determination of vicine and convicine from faba bean with an optimized high-performance liquid chromatographic method. *Food Research International, 76*, 168–177. https://doi.org/10.1016/j.foodres.2015.05.031.

Rackis, J. J., Sessa, D. J., & Honig, D. H. (1979). Flavor problems of vegetable food proteins. *Journal of the American Oil Chemists' Society, 56*(3), 262–271. https://doi.org/10.1007/BF02671470.

Rahman, M., Rehman, A., Chuanqi, X., Xing Long, Z., Binghai, C., Linbao, J., & Su, H. (2015). Extrusion of feed/feed ingredients and its effect on digestibility and performance of poultry: A review. *International Journal of Current Microbiology and Applied Sciences, 4*(4), 48–61.

Raikos, V., Neacsu, M., Russell, W., & Duthie, G. (2014). Comparative study of the functional properties of lupin, green pea, fava bean, hemp, and buckwheat flours as affected by pH. *Food Science and Nutrition, 2*(6), 802–810. https://doi.org/10.1002/fsn3.143.

Revilla, I., & Vivar-Quintana, A. M. (2008). Effect of canning process on texture of faba beans (Vicia faba). *Food Chemistry, 106*(1), 310–314. https://doi.org/10.1016/j.foodchem.2007.02.046.

Rizzello, C. G., Losito, I., Facchini, L., Katina, K., Palmisano, F., Gobbetti, M., & Coda, R. (2016). Degradation of vicine, convicine and their aglycones during fermentation of faba bean flour. *Scientific Reports, 6*, 32452–32452. https://doi.org/10.1038/srep32452.

Rosa-Sibakov, N., Heiniö, R.-L., Cassan, D., Holopainen-Mantila, U., Micard, V., Lantto, R., & Sozer, N. (2016). Effect of bioprocessing and fractionation on the structural, textural and sensory properties of gluten-free faba bean pasta. *LWT - Food Science & Technology, 67*, 27–36. https://doi.org/10.1016/j.lwt.2015.11.032.

Rubio, L. A., Grant, G., Dewey, P., Bremner, I., & Pusztai, A. (1994). The intestinal true absorption of 65Zn in rats is adversely affected by diets containing a faba bean (Vicia faba L.) nonstarch polysaccharide fraction. *The Journal of Nutrition, 124*(11), 2204–2211. https://doi.org/10.1093/jn/124.11.2204.

Santos, C. S., Carbas, B., Castanho, A., Bronze, M. R., Serrano, C., Vasconcelos, M. W., Patto, M. C. V., & Brites, C. (2018). Relationship between seed traits and pasting and cooking behaviour in a pulse germplasm collection. *Crop and Pasture Science, 69*(9), 892–903. https://doi.org/10.1071/CP18205.

Sari, Y. W., Bruins, M. E., & Sanders, J. P. M. (2013). Enzyme assisted protein extraction from rapeseed, soybean, and microalgae meals. *Industrial Crops and Products, 43*, 78–83. https://doi.org/10.1016/j.indcrop.2012.07.014.

Schutyser, M. A. I., Pelgrom, P. J. M., van der Goot, A. J., & Boom, R. M. (2015). Dry fractionation for sustainable production of functional legume protein concentrates. *Trends in Food Science and Technology, 45*(2), 327–335. https://doi.org/10.1016/j.tifs.2015.04.013.

Schuurman, M., van Waardenburg, D., Costa, J. D., Niemarkt, H., & Leroy, P. (2009). Severe hemolysis and methemoglobinemia following fava beans ingestion in glucose-6-phosphatase dehydrogenase deficiency—Case report and literature review. *European Journal of Pediatrics, 168*(7), 779–782. https://doi.org/10.1007/s00431-009-0952-x.

Senberga, A., Dubova, L., Alsina, I., & Strauta, L. (2017). Rhizobium sp. - a potential tool for improving protein content in peas and faba beans. *Rural Sustainability Research, 37*(332), 2–9. https://doi.org/10.1515/plua-2017-0001.

Setia, R., Dai, Z., Nickerson, M. T., Sopiwnyk, E., Malcolmson, L., & Ai, Y. (2019). Impacts of short-term germination on the chemical compositions, technological characteristics and

nutritional quality of yellow pea and faba bean flours. *Food Research International, 122*, 263. https://doi.org/10.1016/j.foodres.2019.04.021.
Sharma, A., & Sehgal, S. (1992). Effect of domestic processing, cooking and germination on the trypsin inhibitor activity and tannin content of faba bean (Vicia faba). *Plant Foods for Human Nutrition, 42*(2), 127–133. https://doi.org/10.1007/BF02196465.
Shi, L., Arntfield, S. D., & Nickerson, M. (2018). Changes in levels of phytic acid, lectins and oxalates during soaking and cooking of Canadian pulses. *Food Research International, 107*, 660–668. https://doi.org/10.1016/j.foodres.2018.02.056.
Siah, S., Konczak, I., Wood, J. A., Agboola, S., & Blanchard, C. L. (2014). Effects of roasting on phenolic composition and in vitro antioxidant capacity of Australian grown faba beans (Vicia faba L.). *Plant Foods for Human Nutrition, 69*(1), 85–91. https://doi.org/10.1007/s11130-013-0400-y.
Singh, R. K., Bohra, N., & Sharma, L. (2019). Valorizing faba bean for animal feed supplements via biotechnological approach: Opinion. *Biocatalysis and Agricultural Biotechnology, 17*, 366–368. https://doi.org/10.1016/j.bcab.2018.12.020.
Singhal, A., Stone, A. K., Vandenberg, A., Tyler, R., & Nickerson, M. T. (2016). Effect of genotype on the physicochemical and functional attributes of faba bean (Vicia faba L.) protein isolates. *Food Science and Biotechnology, 25*(6), 1513–1522. https://doi.org/10.1007/s10068-016-0235-z.
Sözer, N. (2014). *Gluten-free faba bean for bread and pasta*. Retrieved May 19, 2019, from https://www.vttresearch.com/media/news/gluten-free-faba-bean-for-bread-and-pasta
Tabtabaei, S., Vitelli, M., Rajabzadeh, A. R., & Legge, R. L. (2017). Analysis of protein enrichment during single- and multi-stage tribo-electrostatic bioseparation processes for dry fractionation of legume flour. *Separation and Purification Technology, 176*, 48–58. https://doi.org/10.1016/j.seppur.2016.11.050.
Tazrart, K., Lamacchia, C., Zaidi, F., & Haros, M. (2016). Nutrient composition and in vitro digestibility of fresh pasta enriched with Vicia faba. *Journal of Food Composition and Analysis, 47*, 8–15. https://doi.org/10.1016/j.jfca.2015.12.007.
Tiwari, A., & Jha, S. K. (2017). Extrusion cooking technology: Principal mechanism and effect on direct expanded snacks - An overview. *International Journal of Food Studies, 6*, 113–128. https://doi.org/10.7455/ijfs/6.1.2017.a10.
Tiwari, B. K. (2015). Ultrasound: A clean, green extraction technology. *TrAC, Trends in Analytical Chemistry, 71*, 100–109. https://doi.org/10.1016/j.trac.2015.04.013.
Torres, A. M., Avila, C. M., Gutierrez, N., Palomino, C., Moreno, M. T., & Cubero, J. I. (2010). Marker-assisted selection in faba bean (Vicia faba L.). *Field Crops Research, 115*(3), 243–252. https://doi.org/10.1016/j.fcr.2008.12.002.
Turco, I., Ferretti, G., & Bacchetti, T. (2016). Review of the health benefits of faba bean (Vicia faba L.) polyphenols. *Journal of Food and Nutrition Research, 55*(4), 283–293.
Vilariño, M., Métayer, J. P., Crépon, K., & Duc, G. (2009). Effects of varying vicine, convicine and tannin contents of faba bean seeds (Vicia faba L.) on nutritional values for broiler chicken. *Animal Feed Science and Technology, 150*(1), 114–121. https://doi.org/10.1016/j.anifeedsci.2008.08.001.
Vioque, J., Alaiz, M., & Girón-Calle, J. (2012). Nutritional and functional properties of Vicia faba protein isolates and related fractions. *Food Chemistry, 132*(1), 67–72. https://doi.org/10.1016/j.foodchem.2011.10.033.
Waller, D. G., & Sampson, A. P. (2018). 24—Extrapyramidal movement disorders and spasticity. In D. G. Waller & A. P. Sampson (Eds.), *Medical pharmacology and therapeutics* (5th ed., pp. 325–336). Elsevier. https://doi.org/10.1016/B978-0-7020-7167-6.00024-5.
Wang, J., Zhao, J., de Wit, M., Boom, R. M., & Schutyser, M. A. I. (2016). Lupine protein enrichment by milling and electrostatic separation. *Innovative Food Science and Emerging Technologies, 33*, 596–602. https://doi.org/10.1016/j.ifset.2015.12.020.
Wang, Y., Sorvali, P., Laitila, A., Maina, N. H., Coda, R., & Katina, K. (2018). Dextran produced in situ as a tool to improve the quality of wheat-faba bean composite bread. *Food Hydrocolloids, 84*, 396–405. https://doi.org/10.1016/j.foodhyd.2018.05.042.

Warsame, A. O., O'Sullivan, D. M., & Tosi, P. (2018). Seed storage proteins of faba bean (Vicia faba L): Current status and prospects for genetic improvement. *Journal of Agricultural and Food Chemistry, 66*(48), 12617–12626. https://doi.org/10.1021/acs.jafc.8b04992.

Wei, X. (2019). *Effects of short-term germination and autoclaving on selected compounds in faba bean and faba bean application in low-fat pork bologna* (M.Sc. Thesis, University of Saskatchewan).

Weihua, X., Miao, Z., Jing, L., Chuanxiu, X., & Yuwei, L. (2015). Effects of phytase and tannase on in vivo nutritive utilisation of faba bean (Vicia faba L.) flour. *International Food Research Journal, 22*(4), 1550–1556.

Xie, W.-H., Jin, X.-X., Wang, Q., & He, Y.-J. (2014). Effects of germination on iron, zinc, calcium, manganese, and copper availability from cereals and legumes. *CyTA - Journal of Food, 12*(1), 22–26. https://doi.org/10.1080/19476337.2013.782071.

Xu, Y., Coda, R., Shi, Q., Tuomainen, P., Katina, K., & Tenkanen, M. (2017). Exopolysaccharides production during the fermentation of soybean and fava bean flours by Leuconostoc mesenteroides DSM 20343. *Journal of Agricultural and Food Chemistry, 65*(13), 2805–2815. https://doi.org/10.1021/acs.jafc.6b05495.

Yang, J., Liu, G., Zeng, H., & Chen, L. (2018). Effects of high pressure homogenization on faba bean protein aggregation in relation to solubility and interfacial properties. *Food Hydrocolloids, 83*, 275–286. https://doi.org/10.1016/j.foodhyd.2018.05.020.

Youssef, M. M., & Bushuk, W. (1986). Breadmaking properties of composite flours of wheat and faba bean protein preparations. *Cereal Chem, 63*(4), 357–361.

Yu-Wei, L., & Wei-Hua, X. (2013). Effect of different processing methods on certain antinutritional factors and protein digestibility in green and white faba bean (Vicia faba L.). *CyTA - Journal of Food, 11*(1), 43–49.

Chickpea

Sajad Ahmad Sofi, Khalid Muzaffar, Shafia Ashraf, Isha Gupta, and Shabir Ahmad Mir

Introduction

Chickpea (*Cicer arietinum* L.), a legume crop belonging to the Fabaceae family, is consumed all over the world, especially in Afro-Asian countries. Chickpea is the third pulse crop in production in the world after dry beans and field beans. There is an increasing demand for chickpea because it has a high nutritional value and is a cheap source of protein with potentially health-beneficial phytochemicals (Wood and Grusak 2007). Chickpea seeds are used in the form of mature dry seeds as a popular snack food, whether whole or split (dhal), or as a flour that is used as a supplement in weaning food mixes, breads, and biscuits (Alajaji and El Adawy 2006).

Overview of Production

Chickpea (*Cicer arietinum* L.), also called garbanzo bean or Bengal gram, is an Old World pulse grown in more than 50 countries across the Indian subcontinent, North Africa, the Middle East, southern Europe, the Americas, and Australia. Among pulse crops, chickpea has consistently maintained a much more significant status, ranking second in area (15.3% of total) and third in production (15.42%) in the

S. A. Sofi · S. Ashraf · I. Gupta
Division of Food Science and Technology, Sher-e-Kashmir University of Agricultural Science and Technology, Jammu, India

K. Muzaffar
Department of Food Science and Technology, Government Degree College Boys, Baramulla, India

S. A. Mir (✉)
Department of Food Science and Technology, Government College for Women, M. A. Road, Srinagar, India

A. Manickavasagan, P. Thirunathan (eds.), *Pulses*,
https://doi.org/10.1007/978-3-030-41376-7_4

world. During 2018, world chickpea production increased to 14.7 million metric tonnes, with Asian countries contributing 85.5% of total production (FAOSTAT 2018). The major chickpea-producing countries are India with 9.07 million metric tonnes, followed by Turkey (0.47 million metric tonnes), Ethopia (0.47 million metric tonnes), Pakistan (0.33 million metric tonnes), USA (0.33 million metric tonnes), Iran (0.27 million metric tonnes), Mexico (0.18 million metric tonnes), and Canada (0.09 million metric tonnes).

Chemical and Nutritional Qualities

Chickpea is an economical source of vegetable protein, essential amino acids, and other nutrients. The major portion of the chickpea seed is carbohydrate, which constitutes about 60% of raw dry chickpea seeds. Protein content in raw chickpea and chickpea flour is about 19.3% and 22.4%, respectively, and is regarded as the second major component of chickpea. The protein content of chickpea seeds varies depending upon genetic and environmental factors (de Falco et al. 2010). Albumins, globulins, prolamines, glutelins, and residual protein constitute the various types of protein present in the chickpea. The limiting amino acids of chickpea seeds include sulfur amino acids, followed by valine, threonine, and tryptophan (Iqbal et al. 2006). The lipid content of chickpea seeds and seed flour is about 6.04 and 6.69 g/100 g, respectively. The major constituents of neutral lipids include triglycerides. Lecithin, a phospholipid, is the major constituent of polar lipids. Unsaturated fatty acids (oleic, linoleic, linolenic acids) and saturated fatty acids (palmitic acid, stearic acid) are the important fatty acids of chickpea lipids (Chavan et al. 1989). A good amount of dietary fiber (17.4%) is present in chickpeas that can be used for fat binding and retention, as well as in modifying the texture of processed food products (Tosh and Yada 2010). High dietary fiber content, low lipid content, and a very low glycemic index (28%) make chickpea a healthy food (Nasir and Sidhu 2013). Chickpea seeds are also a good source of minerals, such as Ca, P, Mg, Fe, and K (Bampidisa and Christodouloub 2011). Moreover, chickpea shows the presence of various bioactive compounds with potential health benefits, including biochanin, genistein, trifolirhizin, calycosin, formononetin, sissotrin, and ononin in the seeds (Zhao et al. 2009). These components include antifungal, antioxidant, estrogenic, insecticidal, antimicrobial, and contraceptive properties, which confer potential health benefits (Januario et al. 2005). Chickpea has hypocholesteremic effects and antioxidant properties and also regulates bowel movement (Murty et al. 2010).

Exports and Import

According to FAOSTAT (2019) statistics, about 2.4 million tonnes of chickpea have been entered in the global pulse trade to meet the demands of chickpea consumption. India is the largest producer of chickpea, and the second exporter of chickpea after Australia. India exports 0.27 million tonnes of chickpea to various countries such as Algeria, UAE, Sri Lanka, Turkey, and the USA (APEDA 2019). As there are differences in supply and demand in chickpea production, India also depends upon such outside sources as Australia, Russia, Tanzania, Canada, and USA. Australia is the main exporter of chickpea (0.92 million tonnes), meeting half the global export demand. Australia exports chickpea to major destination countries such as India, Bangladesh, Pakistan, UAE, and the United Kingdom. Mexico, Australia, and Turkey contribute 75% of total chickpea export in the world. Mexico is large producer of high-quality Kabuli chickpea seed to most countries with Algeria, Turkey, and Spain their most important customers. Turkey is also a major exporter with 21,615.2 tonnes annually to neighboring countries such as Iraq, Jordan, and Saudi Arabia. Chickpea from the major producing countries India, Turkey, Ethopia, Pakistan, USA, Iran, Mexico, and Canada is imported by major countries with a greater demand for chickpea consumption than production such as Algeria, Bangladesh, Iraq, Spain, Syria, Saudi Arabia, Oman, and Greece.

Postharvest Processing

Harvesting and Storage

Chickpea matures in 3–7 months and, on achieving maturity, the leaves of the plant turn brown/yellow in color. After attaining maturity, plants are left in the field for a few days, and the crop later is threshed by beating with wooden sticks. The main risks to seed quality from mechanical harvesting are (1) the mechanical damage to the seeds and (2) possible physical admixture with seeds of other varieties. Legume crops should be harvested at full maturity (approximate moisture content is 12%) (Van Gastel et al. 2006). Ellis et al. (1988) reported that delaying chickpea seed harvest by 1 month reduced germination from 99% to 80%, and a further delay of 1 month reduced germination to 30%.

The seeds are stored at farmer, trade, or government level in various types of storage structures until the harvest is available in the next season. Most of the postharvest losses in quality and quantity of the grains occur during storage of the seeds (Haysmans 1970). The important factors affecting the storage of chickpea seeds are storage temperature, moisture content, and infestation by insects, rodents, and molds. Safe moisture content for seed storage ranges from 8% to 11%; however, increase in storage temperature or moisture content has a negative effect on the quality of legume seeds (Kosolofski et al. 1998). Adverse storage conditions reduce the cooking, milling, and nutritional quality of chickpea grains. Storage of seeds at higher temperature and relative humidity causes decreases seed quality because of the Maillard reaction and development of the hard-to-cook phenomenon. The increased cooking time required for stored chickpea seeds results in higher energy costs and decreased protein quality. The hard-to-cook phenomenon of the seeds is related to increased bound protein in the seed coat and aleurone layer and changes in pectins and calcium ions, which results in decreased water uptake capacity of the seed cotyledons. If seeds with high moisture content are stored for a long time at elevated temperature, hardening of the seed coat results, which finally causes quantitative and qualitative losses during milling (Chavan et al. 1987). Chickpea dhal and flour with 10.8% moisture do not undergo detectable changes in flavor in storage, but at 14.0% moisture content chickpeas develop a moldy odor within 8 weeks. These changes occur because of the oxidative and hydrolytic degradation of both free and bound lipids and changes in carotenoid content (Arya 1981). A significant amount of vitamins including thiamine and riboflavin also lost during storage of chickpea seeds at poor storage conditions (Chitre et al. 1955). Chickpea seeds stored in bags are susceptible to greater losses of nutrients than those stored in bottles, possibly because of free air circulation in the bags (Shehnaz and Theophilus 1975). Insects and molds also contribute to the postharvest losses in chickpeas during conventional storage conditions. Pulse beetles or bruchids are the most common insects to attack chickpea seeds. Genotypes of chickpea differ in susceptibility to attack by pulse beetles. Kabuli cultivars are more susceptible to bruchids than desi cultivars (Wadnerkar et al. 1978). Insect infestation significantly decreases the protein efficiency ratio (PER) of chickpea dhal. Deterioration in chickpea protein quality is related to changes in the levels of certain amino acids. The amount of available lysine tryptophan and methionine decreases significantly with infestation (Shehnaz and Theophilus 1975). Protein-rich pulses are also contaminated with aflatoxin when exposed to unfavorable environmental storage conditions. Therefore, proper packaging and storage are necessary to prevent postharvest chickpea losses. Preventive measures include storage at optimal temperature, proper aeration, optimal moisture content, use of antifungal sprays, and suitable packaging.

Control of Postharvest Losses

Regular sun drying over several months eliminates bruchids from chickpea grains. Heating chickpea to 58 °C (the thermal death point of *Callosobruchus chinensis*) for 10 min provides freedom from viable infestation. Use of custard apple powder and Sadabahar leaves also protects legume seeds from bruchid infestation. Treating chickpea seeds with vegetable oils such as lemon oil, garlic oil, peanut oil, safflower oil, and neem oil are also effective in controlling insect infestations. Maintaining a storage temperature sufficiently below the temperature required for insect development will also prevent infestation of legume seeds including chickpea. The use of ionizing radiation to control the infestation of stored legumes has been attempted. Fumigation of chickpea seeds using carbon disulfide, carbon tetrachloride, ethylene dichloride, ethylene dibromide, and phosphine can also prevent the growth of insects. Improved storage structures at farm, mill, trader, or government level are important to prevent chickpea infestation (Chavan et al. 1989).

Physical and Chemical Processing

Chickpea seeds are rarely eaten in a raw state. Instead, they are traditionally processed before consumption by various physical, biochemical, or heating methods. Physical processing methods include dehulling, milling, and soaking; sprouting and fermentation are biochemical methods, and boiling/pressure-cooking is an example of heating method (Woodi and Grusak 2007). Traditional processing practices not only remove toxic substances and antinutritional factors but also improve the palatability and digestibility of grain legumes (Singh 1987). Processing generally improves the nutrient profile of legume seed by increasing in vitro digestibility of proteins and carbohydrates from approximately 40% to 98% (El-Adawy 2002; Naveeda and Jamuna 2004). Most of the antinutritional factors such as alpha-galactosides, protease inhibitors, and lectins are heat labile, so cooking can eliminate any potential ill effects before consumption. On the other hand, tannins, saponins, and phytic acids are heat stable but can be reduced by dehulling, soaking, germination, and fermentation (Katyal et al. 2005; Ravi et al. 2011).

Dehulling is the primary technique for converting whole chickpea seeds into *dhal*, or dehulled seed, which is then cooked or milled to flour. Removal of the seed coat during dehulling results in the reduction of antinutritional factors such as tannins that are present in the seed coat (Rao and Deosthale 1982). Cooking time of chickpeas is significantly reduced by dehulling. Compared to whole chickpea seeds, dehulled seeds contain higher amounts of carbohydrate, protein, and vitamins with a lower concentration of dietary fiber and thus higher overall nutritional value. Dehulling of desi chickpea increases its protein content by 116–118% (17.7–25.9% protein for whole seed and 20.6–30.5% for dehulled seed), whereas crude fiber content decreases significantly, from 7.1–10.8% for whole seed to 0.7–1.3% for dehulled seed (Jambunathan and Singh 1981). Similar results were found for kabuli chickpea, but the changes were less because the seed coat content is smaller. However, dehulling of chickpea seeds does not eliminate the effects of oligosaccharides, resistant starch, or protease inhibitors, because these antinutritional factors are present in the seed cotyledons. Various pretreatments used to facilitate the removal of the seed coat include wetting, equilibrating, and drying. The wet dehulling method involves soaking of seeds in water for 1–4 h, and the hull is then removed by the floatation technique. However, if the seed hull is very resistant to soaking, then seeds are subdivided before the actual soaking process (Muehlbauer and Singh 1987).

Milling of dehulled chickpea seeds is carried out for flour preparation. In Pakistan and India, milled chickpea flour is known as gram flour or besan. For preparation of fine flour, the milling operation is performed twice for some pulses (Tabil et al. 1995). The milled flour is of various types such as very fine, medium fine, and coarse and is used for different purposes depending upon the particle size of the flour. The fine and light particles contain more protein, whereas the coarse and heavier particles mostly contain starch granules. Dry milling and air classification of chickpeas is a typical method for fractionation of proteins and starch components. Chickpeas subjected to air classification during milling are used to produce chickpea protein concentrates (Owusu-Ansah and McCurdy 1991). During the milling of dehulled chickpea seeds the starch granules are damaged so that they become more susceptible to enzymic degradation on consumption. In addition, mineral content such as calcium iron and zinc significantly decreases in flour products. The milled chickpea flour is used in for many preparations such as snacks (pakoras), sweets (ladoos and burfis), and driads (bhoojya) (Nasir and Sidhu 2013).

Soaking is a common technique practiced before, or in conjunction with, other processing methods such as germination, fermenting, cooking, and canning. Imbibition of water that occurs during cooking causes the cells to swell as hydration proceeds, thereby reducing the cooking time. Soaking reduces the content of certain antinutritional factors such as oligosaccharides, protease inhibitors, and some tannin, which leach into the soaking medium (Saxena et al. 2003). Rao and Deosthale (1982) stated that overnight soaking of chickpea seeds causes 53% reduction in the tannin content. During soaking, leaching of soluble molecules, such as monosaccharides and disaccharides, oligosaccharides, soluble polyphenolic compounds, and phytic acid also occurs. The amount of leaching depends on the nature of the soaking medium (water, salt solution, or bicarbonate solution) and soaking time (Elham and Boye 2013).

Germination of seeds is the most appropriate technique whereby hydrolytic enzymes are activated, thereby changing the physical and functional properties of seed components (Czukor et al. 2001). El-Adawy (2002) found that germination treatment of chickpea seeds considerably reduces the phytic acid, stachyose, and raffinose content of the seeds. Jood et al. (1985) also stated that germination for 24 h markedly reduced the raffinose, stachyose, and verbascose content in chickpea seeds, and these components were totally eliminated after a germination period of 48 h. Veena et al. (1995) and Mahadevamma and Tharanathan (2004) found that germination of chickpea seeds reduces the resistant starch content from 10.3% to 6.8% but increases the dietary fiber. Reduction in these components would significantly increase carbohydrate digestibility and reduce intestinal discomfort.

Fermentation, as other processing methods, is carried out to reduce the content of antinutritional factors and increase the overall nutritional value of chickpea seeds. Reyes-Moreno et al. (2004) stated that chickpea tempeh flour from fermented chickpea seeds has higher protein and lysine content and in vitro protein digestibility, and is lower in lipid, tannin, and phytic acid content than raw chickpea flour milled from unfermented chickpea seeds. Raffinose content of chickpeas also is reduced significantly during fermentation (Zamora and Fields 1979). The fermentation process increases the dietary fiber and reduces the resistant starch content of chickpea seeds. The fermentation process is an effective method to decrease the saponin content of chickpeas considerably (Veena et al. 1995).

Cooking of pulses causes decrease of most antinutritional factors, thereby improving their nutritional value, and also increases the solubility of many nutrients. Cooking of chickpea was found to be more effective in reducing the level of antinutritional factors (trypsin inhibitors, hemagglutinin activity, tannins, saponins) than germination (El-Adawy 2002). Change in nutritional values depends upon the severity of heat treatment, duration of heating, and method of cooking. ur-Rehman and Shah (1998) studied the effect of different cooking methods (boiling, pressure-cooking, microwaving) on the nutritional value of chickpea. They found that the in vitro starch and protein digestibility increased from 40% (raw chickpea) to 68%, 85%, and 82% by boiling, pressure-cooking, and microwaving, respectively. A greater increase in in vitro starch and protein digestibility during pressure-cooking than boiling was also investigated by Khatoon and Prakash (2004). However excessive heating during cooking also reduces the nutritive value of protein. Khatoon and Prakash (2004) observed that the thiamin content of chickpea decreases after pressure-cooking and microwave heating. Fat is also lost during boiling and pressure-cooking of the chickpea (Parihar et al. 1999).

Protein Isolation

In the food industry, proteins are important ingredients that improve the nutritional, functional, and sensory quality of various developed food products. The chickpea is a protein-rich legume with protein content ranging from 19% to 29% with greater biological value, from 75% to 85% (Boye et al. 2010). Protein concentrates and isolates are protein-dense ingredients extracted from chickpea with wide functions in product formulation. The most commonly used and economic method of extraction of protein isolates from pulses in the food industry is alkaline extraction for starch isolation and subsequent precipitation of protein supernatant at isoelectric pH (Sanchez-Vioque et al. 1999). Utilization of chickpea protein in food formulation depends upon the physicochemical, functional, thermal, and structural properties (Withana-Gamage et al. 2011). Protein isolate extracted from chickpea can be used in product development (bakery industry, extruded products, baby foods), energy-rich supplements, and emulsion formulations.

Product Development

As chickpea is a rich source of carbohydrates, proteins, nutrients, and micronutrients, it can be utilized to develop value-added products, hence making it more economical, affordable, and nutritionally rich. Food-based strategies aim to improve the quality of the overall diet by increasing the availability and consumption of nutrition-rich food products. Chickpea, being a rich source of nutrients, has been incorporated in a variety of products and has shown health benefits in a variety of developed products such as breads, biscuits, muffins, cookies, pasta, infant foods, snacks, spaghetti, and dairy desserts.

Bread

Bread has been used as a human food from ancient times and contributes more than 50% of dietary energy by its high carbohydrate content (Onoja et al. 2011). It is rich in both macro- and micronutrients, especially carbohydrates, proteins, and fiber, as well as magnesium, iron, phosphorus, sodium, and vitamins, mostly B-vitamins. White bread contains 35–43% moisture, 6–16% proteins, 45–58% carbohydrates, 0.5–1.5% lipids, 0.5–1.5% ash, and 1–1.5% salt; 100 g of bread has approximately 250–270 calories (Burcu et al. 2016). Legume flours, as the result of their amino acid composition and fiber content, are ideal ingredients for improving the nutritional value of bread and bakery products. Because protein has a major role in the quality of bread, supplementation of wheat dough with chickpea or other grain legume flours and proteins certainly affects the sensory textural properties of the fortified wheat flour dough and its subsequent finished products (Sathe et al. 1981; Eliasson 1990; Singh and Ram 1990). These effects can be measured by using physical dough testing devices to evaluate the bread-making potential and performance characteristics of the fortified flour. Incorporation of chickpea flour has shown increased levels of aspartic acid and lysine and arginine acids (from 4.7% to 8.79%, from 2.42% to 5.24%, and from 4.6% to 6.47%, respectively). Low- or high-chickpea addition has different effectd on the elasticity and energy of dough. Chickpea flour has been added to enhance the strength of wheat flour to replace 15% and 30% w/w of wheat flour (Hefnawy et al. 2012). The color of crust and crumb progressively becomes darker as the level of chickpea flour substitution increases, as studied by Mohammed et al. (2012a, b). Studies conducted by Simona et al. (2015) indicate that addition of chickpea flour increased total protein content of blends from 14.9% in the case of 10% incorporated chickpea flour to 16.9% for the highest chickpea flour content, 30%. By increasing the addition of chickpea flour from 10% to 30%, the crude fiber content also increased from 3.1% to 6.3%, and the ash content and total lipids content were increased from 1.80% to 2.70% and from 2.3% to 3.4%, respectively. Also, addition of chickpea flour up to 30% increases the amount of total lipids by almost twofold compared to wheat flour; similarly, a fourfold increase was seen in ash content on addition of 30% chickpea flour. However, addition of chickpea flour resulted in decreased loaf volume, increased water retention capacity, and darker crust and crumb (Mohammed et al. 2012a, b). Consumers of chickpea-incorporated bread found the color, texture, and taste appealing. The overall results indicated that the incorporation of chickpea flour up to 20% in bread-making gives parameter values at least as good as in control samples (Mohammed et al. 2012a, b). The scores of general acceptability are found to be 8.1, 7.91, 7.62, and 7.1 in control bread and bread supplemented with 10%, 20%, and 30% chickpea flour, respectively. Abdel et al. (2013) showed that bread with 5% chickpea flour was found to be more acceptable in sensory evaluation compared with wheat flour bread (Simona et al. 2015).

Spaghetti

Spaghetti is a famous pasta-like product with a good nutritional profile. It is a good source of complex carbohydrates and moderate source of protein and vitamin B. It is prepared from a homogeneous mixture of durum wheat semolina and water to form a dough that is extruded into the desired shape at room temperature and atmospheric pressure or under vacuum followed by drying (Hernandez-Nava et al. 2014). Chickpea-fortified spaghetti has been found to be acceptable to consumers and provides an enhanced nutritional status via the amino acid profile. The total protein content (13.82–17.42%) and the content of most amino acids such as lysine (0.36–0.62%) increased with fortification of chickpea at 15–30%, whereas amino acids such as glutamine, proline, cysteine, and methionine decreased (4.48 to 4.22%, 1.57 to 1.51%, 0.34 to 0.33%, and 0.21 to 0.20%, respectively) with increase in chickpea flour at 15–30%. Spaghetti processing and handling characteristics deteriorated as the level of fortification increased. However, spaghetti stickiness improved with increasing fortification and cooking loss was reduced. Chickpea-fortified spaghetti retained firmness much better than durum after refrigeration and the overall sensory quality increased (Wood 2009). In another study, fortification of maize-based spaghetti with chickpea flour at concentrations of 5%, 10%, 15%, 20%, 25%, and 30% continuously increased until the overall sensory quality of pasta reached its sensory threshold. Poor elasticity and increased firmness were seen in spaghetti loaded with 15% chickpea flour, so this concentration represented the highest chickpea flour concentration to be used (Padalino et al. 2014). In another study, it was observed that wheat flour replaced with different processed chickpea flours to produce spaghetti has significant effects on chemical composition, cooking quality, color attributes, and sensory evaluation of spaghetti. Generally, protein solubility values of all processed flours decreases in water and NaCl solution as compared with raw flour (Esmat et al. 2010). The replacement level in spaghetti samples with different processed chickpea flours increases protein content and decreases fiber and total carbohydrates content. The mineral content is expected to be high in spaghetti samples containing microwave-cooked chickpea flour at different levels as compared with spaghetti samples containing traditional cooking and fried chickpea flours (Esmat et al. 2010). The reduction in cooked weight and cooked volume is observed to be high in spaghetti samples replaced with microwave-cooked chickpea flour than in samples replaced with other forms of chickpea flours (Esmat et al. 2010). Cooking loss of replaced spaghetti is increased gradually with increase the level of replacement compared to the control spaghetti. Replacing wheat flour with different processed chickpea flours tend to reduce lightness and yellowness values and increase redness values of spaghetti samples (Esmat et al. 2010). Spaghetti samples replaced with microwave chickpea flour at all levels have better color values than those found in samples replaced with differently processed chickpea flours. Spaghetti samples replaced with microwave-cooked flour at all levels are found to have the highest values for all evaluated sensory characteristics. Goni and Valentine-Gamazo (2003) showed that spaghetti containing 25% chickpea flour had a significantly lower glycemic index (GI) than traditional durum spaghetti; chickpea inclusion also increases the mineral and fat content without affecting the total starch content. Zhao et al. (2005) incorporated 5–30% of different pulse flours into spaghetti and found that trimness and color intensity increases although overall quality decreases.

Snacks

Snacks differ from meals in terms of size, nutritional content, and hunger and thirst sensations before and after the event (Bellisle et al. 2003). Different foods are referred to as meals or snacks, with sweets, cereal bars, biscuits, and fizzy drinks more likely to be termed snacks. Snacking is viewed as food that is eaten in addition to standard meals and therefore perceived as providing extra calories. The most widely consumed extruded snacks are prepared primarily with cereals/grains because of their good expansion characteristics. However, their nutritional value is far from satisfying the needs of health-conscious consumers (Rampersad et al. 2003). Hence, there is an increasing consumer demand for more nutritious snacks (Agriculture and Agri Food Canada 2008). The incorporation of chickpea at a 20% level for preparation of extruded snack foods shows acceptable sensory scores (Singh et al. 2017a, b). During the first month of storage, the moisture content and free fatty acid content values of snacks have been observed as 5.88% and 0.242%, which increases to 7.18% and 0.603%, respectively, whereas the freshness and overall acceptability of snacks decreased from 90.7% and 7.66% to 79.7% and 7.00%, respectively, at the end of 3 months storage, as observed by Singh et al. (2017a, b). Deep-fried snacks were prepared using chickpea flour at different concentrations (36%, 38%, 40%, 42%, 44%) by Ravi and Susheelamma (2004). Utilization of chickpea flour resulted in decreased oil, fat, and moisture content. With the increased chickpea flour concentration, the peak force required for compression also showed significant increase. In another storage study on chickpea flour-incorporated snacks, Singh et al. (2017a, b) reported increased moisture content and decreased hardness from 90.79 to 79.70 N with an acceptable range of free fatty acids after 3 months of storage. Sensory characteristics of chickpea flour snacks resulted in decreased lightness and increased redness and yellowness. Singh et al. (2017a, b) suggested that chickpea flour can be used for preparation of snacks at 20% levels.

Biscuits

Biscuits are ready-to-eat, cheap, and convenient food products that are consumed by all age groups in many countries. Biscuits belong to the flour confectionery. A biscuit is flat, crisp, and may be sweetened or unsweetened according to preferences. Biscuits can be made from hard dough, hard sweet dough, or short or soft dough. Biscuits are produced by mixing various ingredients such as wheat flour, fat, sweetener, and water to form a dough. The dough formed, unlike bread, is not allowed to ferment, and is then baked in an oven (Iwegbue 2012). Biscuits are classified into three broad groups as spongy goods, crackers, and sweet dough, based on the method used for their manufacture. Biscuits are divided into two groups, hard and soft dough. The soft dough biscuits are rich in fat and sugar and include short cakes, shortbread, and frosted biscuits. Other types of biscuits are cream crackers, soda crackers, savory crackers, and water, digestive, and short dough biscuits (Okoli et al. 2010). Biscuits contain carbohydrate 62.5%, crude fat 21.4%, crude protein 8.81%, moisture content 5.8%, ash 0.52%, and crude fiber 0.84% (Usman et al. 2015). Blends of plantain and chickpea flours with different concentrations along with refined wheat flour have been used for the production of biscuits. Evaluation of flours for their chemical and functional properties shows that plantain flour has the highest crude fiber (3.6%) and carbohydrate content (80.8%), whereas chickpea flour has the highest protein content (19.3%) and fat content (4.4%). Plantain flour shows the highest water absorption (167.7%) and the lowest oil absorption capacity (144.6%). Chickpea flour shows the highest foaming capacity and stability. The thickness and diameter of the biscuits do not differ significantly. The spread ratio and percent spread decreases with the addition of plantain and chickpea flours each up to a concentration of 30%. The fracture strength of biscuits increases significantly with addition of plantain and chickpea flours and was highest at 40% concentration (21.1 N). The protein and crude fiber content of biscuits increases significantly from 7.1% to 9.2% and 1.1% to 3.6%, respectively, with increasing extent of chickpea flour and plantain flours in the blends. The sensory properties of biscuits prepared by replacing refined wheat flour up to 20% each with plantain and chickpea flour are more or less similar to those biscuits prepared without adding chickpea flour (Yadav et al. 2012). Pinky et al. (2017) prepared chickpea husk biscuits by replacing wheat flour at levels of 0%, 5%, 10%, 15%, 20%, and 25%. The developed biscuits were subjected to physicochemical analysis that revealed an increase in ash (3.51% from 1.28%), crude fiber (2.15% to 10.48%), carbohydrates, and total dietary fiber (2.15% to 10.48%). Protein and moisture content of the biscuits increased significantly with the increase in chickpea husk level and was highest at 25%. Rababah et al. (2006) fortified the dough, making biscuits with different levels of broad bean flour, chickpea flour (3%, 6%, 9%,12%), and isolated soy protein (3%, 6%, 9%). Physicochemical analysis results showed increase in protein content from 16.82% for the control to 17.83% to 22.84% for 3% to 12% isolated soy proteinate (ISP), 16.93% to 19.64% for 3% to 12% chickpea, and 17.13% to 20.16% for 3% to 12% broad bean. Fat content increased with the increased fortification levels from 14.13% for the control to 14.59%, 14.60%, 14.95%, and 15.31% for 3%, 6%, 9%, and 12% chickpea, respectively. It was increased to 14.38%, 14.44%, 14.50%, and 14.56% for 3%, 6%, 9%, and 12% broad bean. Fat content was increased to 13.43%, 13.51%, and 14.38% for 3%, 6%, and 9% ISP. Baljeet et al. (2014) prepared biscuits from different blends of refined wheat flour, germinated chickpea flour, and carrot pomace in ratios of 100:0:0, 90:5:5, 84:8:8, and 80:10:10. Proximate analysis revealed an increase in ash (0.8–1.20%), protein (7.1–7.90%), and moisture (from 2.5% to 3.1%). Crude fiber increased (0.5–3.2%) with the increased incorporation of carrot pomace and germinated chickpea flour. The increased content of protein and fiber was attributed to the chickpea flour and carrot pomace, respectively. With the increase in chickpea husk, consumer acceptability decreased, with undesirable effects in mouth feel, color, taste, and aroma. Sensory analysis of chickpea husk-incorporated biscuits depicted the moderate overall acceptability for 20% chickpea biscuits (Pinkly et al. 2017). Incorporation of chickpea flour resulted in increased $L*$ and decreased $a*$ value compared to control, followed by broad bean and ISP biscuits; increased yellowness ($b*$) was found in fortified ISP biscuits followed by broad bean, control, and chickpea-fortified biscuits (Rababah et al. 2006). Sensory analysis of developed biscuits revealed that with the increased ratio of ISP and chickpea, acceptable characteristics and JAR ("just about right") values decreased. The same effect was observed with the texture with the increase in fortification levels. A fortification ratio of ISP at 3% and chickpea at 3% gave the best descriptive results of fortification. Descriptive analysis showed that 3% chickpea-, 3% ISP-, and 12% broad bean-fortified biscuits were rated best (Rababah et al. 2006). Sensory characteristics of biscuits developed with 5% carrot pomace and germinated chickpea flour were more or less similar to control biscuits with respect to color, taste, flavor, texture, and overall acceptability, but above the 8% incorporation level a significant decrease was observed in sensory characteristics, whereas incorporating 8% chickpea flour and carrot pomace produced protein- and fiber-enriched biscuits with overall acceptability (Baljeet et al. 2014).

Cookies

Cookies are nutritive snacks produced from unpalatable dough that is transformed into an appetizing product by the application of heat in an oven (Anozie et al. 2014). Cookies are ready to-eat conveniences and an inexpensive food product containing important digestive and dietary principles. Cookies also offer a great way to blend flour replacements and thus an easy and suitable way of improving nutrition. Cookies contain moisture 3.34%, ash 1.41%, protein 11.21%, fat 29.86%, crude fiber 1.32%, carbohydrate 52.79%, and energy value 525.02 (kcal/100 g). Cookies also contribute valuable quantities of iron, calcium, protein, calorie, fiber, and some of the B-vitamins to our diet and daily food requirements (Ikuomola et al. 2017). Cookies prepared from chickpea had significantly increased protein and resistant starch content. In terms of textural measurement, chickpea cookies have increased hardness, crispiness, elasticity, gumminess, and chewiness as compared to cookies prepared without using chickpea flour. For sensory evaluation, chickpea cookies show high differences in flavor, crispiness, and aftertaste attributes. Chickpea cookies give the best flavor and crispiness. Thus, the incorporation of chickpea flour does not change the functional properties but increases the protein and resistant starch (RS) content and acceptability of cookies as observed by Aziah et al. (2012). Cookies containing 25% chickpea flour were ranked as the most acceptable with an overall acceptability of 7.9 among other cookies in which were incorporated mung bean flour, cowpea flour, and pigeon pea flour (Thongram et al. 2016).

Noodles

Noodles are a very thin form of wheat flour, water, egg, and salt. Noodles are an increasingly important food commodity worldwide: 97.5 billion servings of instant noodles were eaten in 2016, so by simple arithmetic as many as 270 million servings are eaten every day (World Instant Noodle Association 2016). The consumption level of noodles has become one of the fastest growing sectors in Asian countries because of their ease of cooking and long shelf life. Noodles in particular are an important basic food widely consumed across the world and among the first foods to be authorized by the Food and Drug Administration as a good vehicle for addition of bioactive compounds. Noodles prepared from wheat flour contain moisture 6.83%, fat 4.30%, protein 13.85%, carbohydrate 69.93%, fiber 1.85%, and ash 1.57% (Shere et al. 2018). Noodles prepared from wheat flour, chickpea flour, and okara flour have been evaluated for proximate analysis, dietary fiber, energy content, and cooking qualities. Noodles prepared incorporating chickpea flour have significantly higher ash, crude protein, and crude fiber content and significantly lower crude fat, carbohydrate, and energy content, are also rich in dietary fiber (10.05 g/100 g), and have higher cooking yield, longer optimal cooking time, lower rehydration rate, and higher cooking loss. Chickpea also increases water activity and decreases pH value, reduces brightness, increases redness and yellowness of noodles, reduces tensile strength, and causes changes in the textural properties. The incorporation of chickpea into instant noodles boosts the nutrient content but results in poorer textural properties (Kuen et al. 2017). Sofi et al. (2019) formulated noodles from rice and germinated chickpea flour with increase in protein content and better cooking and sensory properties than control rice noodles.

Sofi et al. (2019) observed that with increase in level of germinated chickpea flour and protein isolate, protein content (7.22–14.35%), antioxidant activity (22.75–33.79%), and total phenolic content (117.65–223.35 mg GAE/100 g) of the rice-based noodles increased significantly ($p \leq 0.05$), whereas lightness, cooking time (13.35 to 10.13 min), cooking loss (7.38% to 6.78%), cooked weight (41.40 to 33.15 g/100 g), and percentage of starch gelatinization (65.36 to 41.26) and in vitro starch digestibility decreased significantly. Hence, noodles prepared with germinated chickpea flour (20%) and chickpea protein isolate (8%) showed better acceptability on the basis of sensory score and were recommended for making noodles with rice flour with improved noodle quality. In another study, Bilgiçli (2013) investigated some legumes (chickpea and soya), pseudo-cereals (buckwheat and quinoa), and cereal (maize and rice) flour blends used in gluten-free noodle formulations. To improve the dough-forming ability, all flour blends were gelatinized at a level of 25%. Noodles containing chickpea, soya, buckwheat, and quinoa flours (flour blend 1) showed higher levels of protein (194.2 g/kg), ash (27.8 g/kg), lipids (81.2 g/kg), calcium (562.85 mg/kg), copper (9.20 g/kg), iron (56.29 mg/kg), potassium (295.21 mg/kg), magnesium (661.78 mg/kg), manganese (24.07 g/kg), phosphorus (042.88 mg/kg), and zinc (40.24 mg/kg) than other gluten-free noodles and control noodles made with wheat flour. In gluten-free noodles, phytic acid content increased up to 9.2 times as compared to control. Prepared noodles were better in terms of overall acceptability.

Pasta

Pasta is cheap to produce and very convenient to prepare, which makes it one of the most popular high-carbohydrate food products (Wood 2009). Pasta contains carbohydrates (74–77%) and protein (11–15%). However, the quality of pasta protein is low because of the limited amounts of essential amino acids, notably lysine (Shewry 2007). Traditionally, wheat semolina has been the primary ingredient of pasta. However, components that increase nutritional value or exert a beneficial effect on health can be used in its production (Sun-Waterhouse and Wadhwa 2014). Pasta was prepared with durum wheat flour mixed with chickpea flour at two different levels and chemical composition, in vitro starch digestibility, and predicted glycemic index observed. Protein, ash, lipid, and dietary fiber content increased whereas total starch decreased with the chickpea flour level in the composite pasta, all in accordance with the composition of the legume flour. Potentially available starch decreases and resistant starch (RS) increases by adding chickpea flour to the pasta. The main indigestible starch component in composite spaghetti is the fiber-associated RS, representing up to 50% of total RS levels. Starch hydrolysis index (HI) decreased as chickpea flour in the pasta increased, reflecting the slow and low digestion of the starch in the leguminous ingredient. Predicted glycemic index is observed to be lower in spaghetti with added chickpea flour than in durum wheat control pasta. Pasta with added chickpea flour might be a dietetic alternative for people with low calorie requirements, as observed by Osorio-Díaz (2008). In another study, the ability of chickpea flour to enrich pasta products (e.g., lasagna) was studied by Sabanis et al. (2006). In addition, the influence of protein and other components upon the rheological properties of the dough and the cooking quality of the wheat–chickpea blends has also been observed. Supplementing lasagna with 5–20% w/w chickpea flour improves the physical characteristics of the dough, which achieves optimum strength and extensible properties thus allowing the lasagna to maintain a firm and elastic form. Organoleptic properties (color, flavor, and overall acceptability) are improved with a low proportion of chickpea flour, especially for 5% w/w substitution. For supplementation of 30% or more, the content of total protein increases with the level of fortification, and lasagna processing, handling, and cooking characteristics deteriorate proportionally; this could be attributed primarily to the gluten fraction, which decreases upon being diluted by the added chickpea protein. The rheological properties of highly supplemented dough products (30–50% w/w) have low extensograph values and the lasagna obtained has a brown color and a soft mushy taste that is unacceptable to consumers. So, durum wheat flour can carry 5–10% (w/w) of chickpea flour and still meet the specification of pasta products in terms of firmness, cooking quality, and sensory evaluation (Sabanis et al. 2006). Further, sensory, nutritional, and cooking qualities of chickpea-fortified pasta were assessed. The fortification of durum wheat semolina by the combination of chickpea flour and defatted soy flour at levels of 10.6%, 14.10%, and 18.14%, respectively, increased cooking time, water absorption, and stiffness of the samples compared to the control. Cooking time and sensory characteristics of fortified pasta were highly acceptable. On the basis of cooking and sensory quality, pasta when fortified with blends of 14% chickpea flour and 10% defatted soy flour had better quality and nutritional value (carbohydrate content 68.61%, protein content 17.99%, fat content 1.40%, and fiber content 4.19%) (Bashir et al. 2012). The resultant pasta can be used as a nutritious food for low-income groups in developing countries and for patients suffering from lifestyle diseases. In another study, chickpea- and spinach-fortified wheat pasta was found to be the best in nutritional, functional, and cooking qualities. Crude protein was notably affected by fortification: 10.5% in spinach wheat pasta, 18.55% in chickpea- and spinach-fortified wheat pasta, whereas it is 10.8% in wheat pasta. Crude fat was extended from 0.98% to 1.55%. Moisture, ash, and acidity ranged from 8.1% to 8.5%, 0.82% to 1.80%, and 0.23% to 0.59%, respectively. Fortification resulted in increased antioxidant properties with textural properties of wheat pasta and fortified wheat. Free radical scavenging properties, ascorbic acid (20.4 mg/100 g) and total phenols (mg/100 g) were higher in chickpea- and spinach-fortified wheat pasta and the lowest in wheat pasta. Maximum hardness was also highest in chickpea- and spinach-fortified wheat pasta (Ghan et al. 2017). Hence, fortified pasta could be produced and used as an alternative for conventional pasta and can add to variety in food products.

Dairy Desserts

Dairy desserts are complex mixtures and matrices including such main components as milk, sugar, starch, hydrocolloids, colorants, and flavors, with a proteinaceous structure presenting a semisolid consistency. These desserts are widely consumed. It is possible to produce a good alternative for dairy products with good properties and potential higher nutritional value. The incorporation of chickpea flour has been successful; the custard properties are affected by the chickpea flour variety type, level of concentration, and storage time in different degrees (Aguilar-Raymundo & Vélez-Ruiz 2018). The physicochemical and rheological properties of a dairy dessert with the addition of chickpea flour (raw and cooked, at different concentrations) changes with the type of flour, such as luminosity ($L*$) (62.75–83.29), pH (6.35–7.11), and acidity (1.56–3.56). The viscosity of the products increases, which, in turn, contributes to a viscoelastic behavior The flow properties of the custards exhibits a non-Newtonian behavior that fits well by three flow models as observed by Raymundo and Ruiz (2018). The physicochemical and flow properties of the custards changes notably as a function of flour addition and storage time. The dessert texture has also been measured; those formulated with Blanco Noroeste chickpea flour exhibit the highest values of hardness (0.356–0.391 N) throughout the storage period. It can be concluded that those custard systems with the highest content of flour present a very good potential as a new dairy product.

Infant Foods

Infant formula is intended as an effective substitute for breast milk and is formulated to mimic the nutritional composition of breast milk. Infant formulas must include proper amounts of water, carbohydrate, protein, fat, vitamins, and minerals. The composition of infant formula is strictly regulated, and each manufacturer must follow established guidelines set by government agencies. For instance, all the major components added to the formula (protein, lipids, carbohydrates) have a range of minimum and maximum values for their effectiveness. These components must have an established history of safe use (Koletzko et al. 2005). Infant formula prepared as ready for consumption should contain no less than 60 kcal (250 kJ) and no more than 70 kcal (295 kJ) of energy per 100 ml (CAC 1989; Guo 2014). Novel infant follow-on formulas are exclusively made from either cow's milk protein or soy protein isolate that is blended with starch or corn syrup or dextrose, vegetable oil/fat, and mineral and vitamin premixes (Nasirpour et al. 2006). The ratio of protein and carbohydrate in chickpea proximate composition is 0.25, which is similar to that of an infant follow-on formula formulation, suggesting that the whole chickpea grain can be utilized (Zia-Ul-Haq et al. 2007; El-Adawy 2002). Therefore, we can use chickpea as a common source for protein, carbohydrates, and minerals with minimal supplementation of the latter for infant weaning food/follow-on formula. The infant follow-on formula has been formulated to meet the requirements stipulated in the codex standard on infant follow-on formula and Codex Alimentarius Commission (1989) and European Commission Directive (2006). The estimated nutrition value of developed chickpea-based infant follow-on formula without fortification shows the need for calcium, sodium, and potassium supplementation to meet the requirements. Furthermore, we assume a loss of significant amounts of the water-soluble vitamins such that vitamin supplementation should meet the daily infant nutrition needs. Germination for 72 h followed by boiling, drying, and dehulling increases the protein content but does not affect the amino acid profile of chickpea. The formulated chickpea-based follow-on formula preceding micronutrient supplementation meets almost all the nutrient requirements of the European Commission (EC) directive on infant follow-on formula except for calcium, sodium, and potassium. Raymundo et al. (2014) prepared infant food using blends of wheat flour, soy flour, and chickpea with and without the addition of skim milk powder. Addition of chickpea flour resulted in increase in moisture content (from 1.52 to 1.84), protein (up to 13.20 from 12.2), and fat content (from 1.91 to 2.52). The sensory quality attributes of the products revealed that the mean score values for the various sensory attributes of color, flavor, taste, texture, and overall acceptability varied from 6.2 to 8.8. Best scores were obtained for the products with 20% chickpea flour incorporated.

Future Trends in Processing and Product Development

Chickpeas are a nutrient-dense food with good nutritional composition. There is an increasing demand for chickpea because it has high nutritional value and is a cheap source of protein. The chickpea is a good source of protein and carbohydrates, which increases its demand as a functional ingredient in food applications. Various physicochemical processing methods were applied to chickpea to enhance its functional and consumer acceptability. Today's convenience-oriented lifestyle and the development of new technologies for processing can attract the chickpea and enhance its value addition. The use of new technologies such as cold plasma, pulsed electric field, UV, irradiation, and ultrasonification as pretreatments decrease antinutritional factors in chickpea flour and increase its nutrient composition and functional properties. Extraction of bioactive peptides from the chickpea pulse that are not suitable for consumption through enzyme technology are in the future of the chickpea food industry. Proteins and starch isolated from chickpea can be used for coating fresh-cut fruits and vegetables as a preservation technique to prevent spoilage. Chickpea by-products such as protein isolates and hydrolysates can be used in drug delivery and microencapsulation of essential products. Protein supplements for athletes and sportspersons and energy drinks from chickpea can be the next target of the food processing industries. In the new trend of using sprouts with green vegetables as salad, sprouts can be prepared from chickpea with various health benefits. The ready-to-use types of products that are cheap, easily available, and convenient has become the focus of the food processing industry. Chickpea in future can be processed into ready-to-use products such as canned chickpeas and chickpea shakes and drinks. The wide use of extruded products in food industries from chickpea or with other nutrient-dense materials can be prepared with a balanced composition of nutrients. The chickpea in future will have an important role in developing gluten-free food products to combat celiac diseases.

References

Abdel, M., Sulieman, E., Sinada, E. A., & Ali, O. A. (2013). Quality characteristics of wheat bread supplemented with chickpea (*Cicer arietinum*) flour. *International Journal of Food Science and Nutrition Engineering, 3*(5), 85–90.

Agriculture and Agri-Food Canada. (2008). *The Canadian snack food industry*. Retrieved from http://www4.agr.gc.ca/AAFCAAC/displayafficher

Aguilar-Raymundo, V. G., & Vélez-Ruiz, J. F. (2018). Physicochemical and rheological properties of a dairy dessert, enriched with chickpea flour. *Foods, 7*(2), 25.

Alajaji, S. A., & El Adawy, T. A. (2006). Nutritional composition of chickpea (Cicer arietinum L.) as affected by microwave cooking and other traditional cooking methods. *Food Chemistry, 19*(8):806–812.

Anozie, G. O., China, M. A., & Beleya, E. A. (2014). Sensory evaluation and proximate composition of snacks produced from composite flour of *Discorea alata* and *Telfaira occidentalis* seed flours. *Journal of Home Economic Research, 20*, 100–108.

Arya, S. S. (1981). Effect of water activity on storage changes in total carotenoids and lipids in Bengal gram (*Cicer arietinum* L.) dhal and flour. *Journal of Food Science and Technology, 18*, 139–142.

APEDA (2019). https://apeda.gov.in/apedawebsite/Latest_Notification/Trade_Notice.html

Aziah, A. A. N., Noor, A. Y. M., & Ho, L. H. (2012). Physicochemical and organoleptic properties of cookies incorporated with legume flour. *International Food Research Journal, 19*(4), 1539–1543.

Baljeet, S. Y., Ritika, B. Y., & Reena, K. (2014). Effect of incorporation of carrot pomace powder and germinated chickpea flour on the quality characteristics of biscuits. *International Journal of Food Research, 21*(1), 217–222.

Bampidisa, V. A., & Christodouloub, V. (2011). Chickpeas (*Cicer arietinum* L.) in animal nutrition: A review. *Animal Feed Science and Technology, 168*, 1–20.

Bashir, K., Aeri, V., & Masoodi, L. (2012). Physio-chemical and sensory characteristics of pasta fortified with chickpea flour and defatted soy flour. *Journal of Environmental Science, Toxicology and Food Technology, 5*(1), 34–39.

Bellisle, F., Dalix, A. M., Mennen, L., Galan, P., Hercberg, S., De Castro, J. M., & Gausseres, N. (2003). Contribution of snacks and meals in the diet of French adults: A diet-diary study. *Physiology & Behavior, 79*, 183–189.

Bilgiçli, N. (2013). Some chemical and sensory properties of gluten-free noodle prepared with different legume, pseudocereal and cereal flour blends. *Journal of Food and Nutrition Research, 52*(4), 251–255.

Boye, J. I., Aksay, S., Roufik, S., Ribereau, S., Mondor, M., Farnworth, E., & Rajamohamed, S. H. (2010). Comparison of the functional properties of pea, chickpea and lentil protein concentrates processed using ultrafiltration and isoelectric precipitation techniques. *Food Research International, 43*, 537–546.

Burcu, A., Avsaroglu, E., Isık, O., Ozyurt, G., Kafkas, E., & Uslu, L. (2016). Nutritional and physicochemical characteristics of bread enriched with microalgae *Spirulina platensis*. *International Journal of Engineering Research and Applications, 6*(12), 30–38.

Chavan, J. K., Kadam, S. S., Salunkhe, D. K., & Beuchat, L. R. (1987). Biochemistry and technology of chickpea (*Cicer arietinum* L.) seeds. *Critical Reviews in Food Science and Nutrition, 25*(2), 107–158.

Chavan, J. K., Kadam, S. S., & Salunkhe, D. K. (1989). Chickpea. In D. K. Salunkhe & S. S. Kadam (Eds.), *CRC handbook of world food legumes: Nutritional chemistry, processing technology and utilization* (Vol. I, pp. 247–288). Boca Raton: CRC Press.

Chitre, R. G., Desai, D. B., & Raut, V. S. (1955). The effect of storage on thiamine, riboflavin and nictonic acid in cereals and pulses. *Indian Journal of Medical Research, 43*, 585–589.

Codex Alimentarius Commission: Codex Standard for Follow-Up Formula Rome. (1989). *Codex Stan 156–1987*. Rome: Codex Alimentarius Commission.

Czukor, B., Bogracheva, T., Cserhalmi, Z., Fornal, J., Schuster-Gajzago, I., Kovacs, E. T., Lewandowicz, G., & Soral-Smietana, M. (2001). Processing. In C. L. Hedley (Ed.), *Carbohydrates in grain legume seeds: Improving nutritional quality and agronomic characteristics*. Wallingford: CAB International.

de Falco, E., Imperato, R., Landi, G., Nocolais, V., Piccinelli, A. L., & Rastrelli, L. (2010). Nutritional characterization of *Cicer arietinum* L. cultivars with respect to morphological and agronomic parameters. *Emirates Journal of Food and Agriculture, 22*, 377–387.

El-Adawy, T. A. (2002). Nutritional composition and antinutritional factors of chickpeas (*Cicer arietinum* L.) undergoing different cooking methods and germination. *Plant Foods for Human Nutrition, 57*, 83–97.

Elham, A., & Boye, J. I. (2013). Composition of processed dry beans and pulses. In V. M. Siddiq & M. A. Uebersax (Eds.), *Dry beans and pulses: Production, processing and nutrition*. New York: Wiley.

Eliasson, A. C. (1990). Rheological properties of cereal proteins. In H. Faridi & J. Faubion (Eds.), *Dough rheology and baked product texture* (pp. 67–110). New York: Van Nostrand Reinhold.

Ellis, R. H., Agrawal, P. K., & Roos, E. E. (1988). Harvesting and storage factors that affect seed quality in pea, lentil, faba bean and chickpea. In R. J. Summerfield (Ed.), *World crops: Cool season food legumes* (pp. 303–329). Dordrecht: Kluwer Academic.

Esmat, A., Abou Arab, I., Helmy, M. F., & Bareh, G. F. (2010). Nutritional evaluation and functional properties of chickpea (*Cicer arietinum* L.) flour and the improvement of spaghetti produced from its. *Journal of American Science, 6*(10), 1055–1072.

European Commission. (2006) Commission Directive 2006/141/EC of 22 December 2006 on Infant Formulae and Follow-on Formulae and Amending Directive 1999/21/EC. In Directive 2006/141/EC. Brussels: Official Journal of the European Union: The Commission of the European Communities.

FAOSTAT. (2018). Retrieved from http://faostat.fao.org/site/567/DesktopDefault.aspx

FAOSTAT. (2019). http://faostat.fao.org/site/567/DesktopDefault.aspx

Ghan, S. A., Mishra, V., Devina, V., & Sharma, A. (2017). Effect of spinach and chickpea flour fortification on cooking, functional and textural properties of wheat pasta. *Journal on Processing and Energy in Agriculture, 21*(2), 81–85.

Goni, I., & Valentin-Gamazo, C. (2003). Chickpea flour ingredient slows glycemic response to pasta in healthy volunteers. *Food Chemistry, 81*(4), 511–515.

Guo, M. (2014). Human milk biochemistry and infant formula. In *Manufacturing technology*. Cambridge: Elsevier.

Haysmans, A. A. C. (1970). Storage of food grains: Problems and prospects. *Bull Grain and Technology, 8*, 92.

Hefnawy, T. M. H., El-Shourbagy, G. A., & Ramadan, M. F. (2012). Impact of adding chickpea (*Cicer arietinum* L.) flour to wheat flour on the rheological properties of toast bread International *Food Research Journal, 19*(2): 521–525.

Hernandez-Nava, R. G., Berrios, J. D. J., Osorio-Diaz, P. P., & Bello-Perez, L. A. (2014). Development and characterization of spaghetti with high resistant starch content supplemented with banana starch. *Food Science and Technology International, 15*(1), 73–80.

Ikuomola, D. S., Otutu, O. L., & Olunira, D. D. (2017). Quality assessment of cookies produced from wheat flour and malted barley (*Hordeum vulgare*) bran blends. *Cogent Food and Agriculture, 3*, 1–22.

Iqbal, A., Khalil, I. A., Ateeq, N., & Khan, M. S. (2006). Nutritional quality of important food legumes. *Food Chemistry, 97*, 331–335.

Iwegbue, C. M. A (2012). Metal Contents in Some Brands of Biscuits Consumed in Southern Nigeria. *American Journal of Food Technology, 7*(3): 160–167.

Jambunathan, R., & Singh, U. (1981). Studies on desi and kabuli chickpea (*Cicer arietinum* L.) cultivars: Mineral and trace element composition. *Journal of Agricultural and Food Chemistry, 29*, 1091–1093.

Januario, A. H., Lourenco, M. V., Domezio, L. A., Pietro, R. C., Castilho, M. S., & Tomazela, D. M. (2005). Isolation and structure determination of bioactive isoflavones from callus culture of *Dipteryx odorata. Chemical & Pharmaceutical Bulletin, 53*, 740–742.

Jood, S., Mehta, U. Singh, R. & Bhat, C. M. (1985). Effect of processing on flatus producing factors in legumes. *J Agric Food Chem, 33*, 268–271.

Katyal, D., Ghugre, P. S., & Udipi, S. A. (2005). Resistant starch in selected raw and processed legumes. *Journal of Food Science and Technology, 42*, 506–510.

Khatoon, N., & Prakash, J. (2004). Nutritional quality of microwave-cooked and pressure-cooked legumes. *International Journal of Food Sciences and Nutrition, 55*, 441–448.

Koletzko, B., Baker, S., Cleghorn, G., Nete, U. F., Gopalan, F., Hernall, O., Hock, Q. S., Jirapinyo, P., Lonnerdal, B., & Pencharz, P. (2005). Global standard for the composition of infant formula: Recommendations of an ESPGHAN coordinated international expert group. *Journal of Pediatric Gastroenterology and Nutrition, 41*, 584–599.

Kosolofski, A., Sokhansanj, S., & Crowe, T. (1998). *Chickpea: Harvesting, handling and storage: A review*. Saskatoon: Department of Agricultural and Bioresource Engineering, University of Saskatchewan.

Kuen, N. H., Hamid, M. A., Mamat, H., Akanda, J. H., & Ahmad, F. (2017). Effect of chickpea and okara composite flours on the quality of instant noodles. In *First International Conference on Social, Applied Science, and Technology in Home Economics (ICONHOMECS 2017)*. Atlantis Press.

Mahadevamma, S., & Tharanathan, R. N. (2004). Processing of legumes: Resistant starch and dietary fiber contents. *Journal of Food Quality, 27*, 289–303.

Mohammed, I., Ahmed, R. A., & Senge, B. (2012a). Dough rheology and bread quality of wheat–chickpea flour blends. *Industrial Crops and Products, 36*, 196–202.

Mohammed, I., Abdelrehman, R., & Senge, B. (2012b). Effects of chickpea flour on wheat pasting properties and bread making quality. *Journal of Food Science and Technology, 51*(9), 1902–1910.

Muehlbauer, F. J., & Singh, K. B. (1987). Genetics of chickpea. In M. C. Saxena & K. B. Singh (Eds.), *The chickpea* (pp. 99–125). Wallingford: CAB International.

Murty, C. M., Pittaway, J. K., & Ball, M. J. (2010). Chickpea supplementation in an Australian diet affects food choices, satiety and bowel health. *Appetite, 54*, 282–288.

Nasir, M., & Sidhu, J. S. (2013). Common pulses: Chickpea, lentil, mungbean, black gram, pigeon pea and Indian vetch. In V. Siddiq & M. A. Uebersax (Eds.), *Dry beans and pulses production, processing and nutrition*. New York: Wiley.

Nasirpour, A., Scher, J., & Desobry, S. (2006). Baby foods: Formulations and interactions (a review). *Critical Reviews in Food Science and Nutrition, 46*, 665–681.

Naveeda, K., & Jamuna, P. (2004). Nutritional quality of microwave-cooked and pressure-cooked legumes. *International Journal of Food Sciences and Nutrition, 55*, 441–448.

Okoli, E. V., Okolo, B. N., Moneke, A. N., & Ire, F. S. (2010). Effect of cultivar and germination time on amylolytic potential, extract yield and wort fermenting properties of malting sorghum. *Asian Journal of Biotechnology, 2*, 14–26.

Onoja, U. S., Odo, G. E., Dibua, U. M. E., & Eze, J. I. (2011). Physico-chemical properties, energy, mineral, vitamin and sensory evaluation of wheat-based bread supplemented with legume, root, tuber and plantain flour. *Global Journal of Pure and Applied Sciences, 17*(3), 319–327.

Osorio-Díaz, P. (2008). Pasta added with chickpea flour: Chemical composition, in vitro starch digestibility and predicted glycemic index. *Ciência e Tecnologia de Alimentos, 6*(1), 6–12.

Owusu-Ansah, Y. J., & McCurdy, S. M. (1991). Pea proteins: A review of chemistry, technology of production, and utilization. *Food Review International, 7*, 103–134.

Padalino, L., Mastromatteo, M., Lecce, L., Spinelli, S., Conte, A., & Nobile, M. A. D. (2014). Optimization and characterization of gluten-free spaghetti enriched with chickpea flour. *International Journal of Food Science and Nutrition, 12*, 151–158.

Parihar, P., Mishra, A., Gupta, O. P., Rajput, L. P. S., & Singh, A. (1999). Effect of cooking processes on nutritional quality of some common pulses. *Advances in Plant Sciences, 12*, 15–20.

Pinkly, S., Kaur, D., Yadav, N., Kumar, R., & Sanghmitra. (2017). Effect of chick pea husk on quality characteristics of biscuits. *International Journal of Food Nutrition and Science, 6*(2), 123–130.

Rababah, T. M., Maidi, A. A., & Khalil, I. E. (2006). Effect of chickpea, broad bean, or isolated soy protein additions on the physicochemical and sensory properties of biscuits. *Journal of Food Science, 76*(1), 438–442.

Rampersad, R., Badrie, N., & Comissiong, E. (2003). Characteristics of flavored snacks from extruded cassava/pigeonpea flour. *Journal of Food Science, 68*, 363–367.

Rao, P. U., & Deosthale, Y. G. (1982). Tannin content of pulses: Varietal differences and effects of germination and cooking. *Journal of the Science of Food and Agriculture, 33*, 1013–1016.

Ravi, R., & Susheelamma, N. S. (2004). The effect of the concentration of batter made from chickpea (*Cicer arietinum* L.) flour on the quality of a deep-fried snack. *International Journal of Food Science and Technology, 39*, 755–762.

Ravi, R., Ajila, C. M., & Prasada Rao, U. J. S. (2011). Role of steaming and toasting on the odor, protein characteristics of chickpeas (*Cicer arietinum* L.) flour, and product quality. *Journal of Food Science, 76*, S148–S155.

Raymundo, A., Fradinho, P., & Nunes, M. C. (2014). Effect of Psyllium fibre content on the textural and rheological characteristics of biscuit and biscuit dough. *Bioactive Carbohydrates Dietary Fibre, 3*, 96–105.

Raymundo, V. G. A., & Ruiz, J. F. V. (2018). Physicochemical and rheological properties of a dairy dessert, enriched with chickpea flour. *Food, 7*, 25–40.

Reyes-Moreno, C., Cuevas-Rodriguez, E. O., Milan-Carrillo, J., Cardenas-Valenzuela, O. G., & Barron-Hoyos, J. (2004). Solid state fermentation process for producing chickpea (*Cicer arietinum* L.) tempeh flour: Physicochemical and nutritional characteristics of the product. *Journal of the Science of Food and Agriculture, 84*, 271–278.

Sabanis, D., Makri, E., & Doxastakis, G. (2006). Effect of durum flour enrichment with chickpea flour on the characteristics of dough and lasagna. *Journal of the Science of Food and Agriculture, 86*, 1938–1944.

Sanchez-Vioque, R., Climente, A., Vioque, J., Bautista, J., & Millan, F. (1999). Protein isolate from chickpea (*Cicer areitinum* L.): Chemical composition, functional properties and protein characterization. *Food Chemistry, 64*, 237–243.

Sathe, S. K., Ponte, J. G., Rangnekar, P. D., & Salunkhe, D. K. (1981). Effects of addition of great northern bean flour and protein concentrates on rheological properties of dough and baking quality of bread. *Cereal Chemistry, 58*, 97–100.

Saxena, A. K., Chadha, M., & Sharma, S. (2003). Nutrients and antinutrients in chickpea (*Cicer arietinum* L.) cultivars after soaking and pressure cooking. *Journal of Food Science and Technology, 40*, 493–497.

Shehnaz, A., & Theophilus, F. (1975). Effect of insect infestation on the chemical composition and nutritive value of Bengal gram and field bean (*Dolichos lablab*). *Journal of Food Science and Technology, 12*, 299–302.

Shere, P. D., Devkatte, A. N., & Pawar, V. N. (2018). Studies on production of functional noodles with incorporation of spinach puree. *International Journal of Current Microbiology and Applied Sciences, 7*(6), 1618–1628.

Shewry, P. R. (2007). Improving the protein content and composition of cereal grain. *Journal of Cereal Science, 46*, 239–250.

Simona, M., Paucean, A., Muste, S., & Pop, A. (2015). Effect of the chickpea (*Cicer arietinum* L.) flour addition on physicochemical properties of wheat bread. *Bulletin UASVM Food Science and Technology, 72*(1), 41–49.

Singh, K. B. (1987). Chickpea breeding. In M. C. Saxena & K. B. Singh (Eds.), *The chickpea* (pp. 127–162). Wallingford: CAB International.

Singh, P. N., & Ram, H. (1990). Effect of phosphorous and sulphur application on protein and amino acid contents in chickpea. *Indian Journal of Pulses Research, 3*, 36–39.

Singh, G., Singh, B., Sharma, S., & Arashdeep, S. (2017a). Development and storage study of maize and chickpea based extruded snacks. *International Journal of Current Microbiology and Applied Sciences, 6*(10), 4798–4804.

Singh, G., Singh, B., Sharma, S., & Singh, A. (2017b). Development and storage study of maize and chickpea based extruded snacks. *International Journal of Current Microbiology and Applied Sciences, 6*(10), 4798–4804.

Sofi, S. A., Singh, J., Chhikara, N., & Panghal, A. (2019). Effect of incorporation of germinated flour and protein isolate from chickpea on different quality characteristics of rice-based noodle. *Cereal Chemistry, 97*(1), 85–94.

Sun-Waterhouse, D., & Wadhwa, S. S. (2014). Industry-relevant approaches for minimising the bitterness of bioactive compounds in functional foods: A review. *Food and Bioprocess Technology, 6*, 607–627.

Tabil, L. G., Jr., Sokhansanj, S., & Tyler, R. T. (1995). Processing of pulses. In *Proceedings of the Pulse Cleaning and Processing Workshop*. Agricultural and Bioresource Engineering and the Extension Division, University of Saskatchewan: Saskatoon.

Thongram, S., Tanwar, B., Chauhan, A., & Kumar, V. (2016). Physicochemical and organoleptic properties of cookies incorporated with legume flours. *Cogent Food & Agriculture, 2*(1), 1172–1198.

Tosh, S. M., & Yada, S. (2010). Dietary fibers in pulse seeds and fractions: Characterization, functional attributes, and applications. *Food Research International, 43*, 450–460.

Ur-Rehman, Z., & Shah, W. H. (1998). Effect of microwave and conventional cooking on some nutrients, antinutrient and protein digestibility of chick-peas (*Cicer arietinum* L.). *Bangladesh Journal of Scientific and Industrial Research, 33*, 200–205.

Usman, G. O., Ameh, U. E., Alifa, O. N., & Babatunde, R. M. (2015). Proximate composition of biscuits produced from wheat flour and maize bran composite flour fortified with carrot extract. *Journal of Nutrition & Food Sciences, 5*, 395–397.

Van Gastel, A. J. G., Bishaw, Z., Niane, A. A., Gregg, B. R., & Gan, Y. (2006). Chickpea seed production. In S. S. Yadav, R. J. Redden, W. Chen, & B. Sharma (Eds.), *Chickpea breeding and management* (pp. 417–444). Wallingford: CAB International.

Veena, A., Urooj, A., & Puttaraj, S. (1995). Effect of processing on the composition of dietary fibre and starch in some legumes. *Nahrung, 39*, 132–138.

Wadnerkar, D. W., Kaunsale, P. P., & Pawar, Y. M. (1978). Studies on preference of pulse beetle (*Callosobruchus maculatus* fab.) to some varieties of arhar and gram. *Bull Grain and Technology, 16*, 122–124.

WINA. (2016). *World Instant Noodles Association*. Retrieved from http://instantnoodles.org/noodles/report.htm

Withana-Gamage, T. S., Wanasundara, J. P. D., Pietrasika, Z., & Shand, P. J. (2011). Physicochemical, thermal and functional characterization of protein isolates from kabuli and desi chickpea (*Cicer arietinum* L.): A comparative study with soy (Glycine max) and pea (*Pisum sativum* L.). *Journal of Science of Food and Agriculture, 91*, 1022–1031.

Wood, J. A. (2009). Texture, processing and organoleptic properties of chickpea-fortified spaghetti with insights to the underlying mechanisms of traditional durum pasta quality. *Journal of Cereal Science, 49*, 128–133.

Wood, J. A., & Grusak, M. A. (2007). Nutritional value of chickpea. In S. S. Yadav, R. J. Redden, W. Chen, & B. Sharma (Eds.), *Chickpea breeding and management* (pp. 101–142). Wallingford: CAB International.

Yadav, R. B., Yadav, B. S., & Dhull, N. (2012). Effect of incorporation of plantain and chickpea flours on the quality characteristics of biscuits. *Journal of Food Science and Technology, 49*(2), 207–213.

Zamora, A. F., & Fields, M. L. (1979). Nutritive quality of fermented cowpeas and chickpeas. *Journal of Food Science, 44*, 234–236.

Zhao, Y. H., Manthey, F. A., Chang, S. K. C., Hou, H.-J., & Yuan, S. H. (2005). Quality characteristics of spaghetti as affected by green and yellow pea, lentil, and chickpea flours. *Journal of Food Science, 70*(6), 371–376.

Zhao, S., Zhang, L., Gao, P., & Shao, Z. (2009). Isolation and characterization of the isoflavones from sprouted chickpea seeds. *Food Chemistry, 114*, 869–873.

Zia-Ul-Haq, M., Iqbal, S., Ahmad, S., Imran, M., Niaz, A., & Bhanger, M. I. (2007). Nutritional and compositional study of desi chickpea (*Cicer arietinum* L.) cultivars grown in Punjab, Pakistan. *Food Chemistry, 105*, 1357–1363.

Common Bean

T. S. Rathna Priya and A. Manickavasagan

Introduction

Common bean is a legume that belongs to the family Leguminosae, a native to America even though its exact place of origin is unknown (Salcedo 2008; Gentry 1969). It is one of the oldest crops cultivated (Geil and Anderson 1994). Common bean, because of its easy availability and nutritional properties, has now become one of the most important crops consumed throughout the world. These pulses belong to *Phaseolus vulgaris* species and vary greatly due to their genetic diversity. In 2001, around 26,500 varieties of *Phaseolus vulgaris* beans have been included in the FAO (Food and Agriculture Organization) pulse list (CGIAR-Beanfocus).

Common bean usually grows in the warm season and is not well suited to withstand freezing temperatures during their growth. Adequate rainfall is needed for its growth especially during the flowering stage. Maturation in most of the varieties will take around 65–110 days, but some varieties may take more than 200 days (Katungi et al. 2009). The common bean (*Phaseolus vulgaris*) comprises four main varieties, namely, black turtle bean, kidney bean, navy bean and cranberry bean, as shown in Fig. 1. Other varieties include pinto bean, pea bean, pink bean, yellow beans and horto beans that are native to specific parts of the world. The black bean has its origin in South America and serves as a staple food to many indigenous cuisines of Mexico, Caribbean and Latin America. The kidney bean usually appears in varying shades (from light to dark) of red and is kidney shaped, hence the name. These kidney beans are used in a variety of dishes, and they add a special appeal to food in terms of colour and taste. The navy beans will be white in colour and are obtained as dried seeds from the mature green bean plant. The navy bean is the most largely consumed among other common bean varieties (Geil and Anderson 1994). The cranberry bean, native to Nigeria, is commonly referred based on its seed coat

T. S. Rathna Priya · A. Manickavasagan (✉)
School of Engineering, University of Guelph, Guelph, ON, Canada
e-mail: mannamal@uoguelph.ca

A. Manickavasagan, P. Thirunathan (eds.), *Pulses*,
https://doi.org/10.1007/978-3-030-41376-7_5

Fig. 1 Common bean varieties of *Phaseolus vulgaris* are (**a**) black turtle bean, (**b**) red kidney bean, (**c**) white navy bean and (**d**) cranberry or Romano bean

cranberry red colour with ivory pinto markings. The cranberry bean on cooking loses its specks and gives a creamy texture which is liked in Northern Italy and Spain.

Production, Export and Import

The highest production of bean takes place in the Latin American region, where around eight million hectares of land is used for bean production and harvesting (CGIAR-Beanfocus). According to the FAO statistics data of 2017, common bean is harvested in around 23 million hectares of land, and production stands at 12 million tonnes annually throughout the world (FAOSTAT 2019). Latin America and Africa alone contributes to around eight million tonnes (CGIAR – Common Bean 2019).

In North America, bean cultivation and production have increased drastically in the last few decades, but the corresponding consumption, to an extent, is affected owing to the presence of antinutritional factors like phytic acid and trypsin inhibitors that severely affects human and animal nutrition (Martín-Cabrejas et al. 2004).

The production yield of beans is highest in the subtropical regions of the United States, Argentina, the Pacific coast of Mexico, Chile and some Asian and European countries, where a warm environment prevails. The traditional yield obtained in these regions is more than 1000 kg/ha. But this contributes to less than 10% of the global bean production. The remaining 90% of the beans are produced in regions under stressed environment conditions where the yield is less than 600 kg/ha. These stressed areas belong to the tropical and subtropical regions of Latin America and Africa. Adverse climatic factors and frequent incidences of diseases and insects cause severe yield loss in these regions (Jones 1999).

Although the bean production has increased to some extent, a tremendous growth is not seen as in the case of other cereal grains. This change can be attributed to the following reasons: (1) shift in consumer preferences from plant protein to animal protein like milk and meat, (2) longer cooking time, (3) antinutritional and flatulence factors that cause discomfort after consumption and (4) the pulse production does not meet the increasing demands of the exceeding population. Because of the increasing demand, many countries, especially China and India, are currently importing pulses thereby increasing the international trade of pulses in a rapid rate (Estévez et al. 1991; FAO 2016).

As per the FAO global pulse trade data, dry beans are the second most exported product next to dry peas but fetch a higher monetary value than dry peas. The global dry bean export stood at four million tonnes in 2013 with a monetary value of 3.8 billion dollars. Also, the total import quantity of dry bean was at 3.4 million tonnes with an estimated value of 3.5 billion dollars in the same year (Miller Magazine 2016).

Chemical and Nutritional Properties

Common bean is nearly considered 'a complete food' because of its rich nutritive composition (Anton et al. 2008b). It is consumed in various parts of the world in different forms and cuisines. The versatile property of these beans in cooking and processing makes it one of the most valuable and commonly consumed crops. Common beans have been a staple food and an integral part of human diet for many centuries. In the present day, these species pave the way for international food security by improving human health and agricultural sustainability (FAO 2016). The nutritional composition of the beans reported by many authors is compiled in Table 1.

Protein

Beans are considered as one of the most important and cheapest sources of protein available for human consumption (Olaofe et al. 1993). Common bean has a dense protein content and therefore supplies a good amount of energy in addition to dietary fibre and some complex carbohydrates. The average crude protein content of

Table 1 Nutritional composition of common bean varieties

Common bean	Protein	Carbohydrate	Lipid	Ash	Reference
Black turtle bean	25.6–25.93	67.83–68.09	1.52–1.59	4.65–44.71	Berrios et al. (1999)
Red kidney bean	25.3	53.2	2	3.3	Olang'o et al. (2000)
White kidney bean	25.63	59.62	1.44	3.55	Güzel and Sayar (2012)
Navy bean	18.15	54.30	2.63	4.14	Sai-Ut et al. (2009)
Cranberry bean	24.04	39.45	1.3	4.05	Wang et al. (2010)
Pinto bean	18.0	59.7	14.4	2.5	Audu and Aremu (2011)
White pea bean	24.49	38.96	1.53	4.29	Wang et al. (2010)

common bean typically ranges from 21 to 25%. The dry bean is attracting special attention worldwide owing to the health impacts of protein from high-fat animals. The amino acid lysine is especially higher in the common bean than the other staple diets such as rice and maize. However, the bean proteins are reasonably deficient in sulphur-containing amino acids, especially methionine and tryptophan (Geil and Anderson 1994; Gupta 1983).

Owing to this, a common practice of consuming the beans with a complementary source of sulphur-containing amino acids is prevalent in many parts of the world. The complementary protein source will usually be cereal grains or flours particularly rice and corn-based products. The complementary sources vary among different nations and their ethnic cuisines. In addition to this, the common bean contains proteinase inhibitors which are responsible for reducing digestibility, and these are only partially deactivated during cooking (Geil and Anderson 1994).

Carbohydrate

The carbohydrate content of the common beans varies between 60 and 65%. Common bean typically consists of complex polysaccharides. Starch is the primary carbohydrate present in all varieties of common beans. Apart from this, small amounts of monosaccharides, disaccharides and oligosaccharides are also present in varying composition depending upon the variety. The common oligosaccharides present are raffinose, stachyose and verbascose. These oligosaccharides remain undigested, because the human digestive system does not have the alpha-galactosidase enzyme required for their hydrolysis. The undigested oligosaccharides cause anaerobic microbial fermentation and result in flatulence. Soaking and cooking the beans prior to cooking effectively removes these oligosaccharides and reduces the flatulence (Cristofaro et al. 1974).

Fibre

Common bean contains considerable amount of carbohydrate as fibres, particularly in the form of cellulose and hemicellulose. The fibre content (both soluble and insoluble fibres) usually ranges between 3 and 7% between different varieties. The rich soluble fibres attribute to the reduction in blood glucose and blood cholesterol levels in humans. Also, the dry beans play a significant role in the human diets because of its good hydration capacity, bulkiness, fermentability and binding nature aiding in smooth functioning of the gastrointestinal tract in humans. Owing to these properties, the common beans are processed in many different traditional ways in each culture (Geil and Anderson 1994).

Fat

Common beans consist of a very low amount of fat and usually range between 0.8 and 1.5%. The fat composition varies greatly depending upon the bean variety and growth conditions. The unsaturated fatty acids, particularly linolenic acid, are the predominant fatty acid in most of the varieties. Also, these are cholesterol-free making them the highly recommended diet for reducing risks of chronic illnesses (Geil and Anderson 1994).

Other Nutrients and Minerals

The common bean contains nutrients and minerals like iron, copper, magnesium, potassium and zinc, which are part of the required daily intake. A single serving cup of beans (a cup) provides more than half of the RDA of folic acid as advised by the US Department of Agriculture. It also meets 25–30% RDA of iron, 25% of magnesium and copper and 15% of potassium and zinc (CGIAR-Beanfocus). Water-soluble vitamins like thiamine, riboflavin, niacin and folacin are present in beans (Bourne 1989). During industrial processing methods, particularly canning, these water-soluble vitamins are significantly lost, thereby reducing the nutrient content. However, normal cooking temperatures do not seem to affect these components greatly (Deshpande et al. 1984). Coloured beans possess polyphenols that are rich in antioxidants, thereby reducing the oxidative stress (Anton et al. 2008a). These phenolic compounds are responsible for the different seed coat colours in common beans (Beninger and Hosfield 2003). Owing to the beneficial and health-promoting factors, pulses especially beans are finding increased applications in the nutraceutical industries for the development of value-added bean-based functional foods.

Although common bean varieties contain high amounts of vitamins and minerals, these cannot be correlated with high bioavailability of these elements. This is because these elements in most cases interact with other nutrients present in the bean like phenolic compounds and phytic acids, thereby reducing the bioavailability. Minerals obtained from plant sources are in general less bioavailable than those obtained from animal sources (Sgarbieri 1984).

Apart from this, consumption of beans has been proven to reduce the risks of heart diseases, obesity and cancer (Anton et al. 2008b). Common bean takes a longer time for digestion and suppresses the sudden rise in blood sugar levels making it a good appetite suppressant (Katungi et al. 2009).

Antinutritional Factors

Apart from the rich nutritive composition, common beans also possess significant amount of antinutritional factors which interfere with their nutritional benefits. These include protease inhibitors, particularly trypsin and chymotrypsin, alpha amylase inhibitors, tannins, phytates, polyphenols, antivitamin factors, goitrogens, cyanogen, oxalate, lectin, raffinose family of oligosaccharides and saponins (Geil and Anderson 1994; Haileslassie et al. 2016). The protease inhibitors and other antinutrients that interfere with the protein digestibility are sensitive to heat and are destroyed mostly by cooking the beans. Therefore, cooking is done to improve the protein digestibility of the beans. But tannins, phytates and polyphenols are highly heat stable and bind with the minerals like iron, calcium and zinc, thus reducing their bioavailability upon metabolism. It is also proven that polyphenols and phytates affect iron absorption especially in women (Haileslassie et al. 2016; Petry et al. 2010) Traditionally, soaking and germination are done to increase the nutrient and mineral availability. Later, studies have also reported that soaking and germination result in diffusion of water-soluble antinutrients and hydrolysis due to enzyme activation. These processes significantly reduce the antinutritional factors in varying levels (Haileslassie et al. 2016; Hotz and Gibson 2007). On the contrary, some studies show that polyphenols play an important role in the prevention of cardiovascular diseases and phytates in the prevention of cancer (Haileslassie et al. 2016).

Postharvest Processing

Stages in Postharvest Processing

The common beans of *Phaseolus vulgaris* belong to the internationally recognised class of beans because of their vast production and consumption throughout the world. These are recognised in the developing countries because of their rich nutritional contents, good shelf life and easy storage conditions. Despite the beneficial properties, these beans have some limitations like the presence of antinutritional factors, longer cooking time and low sulphur-containing amino acids which affect the processing conditions. Figure 2 depicts the different stages involved in postharvest processing of common beans to reduce the limitations.

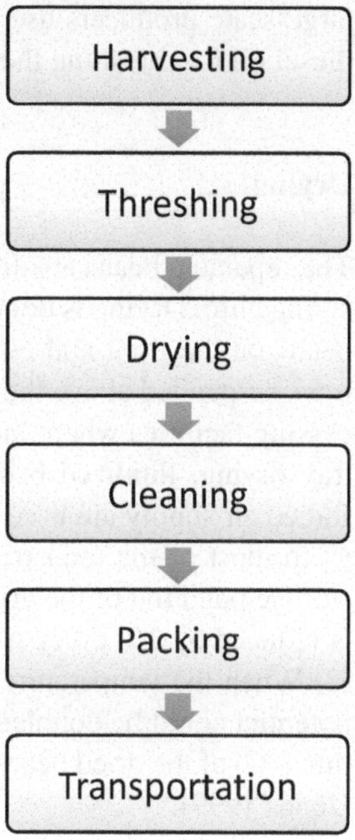

Fig. 2 Postharvest processing methods of common beans

Harvesting

During harvesting, factors like humidity, temperature and climatic conditions must be considered as they may have adverse effects on quality of bean. Even when the bean is harvested from the crop, the bean seed respires and continues to ripen. When the environmental conditions accelerate this process, the rate of biochemical reactions would increase causing deterioration in bean quality. Therefore, the moisture levels of the bean and the humidity should range between 13–15% and 12–13%, respectively, during harvesting. Harvesting is done manually in most of the countries. But in some developed countries, mechanised harvesting is done using combined harvesters (Jones 1999).

Threshing

After harvesting, the bean plants are dried to reduce the moisture content. The small-scale farmers leave the harvested plants on the fields to dry under sunlight, whereas large-scale producers use drying silos. The bean plants are threshed to separate the bean from the plants. Traditional method involves beating the dried plants piled on plastic bags to separate the bean. It is either beaten by sticks or spread on the floor and run over by tractors or other vehicles. Also, the beans that are used for seeds are carefully threshed by hand to avoid damage to the beans. The commercial large-scale producers use standing or moving machineries that run on gasoline or diesel oil for threshing the bean (Jones 1999).

Drying

The separated beans are dried again to achieve a final humidity of 11–12% for better storage life. Drying is done naturally in most developing countries by spreading the beans on the floor and covering it with cloth to absorb the humidity or on netted trays suspended above the ground in the direction of the wind and sun. In bean processing facilities where large volumes are involved, artificial drying processes like tray drying, fluidized bed drying and tunnel drying with high temperatures and forced air supply are used to rapidly dry the beans.

In most of the industrial units, three practical rules are observed to decide the storage potential of the dried beans. They are the following: (1) When the humidity is reduced by 1% every time, the storage potential could be doubled subsequently. (2) When the temperature of the seeds is reduced by 5 °C every time, the storage potential could be doubled. (3) When the sum of humidity level (wb) and temperature (°C) of the dried beans is less than 45, adequate storage conditions are achieved (Jones 1999).

Cleaning

Cleaning is done to separate the bean from any undesired particles like stones, dirt, chaffs or any other lighter materials. The small-scale plants use air currents and sieves to segregate the bean from other particles followed by a manual process of removing the broken or discoloured seeds. The seeds will be allowed to fall on the ground from a height in a windy area. The strong air currents carry away all the lighter materials and the heavy seeds will fall. As an alternative wind source, electric fans or pumps are also used for cleaning. Sieves are used to remove the heavier particles. The mesh size of the sieves is adjusted to that of the bean size, so that the particles smaller than the bean would fall through the mesh. Final sorting is done manually by spreading the bean on a table painted blue for contrast and picking up all the damaged and discoloured seeds.

Large-scale facilities use specialised air and sieving machinery to clean and sort the beans. The distance of the fall from the sieving mesh to the ground is adjusted such that the seed damage would be minimum. The sorted beans are then sent for packing and distribution (Jones 1999).

Packaging

The clean and sorted beans are then packed in polypropylene bags, laminated paper bags or burlap bags and are transported to domestic and international markets.

Transportation

In small-scale farms, transportation is not really a problem as the crops are either stored for seed or for domestic consumption. Also, transportation conditions do not affect the quality significantly as in the case of greens. If the distance is large, the transportation cost would reflect on the price of the commodity, and the intermediate sellers would make profit on it. In case of domestic retail shops or markets, small trucks or trailers are used for transporting the harvested beans. When huge volumes need to be exported, then bulk containers are used. In both cases, the physical handling of beans must be kept to a minimum, and moisture and temperature controls should be monitored throughout. The containers and trucks must be sanitised prior to loading and must be inspected regularly (Jones 1999).

Storage

Storage critically affects the bean quality which in turn affects the final product quality. Storage conditions like temperature, relative humidity and bean moisture level must be monitored regularly. The humidity and moisture levels are usually kept below 12% to preserve the bean quality. The beans are stored over a period to speculate a higher selling price during off-seasons and for a safe consumption. Also, farmers store the seeds for the next harvest season. When the storage conditions are not maintained properly, it will lead to bean defects that may even alter the final grain texture. For example, 'bin burn' is a defect caused by high moisture and high temperature during storage that leads to brown discolouration and off-flavour development in beans. Low moisture, low humidity and high temperature during storage result in a phenomenon called 'hard shell' or 'hard-to-shell', wherein the beans fail to absorb water during soaking. The beans slowly soften during cooking but require a longer cooking period. The seed coat of the bean undergoes physical alterations in its cell structure and is caused by both genetic and environmental conditions (Coelho et al. 2007; Jones 1999). This condition is naturally reversible when the relative humidity is high. Another condition called 'hard to cook' occurs in grains under high temperatures and high relative humidity. This occurs when the cotyledons of the bean undergo alterations due to the environmental conditions. These grains become hard due to the cross-linking of tannins with the macromolecules of the cotyledon cell wall and fail to soften during cooking, even though they absorb enough water on soaking. Aguilera and Steinsapir studied the intracellular modifications of the hard-to-cook beans during soaking and cooking. They reported that in 'hard-to-cook' beans, the middle lamella was very tough and did not loosen causing intracellular separation, whereas in normal beans, the middle lamella loosened sufficiently allowing individual cells to separate without breaking the cell wall (Aguilera and Steinsapir 2013). The 'hard-to-cook' condition is irreversible, and the development mechanism and hereditary conditions are still not clearly understood (Coelho et al. 2007; Jones 1999).

Table 2 Storage facilities for dry beans

Storage type	Storage containers	Storage period	Storage conditions
>1000 kg	Large silos made of wood, steel or concrete	Greater than 6 months	12–15% humidity
200–1000 kg	Containers made of plastic, metal, wood or aluminium	3–6 months	11–12% humidity
<200 kg	Earthenware pots, jute sacks, burlaps, gallon tins or straw baskets	Less than 3 months	Less than 12% humidity

Source: Phaseolus Bean: Post-harvest Operations, Food and Agriculture Organisation of the United Nation. INPhO-Post-harvest Compendium (Jones 1999)

The dry beans are usually stored for less than 6 months in small-scale industries, except those that are used as seeds which are stored for a relatively longer period after treating with insecticides. The different materials used for the storing beans along with the proper storage conditions are given in Table 2.

Processing Methods to Reduce Antinutritional Factors

Even though common beans have antinutritional properties, low sulphur-containing amino acids, flatulence factors that would account for low protein digestibility, development of an easily consumable product with good shelf life, better nutritional properties and reduced antinutritional compounds are feasible owing to the different processing methods. The nutritional properties of the common beans can be effectively improved by different processing methods discussed below.

Soaking

Soaking is the most common household practice done to improve digestibility. During soaking, water-soluble minerals like Na, K and Mg phytate will diffuse into the soaking medium and reduce the phytate levels. Also, some polyphenols and oxalates that retard the absorption of minerals like calcium and iron are lost during soaking, thereby improving their absorption (Hotz and Gibson 2007; Patterson et al. 2017). Beans soaked for 16 h (in 0.03% Na_2 EDTA) followed by cooking for 60 min showed reduction in phytic acid, trypsin inhibitor activity and improved protein digestibility (Estévez et al. 1991). Also, it has been proved that the bioavailability of zinc increased when beans are soaked in EDTA solution. Zinc forms soluble complexes with phytic acid, and these complexes are easily absorbed by the intestine (Estévez et al. 1991; Reddy et al. 1984). Other studies had analysed the effects of soaking with or without cooking on the antinutrient content of the beans and had obtained varying results. A minimum of 4 h of soaking is observed in several studies to reduce the phytate, tannin and polyphenol content of the beans (Haileslassie et al.

2016). The bean variety and soaking conditions such as the seed to water ratio, soaking medium, soaking time and soaking temperature are important factors that influence the extent of antinutrient reduction. The seed to water ratio varies between 1:2 and 1:10 at room temperature with water (distilled, tap or deionised) as the soaking medium for most of the studies, and the soaking time ranges from 4 to 48 h (Haileslassie et al. 2016).

Dehulling

Dehulling, also called as decortication, is the process of removing the seed coat in pulses. Some studies report that dehulling facilitates the removal of some of the antinutrients present in the beans, reduces the cooking time and improves the quality and digestibility of bean proteins. However, most of the antinutrients are present only in the cotyledons, except polyphenols which are present in the seed coat. Therefore, many other studies suggest that dehulling results in an overall increase in the antinutrient content of the beans. Dehulling also causes an increase in the enzyme inhibitors commonly present in the beans. The dehulled red kidney bean variety showed a 2.7% increase in the chymotrypsin inhibitors (Patterson et al. 2017).

The polyphenols present in the dry beans are mostly concentrated on their seed coats, and dehulling results in a decrease in phenolic compounds and tannins. Several studies showed a decrease in the total phenolic content (TPC) of various pulse varieties after decortication. In accordance with this, Deshpande et al. (1982) found a 68–94% reduction in the tannin content of pigmented bean cotyledons after dehulling. However, dehulling has no effects on the lectin content of the dry beans (Deshpande et al. 1982; Dhurandhar and Chang 1990; Patterson et al. 2017; Shimelis and Rakshit 2007).

Germination

Germination is one of the common traditional processes that is done to enhance the nutritional profile of the beans. It also improves the digestibility and acceptability of the final product. The germination conditions such as moisture, pH, duration and light or dark conditions vary for each pulse type and variety. Therefore, it becomes difficult to completely understand and compare the effect of germination on the levels of antinutrient reduction. However, methodical studies were conducted on the edible beans by various authors, and they reported varying levels of reduction in the antinutrient levels. The germination process was found to be more effective than the thermal processing methods to reduce the phytate content of beans. The germination process increases the activity of the phytase enzyme which is responsible for the hydrolysis of phytate thereby indirectly reducing the phytate content in the final product. There are many factors that influence the rate of phytate hydrolysis. These include species, variety, germination conditions (pH, moisture, temperature), germination stage, phytate solubility and presence of specific inhibitors (Hotz and

Gibson 2007; Patterson et al. 2017). Vidal-Valverde and his co-workers (1994) conducted many studies to optimise the germination conditions and to reduce the phytate and tannin levels in lentils and edible beans. They reported that the maximum level of phytate reduction to lower inositol phosphates in beans was observed after 6 days of germination in the presence of light. The white kidney beans when germinated for 7 days in light showed a 76% reduction in the TI activity and 85% reduction in the lectin content (Savelkoul et al. 1994; Patterson et al. 2017). Germination has also been shown to reduce the polyphenols by 64% in kidney beans, and mixed observations have been reported for reduction of TI and tannin content in pulses (Patterson et al. 2017).

Fermentation

Fermentation is one of the oldest methods of preserving food. In beans, it improves the nutritional and sensory properties and reduces the antinutrient content significantly. Fermentation in beans either can occur spontaneously by its native microflora or can be artificially done by starter culture inoculation. It involves the activity of microbial phytase enzymes that hydrolyse phytates to its lower inositol phosphates. In contrast to the phytates, these hydrolysed phosphates do not inhibit calcium and iron absorption. Also, the protein quality, digestibility and vitamin B content of the beans have been shown to improve during fermentation. In addition, it has been proven that fermentation improves the shelf life and the microbiological safety of the product (Hotz and Gibson 2007). Also, low molecular weight organic acids such as citric acid, malic acid and lactic acid are also produced during fermentation, and they are found to increase the iron and zinc absorption. However, these results are only based on in vitro studies, and they further require in vivo confirmatory studies (Hotz and Gibson 2007).

Khattab et al. (2009) investigated two varieties of kidney beans and found that there is an increase in the in vitro protein digestibility (IVPR) when the beans are soaked and fermented (Khattab et al. 2009; Oghbaei and Prakash 2016). When beans are fermented with lactic acid bacteria (inoculated fermentation), 20.2% and 31.6% reductions in the TI activity were observed in the raw and the cooked beans, respectively. Also, 38, 50 and 95% reductions in the TI activity were observed in LAB-fermented raw bean flours, 12-hour-soaked flours and 12-hour-soaked and 90-min-cooked flours, respectively.

During fermentation, the tannins present in the beans either diffuse into the water or form insoluble complexes with the bean proteins, which reduces the tannin content. A maximum reduction of around 89% in the tannin content of soaked, cooked and fermented beans was reported by Barampama and Simard (1994). Khattab and Arntfield (2009) also reported notable reduction in the tannin content of red kidney bean flour fermented for 24 h with *Saccharomyces cerevisiae* starter culture (Barampama and Simard 1994; Khattab and Arntfield 2009; Patterson et al. 2017). Fermentation with or without other pretreatment methods effectively reduces the antinutrients and improves the nutritional and flavour profile of beans.

Thermal Processing Methods

Thermal processing of common beans facilitates the inactivation and destruction of the heat labile antinutritional factors and significantly enhances the protein quality and digestibility of beans (Tiwari and Singh 2012). There are various thermal treatment methods that have various effects on the antinutrients present in the common beans. These are discussed below.

Cooking

Cooking beans is a common household practice done in many parts of the world to improve the palatability and digestibility of the beans. Cooking is done in two ways—either by directly boiling the beans in an open pan or by cooking it under pressure in a closed vessel. The effects of cooking on the antinutrient factors vary with the bean type and the cooking conditions. Studies report the boiling conditions to range between 90 and 100 °C for 15–35 min with a 1:4 seed to water ratio and pressure-cooking conditions at 121 °C at 15 psi for 15–30 min. The common bean varieties, especially the black turtle bean, cranberry bean and dark red kidney bean, contain tannins which are concentrated on their seed coat. Cooking significantly reduces the tannin and trypsin inhibitor activity in the common beans but causes less significant effects on the phytate reduction since phytates are relatively heat stable. The TI activity was reduced by around 93% in common beans by conventional cooking process (Patterson et al. 2017).

Cooking beans has also been found to inactivate toxic elements like lectins (hemagglutinins) and glycoproteins. Grant et al. (1982) reported that the beans hydrated and boiled for 10 min had their toxic elements inactivated. Bender and Reaidi (1982) found that the hemagglutinin activity of partially cooked common beans (10–15 min at 80 °C) increased around 5 times and hence reported that partially cooked beans are more toxic than the raw beans (Koehler et al. 1986).

Microwave Heating

Microwave heating is a type of thermal processing that uses magnetic waves to heat food materials that are non-conductive and provides uniform heating. Microwave heating significantly reduces cooking time and requires low maintenance with easy operations when compared to conventional cooking methods (Chandrasekaran et al. 2013; Patterson et al. 2017). During microwave processing, changes in the functional, rheological and morphological properties of the pulse starches occur. These changes improve the nutritional profile through destruction of the antinutritional factors present in the pulse (González and Pérez 2002; Dronachari and Yadav 2015). The trypsin inhibitor (TI) activity was found to be reduced by approximately 60% in the common bean varieties during microwave processing. The reduction level is lower than that of the conventional cooking process. Also, the lectin content of the common beans is not greatly reduced by microwave processing. This method may need further research on the optimisation of microwaving conditions to improve the efficiency and to increase the elimination of antinutrients in pulses (Patterson et al. 2017).

Micronisation

Micronisation, an infrared heating method, is used for processing pulses, cereal grains and other plant materials to reduce the cooking time and to achieve a higher process efficiency. This method also reduces the processing costs when compared to the conventional heating methods. Bellido et al. (2003) studied the effectiveness of micronisation in navy beans and found that the cooking time effectively reduced when suitable pretreatment methods are used. This is because, during micronisation, heating temperatures around 90 °C could be attained within processing times as short as 1 min. The optimisation of suitable pretreatment method for tempering the bean is the most crucial stage in micronisation. The tempering method affects the moisture content of beans and therefore influences the desirability of the micronised product. Tempering increases the moisture content of the beans and improves the gelatinisation and protein denaturation processes during heating, thereby reducing the cooking time. However, excess moisture is not desirable for micronisation, and hence optimum moisture content must be maintained (Bellido et al. 20032006).

Micronisation also improves the digestibility by reducing the phytates and TI activity in pulses. Fasina et al. (2001) observed a 50% reduction in the TI activity of black beans and less than 10% reduction in kidney beans during micronisation and reported that the reduced efficiency in kidney beans is due to the large size of the seed that arrested heat penetration inside the bean. On the contrary, significant reduction in the antinutrient levels of red kidney beans (more than 88% reduction in TI activity and around 35% reduction in phytate levels) when micronised to 90 °C has been reported by Khattab and Arntfield (2009). Also varying levels of reduction in tannin content of kidney bean grown in Canada (6.3%) and Egypt (82.5%) have been observed by them. Thus micronisation is reported to have shown distinct differences in antinutrient reduction levels between different pulse types and varieties, and the exact mechanism of action is still unclear (Fasina et al. 2001; Khattab and Arntfield 2009; Patterson et al. 2017).

Extrusion

Though thermal processing of beans affects the phenolic and antioxidant properties of beans, the exact effect of the extrusion processing of beans on these properties is not clearly understood (Anton et al. 2009). Also, extrusion cooking of beans inactivates several antinutrients that hinder their widespread applications. When the corn starch was fortified with different percentages (15–45%) of bean flour, the protein content was found to initially increase 12-fold, and also a significant increase in the levels of polyphenols and antioxidant activity was also found (Anton et al. 2009). Although these levels reduced after extrusion, antinutritional factors like phytic acid and trypsin inhibitors were reduced to almost 50 and 100%, respectively, indicating that the extruded bean-based products are safe for human consumption (Anton et al. 2009). Alonso and Marzo reported that extrusion cooking was the best method to abolish the antinutritional compounds like trypsin, chymotrypsin and α-amylase inhibitors present in common beans without altering its protein content (Alonso et al. 2000).

Irradiation

Irradiation is the process of using ionising radiations to internally heat and cause changes in the nutritional properties of the food. There are very few studies on irradiation of common beans, and the potential of irradiation in reducing the antinutrients needs further research. El-Niely (2007) studied the effect of varying doses of gamma irradiation in reducing the phytate and tannin content of kidney beans. He reported that radiation doses at 5, 7.5 and 10 kGy reduced the phytic acid content by 7.5, 14.2 and 26.9%, respectively. Also, a 25% reduction in the tannin content of kidney bean was reported at 10 kGy dose of gamma irradiation. It was found that the reduction levels of both phytic acid and tannin content increased linearly with increase in the radiation dose (El-Niely 2007; Patterson et al. 2017).

Product Development

Traditional Pulse Products

The common beans have been harvested and consumed by traditional methods for many decades. The Mediterranean, Asian and Middle Eastern cuisines include common beans in their traditional salads, soups, patties, pilaf and kofte. It is often mixed with other foods such as meat and vegetables. Beans also remain as the staple food for certain regions of the Middle Eastern, Kenyan and Mediterranean countries.

Bean-Based Products

The development of bean-based products is gaining momentum in the western countries because of the increasing popularity of the health benefits of pulses. Since the beans contain twice the amount of protein present in cereals, the protein fraction is used to develop protein snack bars and snack foods.

Functional Ingredient

The common beans, because of its rich nutrient profile, often find application in a variety of products as a functional ingredient. The bean flour is incorporated in many bakery products such as breads, cakes, buns, muffins, donuts, chips and snack bars to enhance the fibre content of the final product. The bean flour is also added in soup formulations to develop ready to cook and instant soup mixes. The protein fraction of the bean is extracted to prepare functional ingredients like protein concentrates and protein isolates. These protein isolates and concentrates are used for their properties to manufacture ingredients such as emulsifiers, foaming agents, flavour binders, lipid binders, gelling agents and thickeners (Taylor et al. 1984). In particular, the cranberry bean flour and small red kidney bean flour have high foaming and emulsion properties along with a good stability when compared with black turtle bean flour. But the overall foaming and emulsion capacities of the common bean flours are reported to be relatively higher than that of the commercial cereal flours (Siddiq et al. 2010).

Ready to Cook and Extruded Products

The whole bean flour is used to replace the cereal flours to improve the nutritional profile of ready to cook and extruded foods like pasta, spaghetti, breakfast cereals, croutons, bread sticks, flat breads, salty snacks and meat substitutes. Anton et al. (2009) studied the nutritional impacts of the addition of navy and red bean flours to corn starch at different levels (15, 30 and 45%) to produce a fortified puffed snack through extrusion cooking. The study revealed that 30% substitution of bean flours was feasible to produce an extruded snack with increased nutritional functionality. The crude protein level was found to be increased by 12-fold along with a reduction in antinutritional compounds like phenolics, trypsin inhibitors and phytic acid (Anton et al. 2009). Rocha-Guzman et al. (2008) investigated the physical properties of extruded products from three common bean varieties of Mexico and reported decrease in oil absorption capacity and increase in the emulsifying capacity of the extruded products. The study concluded that common bean could be possibly used for the production of extruded products with good physical properties (Rocha-Guzman et al. 2008).

The effect of the addition of common bean flour to semolina in the production of spaghetti pasta was studied by Gallegos-Infante et al. (2010). The cooking time and the water absorption capacity of the flour were found to decrease after the addition of bean flour. The study also reported a linear relationship between the colour change and bean flour content in the developed pasta (Gallegos-Infante et al. 2010).

Ready-to-Eat Foods

Ready-to-eat foods are a widely consumed category of foods in the present day, especially cookies and biscuits. These foods are easy to prepare and consist of mainly carbohydrates, fats and sugars. The refined cereal flour used in their preparation is now being replaced with the nutritionally rich bean flour. This increases the fibre and protein content of the final product and improves satiety while maintaining the blood glucose levels. A variety of homemade and industrially processed ready-to-eat foods are now available in the international market (Nesli et al. 2016).

Health Conscious Formulations

In the present scenario, owing to the increased awareness on health, whole bean ingredients are dehydrated, milled and precooked and developed into a ready to cook ingredient. These can be easily added to prepare an instant health conscious formulation. These pretreated ingredients could be reconstituted within minutes. The major advantage of these products is that the nutritional and sensory attributes are least affected during preparation.

Future Trends in Processing and Product Development

Common bean is found to be a healthy alternative to starch-based foods by people all over the world. The growing awareness is mainly due to the rich nutritional composition of the beans. The higher amounts of proteins and complex carbohydrates, especially dietary fibres and resistant starches, enhance the health of the digestive system and provide good energy for the human metabolism. Also, the low-fat content and presence of vitamins and minerals such as iron, calcium and folate make them a very healthy choice for health management and disease prevention. The common bean varieties are gluten-free and could be consumed by people who are gluten intolerant or have celiac disease. So, beans or bean flours could be used to replace cereal flours which contain gluten to develop product with special dietary needs. Also, since beans are rich in protein, they would play a major role in the vegetarian or vegan diets. The common bean varieties are high in fibre, protein and complex carbohydrates and low in fat and caloric density. Also, the glycaemic index of the beans is low, and they maintain the blood glucose levels without sharp fluctuations. This results in a more stabilised insulin release. The resistant starches present in the beans are digested slowly and are responsible for satiety when consumed. These facts suggest that the bean varieties could be possibly used as a healthy food choice for diabetic people and for weight management.

The contribution of bean-based diet consumption towards reducing the risk of cancer is being widely studied in the present. It has been reported that the phenolic compounds, antioxidants and other bioactive compounds present in beans reduce the risks of developing certain types of cancers such as breast, stomach, kidney and prostate cancer. Also, the fibre content of the beans is reported to reduce the risk of colorectal cancer. Upon further studies, bean-based products could further be developed to significantly reduce the risks of cancer (Dahm et al. 2010; Lanza et al. 2006).

The pulse industry has been growing steadily in the recent years owing to many factors such as the bean nutritional properties, increasing demand, growing population and rapid urbanisation. The bean application in food and feed markets is also increasing at a rapid pace, and the increasing awareness about human health makes common bean as one of the healthiest alternatives for disease prevention and other serious health concerns. Also, the current research focusses more on the application of pulses, pulse flours and their individual fractions in developing a more nutritionally rich product. Further research and a deeper understanding of the beneficial properties of different common bean varieties would pave the way for the development of a whole new range of bean-based products in domestic and international markets.

References

Aguilera, J. M., & Steinsapir, A. (2013). Dry processes to retard quality losses of beans (Phaseolus vulgaris) during storage. *Canadian Institute of Food Technology Journal, 18*, 72–78. https://doi.org/10.1016/s0315-5463(85)71723-3.

Alonso, R., Aguirre, A., & Marzo, F. (2000). Effects of extrusion and traditional processing methods on antinutrients and in vitro digestibility of protein and starch in faba and kidney beans. *Food Chemistry, 68*, 159–165. https://doi.org/10.1016/S0308-8146(99)00169.

Anton, A. A., Gary Fulcher, R., & Arntfield, S. D. (2009). Physical and nutritional impact of fortification of corn starch-based extruded snacks with common bean (Phaseolus vulgaris L.) flour: Effects of bean addition and extrusion cooking. *Food Chemistry, 113*, 989–996. https://doi.org/10.1016/j.foodchem.2008.08.050.

Anton, A. A., Ross, K. A., Beta, T., Gary Fulcher, R., & Arntfield, S. D. (2008a). Effect of predehulling treatments on some nutritional and physical properties of navy and pinto beans (Phaseolus vulgaris L.). *LWT - Food Science and Technology, 41*, 771–778. https://doi.org/10.1016/j.lwt.2007.05.014.

Anton, A. A., Ross, K. A., Lukow, O. M., Fulcher, R. G., & Arntfield, S. D. (2008b). Influence of added bean flour (Phaseolus vulgaris L.) on some physical and nutritional properties of wheat flour tortillas. *Food Chemistry, 109*, 33–41. https://doi.org/10.1016/j.foodchem.2007.12.005.

Audu, S. S., & Aremu, M. O. (2011). Nutritional composition of raw and processed pinto bean (Phaseolus vulgaris L.) grown in Nigeria. *Journal of Food, Agriculture and Environment, 9*, 72–80.

Barampama, Z., & Simard, R. E. (1994). Oligosaccharides, antinutritional factors, and protein digestibility of dry beans as affected by processing. *Journal of Food Science, 59*, 833–838. https://doi.org/10.1111/j.1365-2621.1994.tb08139.x.

Bellido, G., Arntfield, S. D., Cenkowski, S., & Scanlon, M. (2006). Effects of micronization pretreatments on the physicochemical properties of navy and black beans (Phaseolus vulgaris L.). *LWT - Food Science and Technology, 39*, 779–787. https://doi.org/10.1016/j.lwt.2005.05.009.

Bellido, G. G., Arntfield, S. D., Scanlon, M. G., & Cenkowski, S. (2003). The effect of micronization operational conditions on the physicochemical properties of navy beans (Phaseolus vulgaris L.). *Journal of Food Science, 68*, 1731–1735. https://doi.org/10.1111/j.1365-2621.2003.tb12320.x.

Bender, A. E., & Reaidi, G. B. (1982). Toxicity of kidney beans (*Phaseolus vulgaris*) with particular reference to lectins. *Journal of Plant Foods, 4*(1), 15–22.

Beninger, C. W., & Hosfield, G. L. (2003). Antioxidant activity of extracts, condensed tannin fractions, and pure flavonoids from Phaseolus vulgaris L. seed coat color genotypes. *Journal of Agricultural and Food Chemistry, 51*, 7879–7883. https://doi.org/10.1021/jf0304324.

Berrios, J. D. J., Swanson, B. G., & Adeline Cheong, W. (1999). Physico-chemical characterization of stored black beans (Phaseolus vulgaris L.). *Food Research International, 32*, 669–676. https://doi.org/10.1016/S0963-9969(99)00144-1.

Bourne, G. H. (1989). *Nutritional value of cereal products, beans and starches*. Basel: Karger.

CGIAR-Beanfocus. (2019) *Common bean: The nearly perfect food—The importance of common bean*. Retrieved from http://ciat-library.ciat.cgiar.org/articulos_ciat/ciatinfocus/beanfocus.pdf

Chandrasekaran, S., Ramanathan, S., & Basak, T. (2013). Microwave food processing—A review. *Food Research International, 52*, 243–261. https://doi.org/10.1016/j.foodres.2013.02.033.

Coelho, C. M. M., de Mattos Bellato, C., Santos, J. C. P., Ortega, E. M. M., & Tsai, S. M. (2007). Effect of phytate and storage conditions on the development of the 'hard-to-cook' phenomenon in common beans. *Journal of Science and Food Agriculture, 87*, 1237–1243.

CGIAR-Common Bean. (2019) *Grain legumes—A CGIAR Research Program*. Retrieved from https://www.grainlegumes.cgiar.org/crops/common-bean/

Cristofaro, E., Mottu, F., & Wuhrmann, J. J. (1974). Involvement of the raffinose family of oligosaccharides in flatulence. In H. L. Sipple & K. W. McNutt (Eds.), *Sugars in nutrition*. New York: Academic.

Dahm, C. C., Keogh, R. H., Spencer, E. A., Greenwood, D. C., Key, T. J., Fentiman, I. S., Shipley, M. J., Brunner, E. J., Cade, J. E., Burley, V. J., Mishra, G., Stephen, A. M., Kuh, D., White, I. R., Luben, R., Lentjes, M. A. H., Khaw, K. T., & Rodwell, S. A. (2010). Dietary fiber and colorectal cancer risk: A nested case-control study using food diaries. *Journal of the National Cancer Institute, 102*, 614–626. https://doi.org/10.1093/jnci/djq092.

Deshpande, S. S., Sathe, S. K., Salunkhe, D. K., & Cornforth, D. P. (1982). Effects of dehulling on phytic acid, polyphenols, and enzyme inhibitors of dry beans (Phaseolus vulgaris L.). *Journal of Food Science, 47*, 1846–1850. https://doi.org/10.1111/j.1365-2621.1982.tb12896.x.

Deshpande, S. S., Sathe, S. K., Salunkhe, D. K., & Rackis, J. J. (1984). Dry beans of *Phaseolus:* A review. Part 3. *C R C Critical Reviews in Food Science and Nutrition, 21*(2), 137–195. https://doi.org/10.1080/10408398409527399.

Dhurandhar, N. V., & Chang, K. C. (1990). Effect of cooking on firmness, trypsin inhibitors, lectins and cystine/cysteine content of navy and red kidney beans (Phaseolus vulgaris). *Journal of Food Science, 55*, 470–474. https://doi.org/10.1111/j.1365-2621.1990.tb06789.x.

El-Niely, H. F. G. (2007). Effect of radiation processing on antinutrients, in-vitro protein digestibility and protein efficiency ratio bioassay of legume seeds. *Radiation Physics and Chemistry, 76*, 1050–1057. https://doi.org/10.1016/j.radphyschem.2006.10.006.

Estévez, A. M., Castillo, E., Figuerola, F., & Yáñez, E. (1991). Effect of processing on some chemical and nutritional characteristics of pre-cooked and dehydrated legumes. *Plant Foods for Human Nutrition, 41*, 193–201. https://doi.org/10.1007/BF02196387.

FAO pulses – International year of pulses. (2016). *Trends in worldwide production, consumption and trade of pulses*. Retrieved from http://www.fao.org/pulses-2016/news/news-detail/en/c/381491/

FAOSTAT. (2019). *Crops—Area harvested, yield and production quantity*. Retrieved from http://www.fao.org/faostat/en/#data/QC

Fasina, O., Tyler, B., Pickard, M., Zheng, G. H., & Wang, N. (2001). Effect of infrared heating on the properties of legume seeds. *International Journal of Food Science and Technology, 36*, 79–90. https://doi.org/10.1046/j.1365-2621.2001.00420.

Gallegos-Infante, J. A., Rocha-Guzman, N. E., Gonzalez-Laredo, R. F., Ochoa-Martínez, L. A., Corzo, N., Bello-Perez, L. A., Medina-Torres, L., & Peralta-Alvarez, L. E. (2010). Quality of spaghetti pasta containing Mexican common bean flour (Phaseolus vulgaris L.). *Food Chemistry, 119*, 1544–1549. https://doi.org/10.1016/j.foodchem.2009.09.040.

Geil, P. B., & Anderson, J. W. (1994). Nutrition and health implications of dry beans: A review. *Journal of the American College of Nutrition, 13*, 549–558. https://doi.org/10.1080/07315724.1994.10718446.

Gentry, H. S. (1969). In H. Scott (Ed.), *Origin of the common bean, Phaseolus vulgaris*. New York: Springer. Retrieved from http://www.jstor.org/stable/4253014.

González, Z., & Pérez, E. (2002). Evaluation of lentil starches modified by microwave irradiation and extrusion cooking. *Food Research International, 35*, 415–420. https://doi.org/10.1016/S0963-9969(01)00135-1.

Grant, G., More, L. J., McKenzie, N. H., & Pusztai, A. (1982). The effect of heating on the haemagglutinating activity and nutritional properties of bean (Phaseolus vulgaris) seeds. *Journal of the Science of Food and Agriculture, 33*(12), 1324–1326.

Gupta, Y. P. (1983). Nutritive value of food legumes. In S. K. Arora (Ed.), *Chemistry and biochemistry of legumes* (pp. 187–327). London: Edward Arnold Publication.

Güzel, D., & Sayar, S. (2012). Effect of cooking methods on selected physicochemical and nutritional properties of barlotto bean, chickpea, faba bean, and white kidney bean. *Journal of Food Science and Technology, 49*, 89–95. https://doi.org/10.1007/s13197-011-0260-0.

Haileslassie, H. A., Henry, C. J., & Tyler, R. T. (2016). Impact of household food processing strategies on antinutrient (phytate, tannin and polyphenol) contents of chickpeas (Cicer arietinum L.) and beans (Phaseolus vulgaris L.): A review. *International Journal of Food Science and Technology, 51*, 1947–1957. https://doi.org/10.1111/ijfs.13166.

Hotz, C., & Gibson, R. S. (2007). Traditional food-processing and preparation practices to enhance the bioavailability of micronutrients in plant-based diets. *The Journal of Nutrition, 137*, 1097–1100. https://doi.org/10.1093/jn/137.4.1097.

Jones, A. L. (1999). *Phaseolus bean: Post-harvest operations*. Rome: Food and Agriculture Organisation of the United Nation. INPhO-post-harvest Compend. P25.

Katungi, E., Farrow, A., Chianu, J., Sperling, L., & Beebe, S. (2009). *Common bean in eastern and southern Africa: A situation and outlook analysis* (pp. 1–44). Cali: International Centre for Tropical Agriculture.

Khattab, R. Y., & Arntfield, S. D. (2009). Nutritional quality of legume seeds as affected by some physical treatments 2. Antinutritional factors. *LWT - Food Science and Technology, 42*, 1113–1118. https://doi.org/10.1016/j.lwt.2009.02.004.

Khattab, R. Y., Arntfield, S. D., & Nyachoti, C. M. (2009). Nutritional quality of legume seeds as affected by some physical treatments, part 1: Protein quality evaluation. *LWT - Food Science and Technology, 42*, 1107–1112. https://doi.org/10.1016/j.lwt.2009.02.008.

Koehler, H. H., Herrick, H. E., & Burke, D. W. (1986). Differentiating the lectin activity in twenty-four cultivars of dry beans (Phaseolus vulgaris L.). *Journal of Food Science, 51*, 1471–1475. https://doi.org/10.1111/j.1365-2621.1986.tb13837.x.

Lanza, E., Hartman, T. J., Albert, P. S., Shields, R., Slattery, M., Caan, B., Paskett, E., Iber, F., Kikendall, J. W., Lance, P., Daston, C., & Schatzkin, A. (2006). High dry bean intake and reduced risk of advanced colorectal adenoma recurrence among participants in the polyp prevention trial. *The Journal of Nutrition, 136*, 1896–1903. https://doi.org/10.1093/jn/136.7.1896.

Dronachari, M., & Yadav, B. K. (2015). Application of microwave heat treatment in processing of pulses 1. *Journal of Academia and Industrial Research, 3*, 401–407.

Martín-Cabrejas, M. A., Sanfiz, B., Vidal, A., Mollá, E., Esteban, R., & López-Andréu, F. J. (2004). Effect of fermentation and autoclaving on dietary Fiber fractions and antinutritional factors of beans (Phaseolus vulgaris L.). *Journal of Agricultural and Food Chemistry, 52*, 261–266. https://doi.org/10.1021/jf034980t.

Miller Magazine. (2016). *World milling and pulses technologies—World dry bean and chickpea market*. Retrieved from https://www.millermagazine.com/english/world-dry-bean-and-chickpea-market/.html

Nesli, S., Ulla, H.-M., & Kaisa, P. (2016). Traditional and new food uses of pulses. *Cereal Chemistry, 94*, 66–73. https://doi.org/10.1094/CCHEM-04-16-0082-FI.

Oghbaei, M., & Prakash, J. (2016). Effect of primary processing of cereals and legumes on its nutritional quality: A comprehensive review. *Cogent Food and Agriculture, 2*, 1–14. https://doi.org/10.1080/23311932.2015.1136015.

Olang'o, N. K., Ojijo, Toshinori, K., & Hirozi, K. (2000). Composition, soaking and softening characteristics of some Kenyan beans (Phaseolus vulgaris L.). *Food Science and Technology Research, 6*, 12–18.

Olaofe, O., Umar, Y. O., & Adediran, G. O. (1993). The effect of nematicides on the nutritive value and functional properties of cowpea seeds (Vigna unguiculata L. Walp). *Food Chemistry, 46*, 337–341. https://doi.org/10.1016/0308-8146(93)90001-V.

Patterson, C. A., Curran, J., & Der, T. (2017). Effect of processing on active compounds in fresh-cut vegetables. *Processing and Impact on Active Components in Food, 94*, 1–10. https://doi.org/10.1016/B978-0-12-404699-3.00001-9.

Petry, N., Egli, I., Zeder, C., Walczyk, T., & Hurrell, R. (2010). Polyphenols and phytic acid contribute to the low iron bioavailability from common beans in young women. *The Journal of Nutrition, 140*, 1977–1982. https://doi.org/10.3945/jn.110.125369.

Reddy, N. R., Pierson, M. D., Sathe, S. K., & Salunkhe, D. K. (1984). Chemical, nutritional and physiological aspects of dry bean carbohydrates—A review. *Food Chemistry, 13*, 25–68. https://doi.org/10.1016/0308-8146(84)90026-8.

Rocha-Guzman, N. E., Gallegos-Infante, J. A., Gonzalez-Laredo, R. F., Bello-Perez, A., Delgado-Licon, E., Ochoa-Martinez, A., & Prado-Ortiz, M. J. (2008). Physical properties of extruded products from three Mexican common beans (Phaseolus vulgaris L.) cultivars. *Plant Foods for Human Nutrition, 63*, 99–104. https://doi.org/10.1007/s11130-008-0076-x.

Sai-Ut, S., Ketnawa, S., Chaiwut, P., & Rawdkuen, S. (2009). Biochemical and functional properties of proteins from red kidney, navy and adzuki beans. *Asian Journal of Food and Agro-Industry, 2*(4), 493–504.

Salcedo, J. M. (2008). Regeneration guidelines: Common bean. In M. E. Dulloo, I. Thormann, M. A. Jorge, & J. Hanson (Eds.), *Crop specific regeneration guidelines [CD-ROM]* (9 pp). Rome: CGIAR System-wide Genetic Resource Programme.

Savelkoul, F. H., Tamminga, S., Leenars, P. P., Schering, J., & Ter Maat, D. W. (1994). The degradation of lectins, phaseolin and trypsin inhibitors during germination of white kidney beans, Phaseolus vulgaris L. *Plant Foods for Human Nutrition, 45*, 213–222.

Sgarbieri, V. C. (1984). Composition and nutritive value of beans (Phaseolus vulgaris L.). In G. H. Bourne (Ed.), *Nutritional value of cereal products, beans and starches. World Rev Nutr Diet* (pp. 132–198). Basel: Karger.

Shimelis, E. A., & Rakshit, S. K. (2007). Effect of processing on antinutrients and in vitro protein digestibility of kidney bean (Phaseolus vulgaris L.) varieties grown in East Africa. *Food Chemistry, 103*, 161–172. https://doi.org/10.1016/j.foodchem.2006.08.005.

Siddiq, M., Ravi, R., Harte, J. B., & Dolan, K. D. (2010). Physical and functional characteristics of selected dry bean (Phaseolus vulgaris L.) flours. *LWT - Food Science and Technology, 43*, 232–237. https://doi.org/10.1016/j.lwt.2009.07.009.

Taylor, P., Sathe, S. K., Deshpande, S. S., Salunkhe, D. K., & Rackis, J. J. (1984). Dry beans of phaseolus. A review. Part 1. Chemical composition. Proteins. *C R C Critical Reviews in Food Science and Nutrition, 20*(1), 37–41.

Tiwari, B. K., & Singh, N. (2012). *Pulse chemistry and technology*. Cambridge: Royal Society of Chemistry.

Vidal-Valverde, C., Frias, J., Estrella, I., Gorospe, M. J., Ruiz, R., & Bacon, J. (1994). Effect of processing on some antinutritional factors of lentils. *Journal of Agricultural and Food Chemistry, 42*(10), 2291–2295.

Wang, N., Hatcher, D. W., Tyler, R. T., Toews, R., & Gawalko, E. J. (2010). Effect of cooking on the composition of beans (Phaseolus vulgaris L.) and chickpeas (Cicer arietinum L.). *Food Research International, 43*, 589–594. https://doi.org/10.1016/j.foodres.2009.07.012.

Cowpea

Subajiny Sivakanthan, Terrence Madhujith, Ashoka Gamage, and Na Zhang

Introduction

Cowpea (*Vigna unguiculata* L. Walp) (Pasquet 1998) also known as black-eyed pea, bachapin bean, southern pea, crowder pea, China pea and cow gram is an herbaceous legume belonging to the family Fabaceae. It is well adapted to harsh arid climates and low fertile soils and more drought- and heat-tolerant than most of its legume relatives (Carvalho et al. 2017; Hall 2004; Timko and Singh 2008). Cowpea has been used for human consumption as well as an animal feed since antiquity. Cowpea plays a vital role in the livelihoods of millions of people in less developed countries of the tropics. For Africans, it plays a pivotal role in the economy and nutrition of their daily life (Houssou et al. 2010).

S. Sivakanthan
Department of Agricultural Chemistry, Faculty of Agriculture, University of Jaffna, Jaffna, Sri Lanka

T. Madhujith (✉)
Department of Food Science and Technology, Faculty of Agriculture, University of Peradeniya, Peradeniya, Sri Lanka

A. Gamage
Department of Chemical and Process Engineering, Faculty of Engineering, University of Peradeniya, Peradeniya, Sri Lanka

N. Zhang
College of Food Engineering, Harbin University of Commerce, Harbin, China

A. Manickavasagan, P. Thirunathan (eds.), *Pulses*,
https://doi.org/10.1007/978-3-030-41376-7_6

Botany

Cowpea is an annual herb with varying growth forms such as erect, semierect, prostrate, trailing, climbing or bushy. It has a taproot system with spreading lateral roots in surface soil. Cowpea roots establish a nitrogen-fixing symbiotic association with nodule-forming bacteria, rhizobia (Leite et al. 2017). The leaves are alternate and trifoliate except the first pair of leaves, which are basic and opposite. Leaves vary in shape from linear-lanceolate to ovate with petiole length of 5–25 cm. Stems are striate, smooth or slightly hairy. The flowers are white, yellowish, pale blue or violet and are born on racemose inflorescences at the ends of 2–20-cm-long peduncles. Pods are coiled, round, crescent or linear. Pods are usually 20–30 cm long and small seeded (Department of Agriculture Forestry and Fisheries, South Africa 2011) (Fig. 1).

World Production and Trade

Cowpea is originated and domesticated in Africa (Goncalves et al. 2016; Lazaridi et al. 2017; Richard 1847). Nowadays, it is widely grown in many other parts of the world as well such as Latin America, Europe, Asia and the United States (FAOSTAT 2019). However, Africa remains the leading producer of cowpea even today. According to FAOSTAT database, in 2017, total area under the cowpea production in the world was around 12.57 million ha, and the total production was 7.40 million MT. Africa is the largest producer in the world accounting more than 95% of the

Fig. 1 Cowpea seeds of different varieties

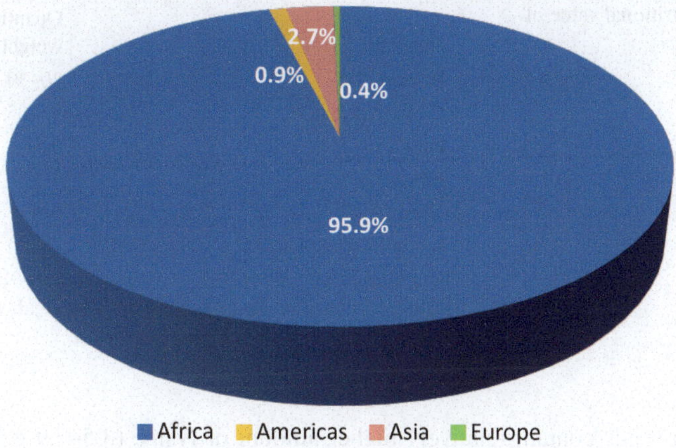

Fig. 2 Production of cowpea in 2017. (Source: FAOSTAT 2019)

world's production of cowpea with Nigeria and Niger the leading producers in Africa. Asia, Americas and Europe contribute less compared to Africa for the world's production as shown in Fig. 2 (FAO 2019). Cowpea is one of the most important agricultural exports from the West and Central Africa (Cowpea Storage Project: Profiles of Progress 2010). Of the developed countries, only the United States is a substantial producer and exporter of cowpea (Gómez 2004). It is reported that over the years, cowpea production increased more compared to other pulses. From 1981 to 2013, share in total pulse crop area increased from 5.7% to 14%, while its production share increased from 3 to 9% (Joshi and Rao 2017).

Nutritional Value

All parts of cowpea plant contain high nutritive value. The seeds (peas or grains) are rich source of good quality protein. The leaves also can be used as a vegetable, and the rest of the plant parts can be used as animal feed. Cowpea serves as an inexpensive source of good quality protein, especially for the poor population. It is a major source of protein in diet of many people in sub-Saharan Africa (Sebetha et al. 2015). Table 1 shows the amount of major nutrients present in cowpea grains.

Cowpea is considered as a nutrient-dense food with low energy density. Compared to cereal grains, cowpea has higher protein and carbohydrate contents with a relatively less fat content and a complementary amino acid pattern (Jayathilake et al. 2018). Cowpea seeds contain significant amount of essential amino acids such as leucine, lysine, phenylalanine, tyrosine, aspartate, glutamate and arginine. Most limiting amino acids in cowpea are S-containing amino acids such as methionine and cysteine, tryptophan and threonine (Farinu and Ingrao 1991; Jirapa et al. 2001), whereas cowpea is an excellent source of lysine (Iqbal et al. 2006).

Table 1 Nutritional value of cowpea grains

Nutrient	Quantity on dry weight basis
Protein (%)	20–39
Carbohydrate (%)	50–65
Fat (%)	0.3–3
Fibre (%)	5–6.9
Minerals (%)	3–4
Essential amino acids (g/100 g protein)	27–33

Jayathilake et al. (2018), Goncalves et al. (2016), Kalogeropoulos et al. (2010), Ahenkora et al. (1998) and Prinyawiwatkul et al. (1996)

Cowpea seeds contain relatively higher amount of lysine (3.5–7.9 g 16 g^{-1} N) (Goncalves et al. 2016) compared to most cereal grains (2.3–4 g 16 g^{-1} N) (Chavan et al. 1989). The bulk of the diet of the Africans mainly consists of starchy food made from cassava, yam, plantain and banana, millet, sorghum and maize. The incorporation of cowpea into these starchy diets could enhance the protein quality via synergistic effect of high protein and high lysine from cowpea and high methionine and high energy from the starchy foods. Thus, it ensures a nutritional security (Simion 2018).

Cowpea is also a good source of minerals and vitamins (Cruz and Aragão 2014), most importantly, vitamin C (Tresina and Mohan 2011; Etokakpan et al. 1983) and carotenoids (Hashim and Pongjata 2000). Calcium, zinc and iron are the important minerals found in cowpea seeds (Thangadurai 2005; Adebooye and Singh 2007; Goncalves et al. 2016). In addition to these basic nutrients, cowpea has soluble and insoluble dietary fibre and phytochemicals and thus possesses therapeutic potentials such as antidiabetic, anticancer, anti-hyperlipidemic, anti-inflammatory and antihypertensive properties (Jayathilake et al. 2018). Cowpea contains an interesting profile of polyphenols, especially the highly glycosylated flavan-3-ols, and the unusual predominance of quercetin glycosides.

Antinutritional Factors

Even though cowpea is a very good source of protein and other nutrients, the presence of antinutritional factors brings about nutritional implications. Major antinutritional factors include tannin, protease inhibitors (trypsin inhibitors and chymotrypsin inhibitors), lectins, phytic acid, oxalic acid and flatulence-causing oligosaccharides (Goncalves et al. 2016; Khattab and Arntfield 2009). Intake of these substances over a long period of time is supposed to cause some adverse health effects because these compounds can interact with macro- and micronutrients, impairing their absorption during digestion, thus reducing the bioavailability of nutrients. However, in recent years, it has been suggested that they also have some beneficial health effects to human (Abizari et al. 2012a; Goncalves et al. 2016; Singh et al. 2017). For instance, tannin and phytic acid possess anticancer properties. Phytic acid is believed to prevent colon cancer by reducing oxidative stress in the lumen of the intestinal tract by the iron chelating effect (Simion 2018).

Tannins are water-soluble phenolic compounds (molecular weights ranging between 500 and 3000 Da). Most of the tannins in cowpea are nonhydrolysable, also known as condensed tannins, flavolans or proanthocyanidins. They have the ability to interact with proteins and, in most cases, precipitate the proteins (Madsen and Brinch-Pedersen 2016). In addition, tannins lower digestibility of protein and carbohydrates via inhibiting the digestive enzymes and forming complexes with them (Khattab and Arntfield 2009; Jain et al. 2019). Tannins also can chelate ferric ions, thus interfering iron absorption (Madsen and Brinch-Pedersen 2016) and may lead to anaemia.

Lectins are carbohydrate-binding proteins of non-immune origin. Their binding with carbohydrates is reversible and highly specific. Thus, lectins may interfere with the digestion and absorption of carbohydrates. However, like other antinutritional factors found in cowpea, lectins also have been reported to possess antitumour activities (Lagarda-Diaz et al. 2017).

Cowpea seeds contain enzyme inhibitors such as protease inhibitors, trypsin and chymotrypsin inhibitors and amylase inhibitors. Trypsin inhibitors impair digestion and absorption of dietary protein by strongly inhibiting trypsin activity (Khattab and Arntfield 2009). Trypsin inhibitors belong to two families such as Bowman–Birk or the Kunitz families. Both are single-chain polypeptides; however, Bowman–Birk inhibitors are smaller (approximately 8 kDa) with seven disulphide bridges and two active domains for the inhibition of trypsin and/or chymotrypsin than Kunitz family (20 kDa), which has only two disulphide bridges. α-Amylase inhibitors interfere with the utilization of starch (Madsen and Brinch-Pedersen 2016).

Phytic acid (myo-inositol-(1,2,3,4,5,6)hexakisphosphate) is the major phosphorus storage compound in plants (Madsen and Brinch-Pedersen 2016). It has the ability to bind essential dietary minerals (zinc, iron, calcium, magnesium, copper and manganese), proteins and starch, thereby reducing their bioavailability (Khattab and Arntfield 2009).

Oxalic acid may form deleterious complexes with metal ions, e.g. calcium oxalate causing renal damage (Maina et al. 2015). High consumption of these compounds leads to flatulence and osmotic effect and interferes with digestion and absorption of other nutrients (Martinez-Villaluenga et al. 2008).

The presence of the α-galactosides such as raffinose, stachyose and verbascose is also reported to have adverse effects on nutritional value of the cowpea (Khattab and Arntfield 2009). α-Galactosides withstand the digestion in the human gastrointestinal tract due to lack of secretion of α-galactosidases. In the caecum, undigested α-galactosides undergo fermentation by anaerobic microflora composed mainly of *Bifidobacterium*, *Bacteroides*, *Fusobacterium* and *Clostridium* spp. Fermentation also leads to the formation of various gases such as hydrogen, carbon dioxide and methane and short-chain fatty acids such as acetate, butyrate and propionate causing flatulence (Madodé et al. 2013).

Ingestion of large amount of α-galactosides usually leads to flatulence in human and monogastric animals. Accumulation of flatus in the intestinal tract results in discomfort, abdominal rumblings, cramps, pain and diarrhoea (Martinez-Villaluenga et al. 2008). However, these flatulence-causing compounds are reported to reduce the risk of colon cancer and cardiovascular diseases in humans (Madodé et al. 2013). α-Galactosides are also reported to have beneficial effect on human health as they can stimulate the growth and activity of living *Bifidobacterium* and *Lactobacillus* in the human colon (prebiotic effect).

Various processing approaches such as soaking, germination, pressure-cooking, fermentation, enzyme treatment and genetic modification of the plant can be used to destroy these antinutritional components (Martinez-Villaluenga et al. 2008).

Postharvest Processing

Postharvest processing of cowpea aims to improve the quality of the grains while minimizing the losses due to insect and other pest infestations and physiological changes. In addition, reducing the effect of antinutritional factors to improve the nutritional value of cowpea is aimed at most processing technologies. Primary postharvest operations include drying, husking, winnowing, separation and storage, whereas secondary processing techniques include soaking, germination, fermentation, frying, milling, cooking and roasting (Subuola et al. 2012; Goncalves et al. 2016). Some conventional methods of postharvest storage and processing techniques have been replaced by new improved technologies. Besides antinutritional factors, consumption of cowpea is also confronted by some other issues such as beany flavour, flatulence and a long cooking time (Goncalves et al. 2016). Thus, cowpea requires extensive processing for their optimal utilization in the diet.

Drying

The pods are usually ready to harvest 60–120 days after planting. Pods should be harvested when 95% in one accession have turned yellowish brown (Dumet et al. 2008). During harvesting, cowpea grains have around 20% moisture. In order to prevent spoilage during storage, the moisture content of the grains should be reduced to between 8 and 12% depending on the intended duration of storage. The grains from pods are removed by hands or mechanically. Sometimes, pods are steeped in water for hours (2–8 h) or treated with oil before sundrying in order to facilitate dehusking process. Sundrying is the conventional method of drying.

Hulling, Winnowing and Splitting

Hulling can be performed by dry method or wet method. Dry method is traditionally practiced in African and Asian countries which involves pounding of the dried grains in mortar with pestles or in hand-operated wooden or stone shellers or power-operated shellers and abrasive hulling machines. Wet grinding process involves conditioning the grain by soaking of the grains before drying. Conditioning techniques through moisture adjustment facilitates easy husking (Department of Agriculture Forestry and Fisheries 2011). The separated husks are removed from the cotyledons by winnowing manually, which is time consuming and laborious. In order to overcome this problem, improved abrasive hulling machines have been developed.

Storage

After harvest, grains are dried, sorted and stored. The storage life of cowpea depends on its moisture content. The lower the moisture content, the better the quality of seeds in storage. In developed countries cold storage is commonly practiced. It is reported that exposure to −18 °C for 6–24 h can reduce the number of pests by more than 99%. If the grain storage is meant for short term, moisture content of around 12% or less is recommended. However, for long-term storage, it should be reduced to 8–9% (Department of Agriculture Forestry and Fisheries 2011).

The major problem with cowpea after harvest is the high susceptibility to insect infestation. The serious insect pest causing major economic loss during storage of cowpea is the cowpea weevil, *Callosobruchus maculatus* (Coleoptera: Bruchidae) (Tiroesele et al. 2015). Thus, appropriate measures should be considered to minimize the losses from insect attack.

During the 1980s, neem oil was widely used to minimize the losses from insect pests in small scale. This method had some advantages such as non-toxic and easily available. However, the use of neem oil was discouraged because of difficulties in extraction and imparted bitter taste to the grains (Gómez 2004). In order to reduce the losses during storage, various control techniques are being used since a long time ago, and some new techniques have been developed. Main control method is the application of chemicals as insecticides. However, it may not be feasible for small-scale farmers due to cost and some technical reasons. Since insecticides have negative impact on humans, environment as well as non-target organisms, there is a need to develop new techniques to overcome these problems. The use of some natural plant products (e.g. garlic and peppermint) which have the ability to produce odours that repel weevils can be used effectively (Tiroesele et al. 2015).

Traditionally, sacks and bamboo baskets have been used extensively to store the cowpea. However, in order to increase the storage period by many months without pest attack, in 2007, Purdue University research team and Bill & Melinda Gates Foundation introduced a new inexpensive packaging called triple-layer bag to

African farmers as a simple solution to protect cowpea from losses during storage. It was named as the Purdue Improved Cowpea Storage (PICS) bag. It is made from two inner high-density polyethylene plastic bags and an outer nylon sack which provide an airtight seal for long-term, pest-free storage. PICS bags addressed the cowpea storage problems without the use of chemicals (Cowpea Storage Project: Profiles of Progress 2010).

Destroying the Antinutritional Factors

Removal of antinutrients in order to improve nutritional quality is necessary for an optimal utilization of cowpea in human nutrition (Preet and Punia 2000). Different processing methods such as dehulling, soaking, low and high pressure-cooking, fermentation and germination can help improve the nutritional quality of food legumes by destroying antinutritional factors to various extents (Khattab and Arntfield 2009; Mubarak 2005). In addition to these processing approaches, breeding techniques also can be used to reduce the amount of antinutritional factors. However, since these compounds are reported to have some beneficial roles in the plant, their manipulation via genetic means should consider their positive roles as well in plants (Simion 2018).

Cooking

Heat processing is as an effective means of inactivating the heat-labile antinutritional factors such as enzyme inhibitors and lectins found in legumes (Akande and Fabiyi 2010). Various cooking methods such as boiling, pressure-cooking, roasting and microwave cooking are reported to reduce the level of antinutritional factors. Several authors have reported that moist heating is more effective than dry heating to inactivate antinutritional factors (Khattab and Arntfield 2009; Akande and Fabiyi 2010). The cooking of pre-soaked grains appeared to be the most effective method for reducing trypsin inhibitor activity (Jain et al. 2019). After cooking in boiling in water, the cooking water may be discarded to reduce the amount of antinutritional compounds; however, other soluble compounds with essential nutritional value also could be removed.

A study has been carried out to determine the effects of cooking and soaking methods on α-amylase inhibitors in cowpea (Piergiovanni and Gatta 1994). Boiling in distilled water and cooking in microwave oven were found to be ineffective, whilst autoclaving was effective in reducing the amount of α-amylase inhibitors. Soaking in alkaline solutions was more effective than soaking in distilled water. Boiling of soaked cowpea seeds gave pronounced effect in the reduction of α-amylase inhibitor.

Soaking

Some antinutritional factors such as α-galactosides and tannins are water soluble. Thus, soaking of grains in water or saline solutions and discarding the soaking solution will remove most of the water-soluble antinutritional factors (Martinez-Villaluenga et al. 2008).

Fermentation

Fermentation has been reported to reduce the antinutritional factors. The fermentation process could be natural (by the microorganism present in the seed) or induced (using a microbial culture) (Martinez-Villaluenga et al. 2008). Phytate content of the cowpea could be reduced during fermentation by endogenous phytase enzyme or by microorganisms such as yeast (Martinez-Villaluenga et al. 2008; Khattab and Arntfield 2009). However, Madodéet al. (2013) have reported that in vivo fermentability index of fermented cowpeas is significantly lower than that of traditionally processed (dehulled, soaked and boiled) cowpeas.

Decortication or Dehulling

Tannins are primarily located in seed coat of cowpea. Thus dehulling of seed can decrease the tannin content of cowpea and improves their nutritional quality. It is reported that dehulling can reduce the tannin content by about 68–99%. For instance, a study reported that decortication significantly reduced tannin content by 85% (Jain et al. 2019).

Germination

Germination is an important traditional processing method practiced before cooking of legumes. Reductions in tannin content of cowpea during germination could be attributed to presence and activity of polyphenol oxidase and enzymatic hydrolysis (Jain et al. 2019). Germination causes breakdown of certain complex compounds into simple compounds, transport of simple compounds to the endosperm and the synthesis of new materials from the breakdown products. During seed germination, storage proteins are hydrolysed, and the amino acids and complex polysaccharides are broken down into simple sugars. Phytic acid is hydrolysed by phytase resulting in the formation of available phosphorus (Akande and Fabiyi 2010).

Product Development

In the present context, there is a tendency of increasing the diversity of foods consumed. A plethora of factors such as changing lifestyles, development of new technologies, changing consumer demands and increasing consumer awareness on healthy foods lead to the rising demand for diverse food products with good nutritional value. Thus, for the food sectors, efforts on new food product development become inevitable.

Preparation of Cowpea

Cowpea is prepared for consumption as whole grain, split or ground forms. Dry grain can be consumed as boiled, fried or steamed in different preparations such as salads, snacks, breakfast cereals and baked goods. Cowpea can play an important role as a functional ingredient in such products, especially with the growing interest in high plant protein diets and 'ancient grains'. Fresh seeds are often served boiled, as well as being consumed fried or fresh (Carvalho et al. 2017; Jayathilake et al. 2018). The seeds can also be cooked with meat, tomatoes and onions into a thick soup and eaten with pancake and bread (Sebetha et al. 2015). Dry mature seeds are also suitable for boiling and canning (Carvalho et al. 2017).

Since cowpea serves as the main protein source for the Africans, a variety of local preparations are prepared and consumed by Africans. Madodé and colleagues did a survey on local cowpea preparations available in Africa, and they have reported them with their nutritional composition (Madodé et al. 2011). According to them approximately 90% of the preparations are prepared using seeds, and the remaining prepared using leaves. Typical processing of cowpea includes steeping, dehulling, whipping, milling, and cooking. Seeds are prepared either alone or in combination with cereals, roots and tubers and/or cooking oils and seasonings such as salt, pepper and roasted shrimp (Madodé et al. 2011).

Preparations Based on Cowpea Grains

Seeds are cooked in different ways such as boiling, roasting and frying or combination of these methods. Usually seasonings are added to enrich the sensory qualities. Cooking raw seeds is usually time-consuming because of high calcium content, making the seeds hard. However, soaking (steeping) overnight in water is usually practiced to shorten the cooking time. Soaking is also helpful in minimizing antinutritional factors. Addition of sodium bicarbonate during cooking is also used to reduce the cooking time. In Africa, *Kanwu*, a kind of rock salt containing carbonate and bicarbonate of minerals such as Ca, Na and Fe, is used as softener (tenderizer).

Cooking in 0.05–0.1% of *Kanwu* can reduce the cooking time to a significant level during open pan boiling. The combination of pressure-cooking with a tenderizer can further reduce the cooking time (Madodé et al. 2011).

Cowpea fritter or snack dish, namely, *Ata* (Benin), *Akara* (Nigeria) or *Koose* (Ghana), is produced using a combination of processing methods. It is traditionally made by steeping and wet-dehulling (manually or mechanically) followed by grinding the dehulled seeds into a batter that is whipped to incorporate air. This dough is seasoned and deep-fat fried. This product takes at least a day to prepare, in order to allow the cowpea to soak and the batter to rest.

Doco or *Ata-doco* is a fried product similar to *Ata.* Whole or dry-sieved cowpea flour is whipped and deep-fried. *Doco* is cooled and fried a second time to obtain hard and dry fritters called *Ataclè*, which are immersed in oil for preservation. *Ata*, *Doco* and *Ataclè* are commonly consumed as a side dish with porridges, yam (*Dioscorea* sp.) or sweet potato (*Ipomoea batatas*) fries.

Yoyoue is a kind of oily flour. Cleaned seeds are roasted, seasoned, ground and finally deep-fried in cooking oil. *Magni-magni*, *Lèlè* or *Alèlè* (Benin), *Moinmoin* (Nigeria) or *Koki* (Cameroon) is a steamed product. Seeds are steeped, wet-dehulled, ground with seasonings to taste, whipped, mixed with palm oil and salt and finally wrapped in banana leaves for steaming.

In Ghana, cowpea flour is used to make a variety of items such as cowpea straw, cowpea pie, fried cowpea paste, cowpea porridge, cowpea twisted cake, stew, cowpea cutlet, cowpea pancake, cowpea chips, cowpea rock buns, cowpea pudding, cowpea doughnuts, cowpea biscuits, etc. (Gómez 2004).

Cowpea soup is another preparation based mainly on seeds. To prepare this soup, seeds are washed, soaked, dehulled, boiled, mashed and sieved. The sieved seeds are then cooked with palm oil along with other ingredients such as pepper, spices and seasoning with or without fresh or dried fish to taste to produce *Gbegiri*. It is eaten with reconstituted yam flour product *Amala* (Subuola et al. 2012).

Cowpea Preparations Mixed with Cereals

Cowpea preparations can be prepared by mixing with cereals such as rice and maize. *Atassi* (Benin) or *Waakye* (Ghana) is a product prepared by cooking parboiled seeds mixed with cleaned rice (usually at the ratio of cowpea/rice, 2–3).

Atchonkouin, *Kossibobo*, *Adalou* or *Aibli* is a combination of cowpea and maize. Maize is parboiled and then boiled together with cowpea. *Abla* is a paste made from maize (about 30%), cowpea (about 30%) and crude palm fruit extract or refined palm oil (about 40%). To prepare this, fine cowpea flour is mixed with fine maize flour and potash filtrate and blended with palm oil to obtain a homogeneous dough, which is wrapped in banana leaves and steam-cooked. By replacing cowpea flour by dehulled seeds, *Kowe* is obtained.

Cowpea pie is used as main dish and it is prepared by incorporating rice. Cowpea and rice are cooked until they become very soft and are mashed together. To this mixture gravy with seasonings is added and baked. *Adalu* is a mixture of cowpeas and maize. It is also called *NiébéetMaïs* or, in English, 'black-eyed peas and corn'. In Africa, it is usually made with dried cowpeas and either fresh or dried maize. It can also be adapted to use canned or frozen black-eyed peas and corn.

Cowpea Preparations Based on Mixtures with Roots or Tubers.

A type of *Magni-magni* processed without oil and seasonings is *Toubani*. Cowpea is partially dehulled, whipped and moulded in recycled tins or leaves before steam cooking. Yam (*Dioscorea* sp.) or cassava (*Manihot esculenta*) flour is usually added as a binder along with sodium bicarbonate.

Cowpea can also be used to enhance the acceptability of water yam (*Dioscorea alata*), which is inappropriate for making pounded yam. *Dioscorea alata* tubers are grated, mixed with cowpea flour (5% of the grated yams' weight), whipped and deep-fried to produce a fritter named *Alounganta. Tche* is a kind of *Abobo* obtained by boiling seeds with peeled and sliced yam.

Products Prepared by Mixing with Other Products

Akpada is a very thick cowpea sauce. A thick tomato sauce is prepared, to which cowpea flour is added, mixed and cooked together. Red-Red is a popular dish in Ghana made from cowpeas prepared by the combination of red pepper and red palm oil. The Red-Red cowpeas stew is usually served with fried plantains.

Nowadays, breakfast cereals as convenient and energy-dense food play a vital role in daily diets of the people of most countries. The typical breakfast cereals are rich in carbohydrates and fibre. There is a potential to prepare breakfast cereals enriched with cowpea (sprouted or non-sprouted) using extrusion technique (Marengo et al. 2017).

Cowpea Protein Isolates

Since cowpea is a rich source of proteins, functional properties of proteins can be exploited. Functional proteins of cowpea can be extracted using different methods and utilized in various products. Proteins in the foods can interact directly or indirectly with other components in the foods and change the properties of foods due to their functional properties. Cowpea protein isolates can be used as a food additive such as emulsifier, texturizer, stabilizer and fat replacer or as a supplement to enrich the nutrient content of other food stuffs. For instance, protein isolates from cowpea can be utilized to improve the properties of gluten-free rice muffins (Shevkani et al.

2015). Cowpea protein isolate can also be incorporated with food products, especially cereal-based products, which contain lysine as a limiting amino acid and are rich in methionine (Frota et al. 2017). Protein isolates can also be used to make textured vegetable proteins. Textured vegetable protein made from soybean is popular among vegetarians in most of the tropical countries. However, textured vegetable proteins made using cowpea proteins are not available in the market so far.

Cowpea-Fortified Foods

Fortification is the practice of deliberately increasing the content of an essential micronutrient, i.e. vitamins and minerals in a food, so as to improve the nutritional quality of the food supply and provide a public health benefit with minimal risk to health (Allen et al. 2006). Cowpea is rich in some vitamins (vitamin C and carotenoids) and minerals (such as calcium, zinc and iron). Thus, it is possible to fortify other foods which are lacking in those particular micronutrients with cowpea. However, consideration should be given to the elimination of antinutritional factors which limit the bioavailability of these micronutrients.

A study has been carried out to develop a recipe for an enriched cheese bread with whole biofortified cowpea flour; its chemical composition and consumer acceptance were evaluated. Cookies fortified with defatted cowpea flour (10% of wheat flour replaced by cowpea flour) contained increased Ca content in addition to increased crude fibre, protein and ash contents (Cavalcante et al. 2016).

Fortified Cowpea Foods

Moreover, cowpea itself can also be fortified to enrich its micronutrient content. For example, a study has been carried out to fortify cowpea meal with NaFeEDTA, and the results indicated that fortification of cowpea flour with NaFeEDTA overcomes the combined inhibitory effect of phytic acid and polyphenols and is effective in reducing the prevalence of iron deficiency and iron deficiency anaemia (Abizari et al. 2012b).

Weaning Foods

Weaning foods are introduced to the infants during the period between breastfeeding and total solid food intake. Hence, easily digestible nutritious and energy-dense foods should be given during the weaning period. Malting is one of the simple traditional techniques to improve the nutritional quality as well as digestibility of cereals and legumes. Malting process consists of three stages such as steeping, germination and drying. During germination, endogenous enzymes, such as α-amylase and α-glucosidase, limit dextrinase, and proteases hydrolyse the polysaccharides into simple sugars (Gupta et al. 2010). In addition, several other favourable changes such as enhanced flavour and increase in essential amino acids (lysine, methionine and tryptophan) and vitamins (riboflavin, niacin and ascorbic acid) occur during germination (Baranwal 2017). Thus, malting can be taken advantage of in the development of weaning foods.

Steeping in water followed by germination for 24–48 h at 25–28 °C can be used as malting conditions. Most of the commercially available weaning foods are extruded or roller-dried which need high capital. Malting is a very simple and inexpensive technique; thus, it could be easily adapted in developing countries. Malted cereals and legumes can be used in weaning foods. For example, weaning foods can be prepared by mixing malted sorghum, green gram, black gram and steamed or germinated cowpea. It is reported that in vitro protein quality and starch digestibility of cowpea protein can be improved through germination (Jirapa et al. 2001). Thus, it is necessary to determine the optimum conditions for malting to be used in weaning foods with good nutritional and sensory characteristics.

In a study to evaluate the possibility of processing a ready-to-eat nutrient-rich weaning food for infants from cooking banana fortified with cowpea and peanut using in vitro digestibility, it is reported that it is feasible to produce precooked weaning food which has the potential to meet the nutritional needs of an infant (Bassey et al. 2013). In order to get more balanced nutritious weaning food from cowpea, other grains such as mung bean, green gram, rice, etc. could be incorporated.

Future Trends in Processing and Product Development

There has been a mounting interest in the use of legumes as a balanced source of nutrient during the past few years. Among legumes, cowpea is considered an important crop because of its wide adaptation to agroclimatic conditions and its nutritional value, importantly, protein at low cost. However, cowpea remains an underexploited crop, and it has received relatively little attention from a research standpoint.

There are some limiting factors for the dietary utilization of cowpea. These include the presence of antinutritional and flatulence-causing factors as well as the long duration of cooking because of hard seed coat. Thus, cowpea requires further developments in processing alternatives allowing the effective utilization of nutrients in cowpea with minimal impact on human health. In this scenario, much focus should be given to the intense research efforts in order to expand the cowpea preparations in new forms with potential contribution to human nutrition. In order to develop new products with good nutritional value, a thorough understanding of the composition of the cowpea is crucial. The development of innovative cowpea preparations and cowpea-enriched preparations is a promising way to exploit the full potential of cowpea as a good nutritional source.

Cowpea grains are reported to contain an array of polyphenols, which exert beneficial health effects. Polyphenols are concentrated in the seed coat; thus, processing technologies that remove the seed coat will almost entirely eliminate the polyphenols, besides eliminating valuable dietary fibre. However, thermal processes such as moist heat cooking used to prepare cowpea preparations reported to have limited effect on the profile of these compounds; thus their benefits are likely retained in such products (Awika and Duodu 2017).

Extrusion cooking is gaining popularity for production of expanded snacks, because there is a huge demand for healthy and nutritious ready-to-eat products from all age groups of consumer all over the world. Extrusion cooking can be applied in cowpea processing in order to produce variety of value-added food products from cowpea with good nutritional value. It has been reported that in vitro protein digestibility of extruded cowpea is significantly higher than that of whole raw cowpea (Jakkanwar et al. 2018). High-protein instant porridge can be prepared using extrusion technology from cowpea incorporated with other grains such as sorghum (Pelembe et al. 2002). In addition, there is a potential to produce textured vegetable proteins from cowpea protein isolates using extrusion cooking. Thus, attempts should be made to produce textured vegetable proteins from cowpea in order to ensure the availability of variety of cowpea-based products and make efficient utilization of cowpea.

Cowpea research has been underway in some African countries for many years. In Nigeria, the Federal Department of Agricultural Research, the Institute for Agricultural Research and Training (IAR&T) at Ibadan, the University of Ife and the Institute of Agricultural Research (IAR) and Centre National de Recherches Agronomiques (CNRA) in Senegal have started cowpea research in the 1960s (Boukar et al. 2018). However, further researches on breeding are needed to exploit the valuable genes to improve cowpea yield and quality. The general target of most breeding programmes is to develop varieties with high yield, improved adaptation to different agroecological zones and improved tolerance to pest attack and other adverse conditions. However, researches on breeding to improve nutritional quality of cowpea grains are lacking. Thus, more emphasis should be given to improve the nutritional value of the cowpea via reducing the antinutritional factors and/or increasing the amounts of essential amino acids and other nutrient contents.

Further research is necessary to evaluate the opportunities to efficiently use cowpea to help protect against nutritional deficiency diseases and economy of the peasant farmers in Africa as well as other developing countries, which eventually could play a role in global food supply helping towards the food and nutritional security.

References

Abizari, A. R., Moretti, D., Schuth, S., Zimmermann, M. B., Armar-Klemesu, M., & Brouwer, I. D. (2012a). Phytic acid-to-iron molar ratio rather than polyphenol concentration determines iron bioavailability in whole-cowpea meal among young women. *The Journal of Nutrition, 142*(11), 1950–1955. https://doi.org/10.3945/jn.112.164095.

Abizari, A. R., Moretti, D., Zimmermann, M. B., Armar-Klemesu, M., & Brouwer, I. D. (2012b). Whole cowpea meal fortified with NaFeEDTA reduces iron deficiency among Ghanaian school children in a malaria endemic area. *The Journal of Nutrition, 142*(10), 1836–1842. https://doi.org/10.3945/jn.112.165753.

Adebooye, O. C., & Singh, V. (2007). Effect of cooking on the profile of phenolics, tannins, phytate, amino acid, fatty acid and mineral nutrients of whole-grain and decorticated vegetable cowpea (*Vigna unguiculata* L. Walp). *Journal of Food Quality, 30*(6), 1101–1120. https://doi.org/10.1111/j.1745-4557.2007.00155.x.

Ahenkora, K., Adu Dapaah, H. K., & Agyemang, A. (1998). Selected nutritional components and sensory attributes of cowpea (Vigna unguiculata [L.] Walp) leaves. *Plant Foods for Human Nutrition (Dordrecht, Netherlands), 52*(3), 221–229.

Akande, K. E., & Fabiyi, E. F. (2010). Effect of processing methods on some antinutritional factors in legume seeds for poultry feeding. *International Journal of Poultry Science, 9*(10), 996–1001.

Allen, L., de Benoist, B., Dary, O., & Hurrell, R. (2006). *Guidelines on food fortification with micronutrients*. Geneva: WHO.

Awika, J. M., & Duodu, K. G. (2017). Bioactive polyphenols and peptides in cowpea (Vigna unguiculata) and their health promoting properties: A review. *Journal of Functional Foods, 38*, 686–697. https://doi.org/10.1016/j.jff.2016.12.002.

Baranwal, D. (2017). Malting: An indigenous technology used for improving the nutritional quality of grains—A review. *Asian Journal of Dairy and Food Research, 36*(3), 179–183. https://doi.org/10.18805/ajdfr.v36i03.8960.

Bassey, F. I., McWatters, K. H., Edem, C. A., & Iwegbue, C. M. A. (2013). Formulation and nutritional evaluation of weaning food processed from cooking banana, supplemented with cowpea and peanut. *Food Science & Nutrition, 1*(5), 384–391. https://doi.org/10.1002/fsn3.51.

Boukar, O., Belko, N., Chamarthi, S., Togola, A., Batieno, J., Owusu, E., Haruna, M., Diallo, S., Umar, M. L., Olufajo, O., & Fatokun, C. (2018). Cowpea (Vigna unguiculata): Genetics, genomics and breeding. *Plant Breeding, 138*, 415. https://doi.org/10.1111/pbr.12589.

Carvalho, M., Lino-Neto, T., Rosa, E., & Carnide, V. (2017). Cowpea: A legume crop for a challenging environment. *Journal of the Science of Food and Agriculture, 97*(13), 4273–4284. https://doi.org/10.1002/jsfa.8250.

Cavalcante, R. B. M., Morgano, M. A., Silva, K. J. D., Rocha, M. M., Araújo, M. A. M., & Moreira-Araújo, R. S. R. (2016). Cheese bread enriched with biofortified cowpea flour. *Ciência e Agrotecnologia, 40*, 97–103.

Chavan, J. K., Kadam, S. S., & Beuchat, L. R. (1989). Nutritional improvement of cereals by sprouting. *Critical Reviews in Food Science and Nutrition, 28*(5), 401–437. https://doi.org/10.1080/10408398909527508.

Cowpea Storage Project: Profiles of Progress. (2010). Bill and Melinda Gates Foundation. 1–2..

Cruz, A. R. R., & Aragão, F. J. L. (2014). RNAi-based enhanced resistance to cowpea severe mosaic virus and cowpea aphid-borne mosaic virus in transgenic cowpea. *Plant Pathology, 63*(4), 831–837. https://doi.org/10.1111/ppa.12178.

Department of Agriculture Forestry and Fisheries. (2011). *Production guidelines for cowpeas*. Republic of South Africa

Dumet, D., Adeleke, R., & Faloye, B. (2008). Regeneration guidelines: Cowpea. In M. E. Dulloo, I. Thormann, M. A. Jorge, & J. Hanson (Eds.), *Crop specific regeneration guidelines*. Rome: CGIAR System-wide Genetic Resource Programme.

Etokakpan, O. U., Eka, O. U., & Ifon, E. T. (1983). Chemical evaluation of the effect of pest infestation on the nutritive value of cowpeas Vigna unguiculata. *Food Chemistry, 12*(3), 149–157. https://doi.org/10.1016/0308-8146(83)90001-8.

FAOSTAT. (2019). *Food and Agriculture Organization of the United Nations*. Retrieved from http://www.fao.org/faostat/en/#data

Farinu, G. O., & Ingrao, G. (1991). Gross composition, amino acid, phytic acid and trace element contents of thirteen cowpea cultivars and their nutritional significance. *Journal of the Science of Food and Agriculture, 55*(3), 401–410. https://doi.org/10.1002/jsfa.2740550308.

Frota, G., Lopes, L. A. R., Silva, I. C. V., Alfredo, J. A. B., & Arêas, J. A. G. (2017). Nutritional quality of the protein of *Vigna unguiculata* L. Walp and its protein isolate. *Revista Ciência Agronômica, 48*(5), 792–798. https://doi.org/10.5935/1806-6690.20170092.

Gómez, C. (2004). *Cowpea: Postharvest operations*. Rome: Food and Agriculture Organization of the United Nations.

Goncalves, A., Goufo, P., Barros, A., Dominguez-Perles, R., Trindade, H., Rosa, E. A., Ferreira, L., & Rodrigues, M. (2016). Cowpea (Vigna unguiculata L. Walp), a renewed multipurpose crop for a more sustainable Agri-food system: Nutritional advantages and constraints. *Journal of the Science of Food and Agriculture, 96*(9), 2941–2951. https://doi.org/10.1002/jsfa.7644.

Gupta, M., Abu-Ghannam, N., & Gallaghar, E. (2010). Barley for brewing: Characteristic changes during malting, brewing and applications of its by-products. *Comprehensive Reviews in Food Science and Food Safety, 9*(3), 318–328. https://doi.org/10.1111/j.1541-4337.2010.00112.x.

Hall, A. E. (2004). Breeding for adaptation to drought and heat in cowpea. *European Journal of Agronomy, 21*(4), 447–454. https://doi.org/10.1016/j.eja.2004.07.005.

Hashim, N., & Pongjata, J. (2000). Vitamin a activity of rice-based weaning foods enriched with germinated cowpea flour, banana, pumpkin and milk powder. *Malaysian Journal of Nutrition, 6*(1), 65–73.

Houssou, A. P. F., Ahohuendo, B. C., Fandohan, P., & Hounhouigan, D. J. (2010). Analysis of pre- and post-harvest practices of cowpea (*Vigna unguiculata* (L.) Walp.) in Benin. *International Journal of Biological and Chemical Sciences, 4*(5), 1730–1741. https://doi.org/10.4314/ijbcs.v4i5.65576.

Iqbal, A., Khalil, I. A., Ateeq, N., & Sayyar Khan, M. (2006). Nutritional quality of important food legumes. *Food Chemistry, 97*(2), 331–335. https://doi.org/10.1016/j.foodchem.2005.05.011.

Jain, A. K., Kumar, S., & Panwar, J. D. S. (2019). Antinutritional factors and their detoxification in pulses-a review. *Agricultural Reviews, 30*(1), 64–70.

Jakkanwar, S., Rathod, R., & Annapure, U. (2018). Development of cowpea-based (Vigna unguiculata) extruded snacks with improved in vitro protein digestibility. *International Food Research Journal, 25*(2), 804–813.

Jayathilake, C., Visvanathan, R., Deen, A., Bangamuwage, R., Jayawardana, B. C., Nammi, S., & Liyanage, R. (2018). Cowpea: An overview on its nutritional facts and health benefits. *Journal of the Science of Food and Agriculture, 98*(13), 4793–4806. https://doi.org/10.1002/jsfa.9074.

Jirapa, P., Normah, H., Zamaliah, M. M., Asmah, R., & Mohamad, K. (2001). Nutritional quality of germinated cowpea flour (Vigna unguiculata) and its application in home prepared powdered weaning foods. *Plant Foods for Human Nutrition, 56*(3), 203–216. https://doi.org/10.1023/A:1011142512750.

Joshi, P. K., & Rao, P. P. (2017). Global pulses scenario: Status and outlook. *Annals of the New York Academy of Sciences, 1392*(1), 6–17. https://doi.org/10.1111/nyas.13298.

Kalogeropoulos, N., Chiou, A., Ioannou, M., Karathanos, V. T., Hassapidou, M., & Andrikopoulos, N. K. (2010). Nutritional evaluation and bioactive microconstituents (phytosterols, tocopherols, polyphenols, triterpenic acids) in cooked dry legumes usually consumed in the Mediterranean countries. *Food Chemistry, 121*(3), 682–690. https://doi.org/10.1016/j.foodchem.2010.01.005.

Khattab, R. Y., & Arntfield, S. D. (2009). Nutritional quality of legume seeds as affected by some physical treatments 2. Antinutritional factors. *LWT - Food Science and Technology, 42*(6), 1113–1118. https://doi.org/10.1016/j.lwt.2009.02.004.

Lagarda-Diaz, I., Guzman-Partida, A. M., & Vazquez-Moreno, L. (2017). Legume lectins: Proteins with diverse applications. *International Journal of Molecular Sciences, 18*(6), 1242. https://doi.org/10.3390/ijms18061242.

Lazaridi, E., Ntatsi, G., Fernández, J. A., Karapanos, I., Carnide, V., Savvas, D., & Bebeli, P. J. (2017). Phenotypic diversity and evaluation of fresh pods of cowpea landraces from southern Europe. *Journal of the Science of Food and Agriculture, 97*(13), 4326–4333. https://doi.org/10.1002/jsfa.8249.

Leite, J., Fischer, D., Rouws, L. F. M., Fernandes-Júnior, P. I., Hofmann, A., Kublik, S., Schloter, M., Xavier, G. R., & Radl, V. (2017). Cowpea nodules harbor non-rhizobial bacterial communities that are shaped by soil type rather than plant genotype. *Frontiers in Plant Science, 7*, 2064–2064. https://doi.org/10.3389/fpls.2016.02064.

Madodé, Y. E., Houssou, P. A., Linnemann, A. R., Hounhouigan, D. J., Nout, M. J. R., & Van Boekel, M. A. J. S. (2011). Preparation, consumption, and nutritional composition of west African cowpea dishes. *Ecology of Food and Nutrition, 50*(2), 115–136. https://doi.org/10.1080/03670244.2011.552371.

Madodé, Y. E., Nout, M. J. R., Bakker, E.-J., Linnemann, A. R., Hounhouigan, D. J., & van Boekel, M. A. J. S. (2013). Enhancing the digestibility of cowpea (*Vigna unguiculata*) by traditional processing and fermentation. *LWT - Food Science and Technology, 54*(1), 186–193. https://doi.org/10.1016/j.lwt.2013.04.010.

Madsen, C. K., & Brinch-Pedersen, H. (2016). The antinutritional components of grains. In *Reference Module in Food Science*. Amsterdam: Elsevier. https://doi.org/10.1016/B978-0-08-100596-5.00111-6.

Maina, A. N., Tchiagam, J. B. N., Gonne, S., Hamadama, Y., Bell, J. M., & Yanou, N. N. (2015). Diallel analysis of polyphenols and phytates content in cowpea (*Vigna unguiculata* L. Walp.). *Scientia Agriculturae, 12*(1), 46–51. https://doi.org/10.15192/PSCP.SA.2015.12.1.4651.

Marengo, M., Baffour, L. C., Buratti, S., Benedetti, S., Saalia, F. K., Carpen, A., Manful, J., Johnson, P.-N. T., Barbiroli, A., Bonomi, F., Pagani, A., Marti, A., & Iametti, S. (2017). Defining the overall quality of cowpea-enriched rice-based breakfast cereals. *Cereal Chemistry, 94*(1), 151–157. https://doi.org/10.1094/CCHEM-04-16-0092-FI.

Martinez-Villaluenga, C., Frias, J., & Vidal-Valverde, C. (2008). Alpha-galactosides: Antinutritional factors or functional ingredients? *Critical Reviews in Food Science and Nutrition, 48*(4), 301–316. https://doi.org/10.1080/10408390701326243.

Mubarak, A. E. (2005). Nutritional composition and antinutritional factors of mung bean seeds (*Phaseolus aureus*) as affected by some home traditional processes. *Food Chemistry, 89*(4), 489–495. https://doi.org/10.1016/j.foodchem.2004.01.007.

Pasquet, R. S. (1998). Morphological study of cultivated cowpea *Vigna unguiculata* (L.) Walp. Importance of ovule number and definition of cv gr Melanophthalmus. *Agronomie, 18*(1), 61–70.

Pelembe, L. A. M., Erasmus, C., & Taylor, J. R. N. (2002). Development of a protein-rich composite Sorghum–cowpea instant porridge by extrusion cooking process. *LWT - Food Science and Technology, 35*(2), 120–127. https://doi.org/10.1006/fstl.2001.0812.

Piergiovanni, A. R., & Gatta, C. D. (1994). α-Amylase inhibitors in cowpea (*Vigna unguiculata*): Effects of soaking and cooking methods. *Food Chemistry, 51*(1), 79–81. https://doi.org/10.1016/0308-8146(94)90051-5.

Preet, K., & Punia, D. (2000). Proximate composition, phytic acid, polyphenols and digestibility (in vitro) of four brown cowpea varieties. *International Journal of Food Sciences and Nutrition, 51*(3), 189–193.

Prinyawiwatkul, W., McWatters, K. H., Beuchat, L. R., Phillips, R. D., & Uebersak, M. A. (1996). Cowpea flour: A potential ingredient in food products. *Critical Reviews in Food Science and Nutrition, 36*(5), 413–436. https://doi.org/10.1080/10408399609527734.

Richard, A. (1847). *Tentamen florae Abyssinicae*. Apud Arthus Bertrand, Editorem, Paris.

Sebetha, E. T., Modi, A. T., & Owoeye, L. G. (2015). Cowpea crude protein as affected by cropping system, site and nitrogen fertilization. *Journal of Agricultural Science, 7*(1), 224–234.

Shevkani, K., Kaur, A., Kumar, S., & Singh, N. (2015). Cowpea protein isolates: Functional properties and application in gluten-free rice muffins. *LWT - Food Science and Technology, 63*(2), 927–933. https://doi.org/10.1016/j.lwt.2015.04.058.

Simion, T. (2018). Breeding cowpea *Vigna unguiculata* l. Walp for quality traits. *Annals of Reviews and Research, 3*(2), 555609.

Singh, B., Singh, J. P., Shevkani, K., Singh, N., & Kaur, A. (2017). Bioactive constituents in pulses and their health benefits. *Journal of Food Science and Technology, 54*(4), 858–870. https://doi.org/10.1007/s13197-016-2391-9.

Subuola, F., Widodo, Y., & Kehinde, T. (2012). Processing and utilization of legumes in the tropics. In A. A. Eissa (Ed.), *Trends in vital food and control engineering*. Rijeka: InTechOpen. https://doi.org/10.5772/36496.

Thangadurai, D. (2005). Chemical composition and nutritional potential of *Vigna unguiculata* ssp. Cylindrica (Fabaceae). *Journal of Food Biochemistry, 29*(1), 88–98. https://doi.org/10.1111/j.1745-4514.2005.00014.x.

Timko, M. P., & Singh, B. (2008). Cowpea: A multifunctional legume. In P. H. Moore & R. Ming (Eds.), *Genomics of tropical crop plants* (Plant genetics and genomics: Crops and models) (Vol. 1, pp. 227–258). New York: Springer.

Tiroesele, B., Thomas, K., & Seketeme, S. (2015). Control of cowpea weevil, *Callosobruchus Maculatus* (F.) (Coleoptera: Bruchidae), using natural plant products. *Insects, 6*(1), 77–84. https://doi.org/10.3390/insects6010077.

Tresina, P. S., & Mohan, V. R. (2011). Effect of gamma irradiation on physicochemical properties, proximate composition, vitamins and antinutritional factors of the tribal pulse Vigna unguiculata subsp. unguiculata. *International Journal of Food Science & Technology, 46*(8), 1739–1746. https://doi.org/10.1111/j.1365-2621.2011.02678.x.

Hyacinth Beans

K. A. Athmaselvi, Aryasree Sukumar, and Sanika Bhokarikar

Introduction

The hyacinth bean (include the scientific name here) is a tropical and subtropical food legume which is of Asian origin. In Europe, Africa, Middle East, America, and Asia, hyacinth beans have been cultivated for many years. Leaves, green beans, and pods of hyacinth beans are edible and consumed widely (Maass et al. 2010). *Lablab purpureus* are climbing plants that can grow up to a length of 5 m, with leaves 3-foliate and pinnate, 6–12 cm by 5–9 cm acute leaflets, purplish pink or white flowers, and green podded fruits with 4–5 seeds, flattened 6 cm long and 2 cm wide. Color change of pods from green to light brown is used as a maturity index (Al-Snafi 2017). Different names of hyacinth beans are hyacinthbean, Egyptian bean, field bean, bonavist bean, and *Dolichos lablab* L., *Lablab niger* Medikus, *Lablab vulgaris* (L.) Savi, *Lablab purpureus* (L.) Sweet, *Vigna aristata* Piper, *Dolichos benghalensis* Jacq., and *Dolichos purpureus* L., are the different scientific names (Sheahan 2012).

Hyacinth bean is a legume which can grow as rainfed crop and also with very low rainfall in the range of 60–90 cm. It is a drought-tolerant crop, which can grow from sea level to high altitudes of up to 2300 m. Yield range of seeds are 450–1460 kg/ha from pure stands (Duke 1983). Hyacinth bean is a good protein source, edible in the form of unripe pods, dry seeds, and green mature beans. There are two different

K. A. Athmaselvi (✉)
Indian Institute of Food Processing Technology, Thanjavur, Tamil Nadu, India
e-mail: athmaselvi@iifpt.edu.in

A. Sukumar · S. Bhokarikar
Department of Food Process Engineering, School of Bioengineering, SRMIST, Kattankulathur, Tamil Nadu, India

A. Manickavasagan, P. Thirunathan (eds.), *Pulses*,
https://doi.org/10.1007/978-3-030-41376-7_7

forms of hyacinth beans: the ones whose flowers, stems, and foliage are deeply pigmented and the other one with white-colored flowers and unpigmented foliage and stems. (Akpapunam 1996). It is used as forage, hay, and silage. Hyacinth bean seed and leaves are one of the most palatable animal legumes (Valenzuela and Smith 2002).

Proximate composition studies showed 20–25% crude protein content, 2.6–4.1% crude lipid content, 4.9–6.9% total dietary fiber, 3.9–4.4% ash, and 60–66% carbohydrates. Minerals include sodium, potassium, calcium, magnesium, phosphorus, iron, and manganese. Vitamins include niacin and ascorbic acid. Except for sulfur-containing and tryptophan amino acid, all essential amino acids are found in seed protein. Its fatty acid composition includes 24% saturated fatty acid, 18% monounsaturated fatty acid, 57% polyunsaturated fatty acid, and 44% linoleic acid. Purified seeds of *Dolichos lablab* contain protein globulins arcelin and dolichin. Stachyose is the oligosaccharide present in all varieties of *Lablab purpureus* (Kala et al. 2010). In *Lablab purpureus*, 262 volatile compounds are identified. Terpenoids, terpenes, and their derivatives are the main volatile compounds present, which account to 46% of all detected compounds. Top exporter of hyacinth bean is Mexico, and top importer is the United States. Its global average market price is $1.43 USD/kg.

Postharvest Processing

Harvesting

As per degree of maturity, harvesting time is determined. Mature pods will change color from green to light brown. Hyacinth beans are manually harvested because of the growth habitat. Beans are removed from plants carefully without any damage to pods and plants. Unprofessional handling of pods will cause extended and visible damage. Harvested pods are loosely packed in containers and should not keep in the sun for a long time. Table 1 shows unit operations included in postharvest processing of beans.

Preharvest Drying

Harvesting has to be started after the moisture on plants has evaporated. Dried pod harvesting helps in low contamination and prevention of postharvest diseases. Increasing preharvest drying period will increase the risk of losing the harvest to insects, birds, and rodents. Harvesting before maturity leads to seed decay and mold growth. Mature harvest transport is riskier because there is a chance of seeds detaching and falling on roads before reaching threshing area. Much care is needed during assortment and initial conveyance of harvest.

Table 1 Unit operations in postharvest processing of beans (Uebersax and Siddiq 2013)

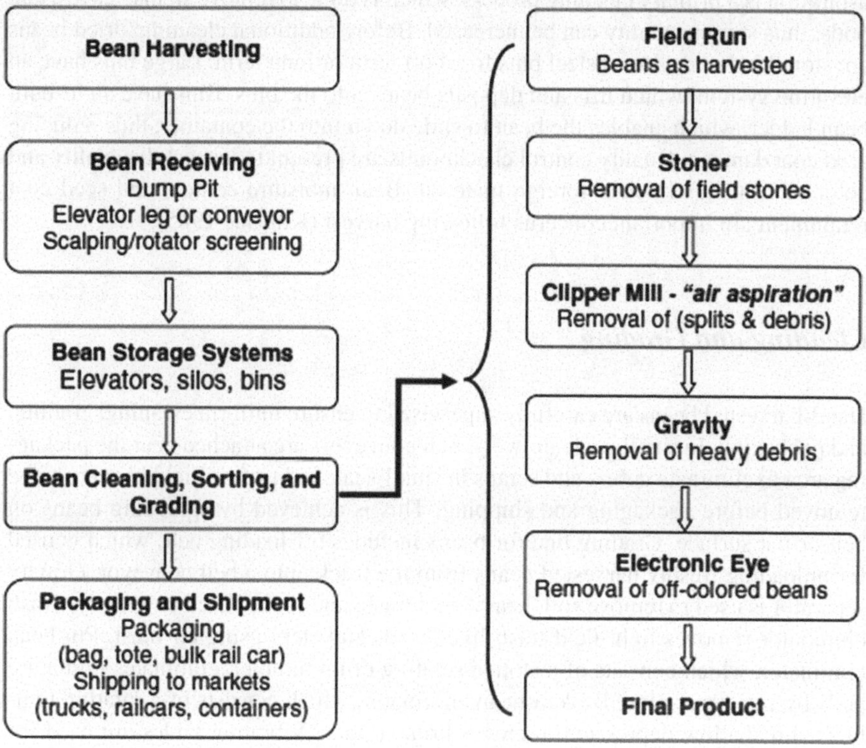

Postharvest Drying

Drying of harvested bean is required to avoid mold growth and to increase storage quality. Time period needed for drying depends on weather and atmospheric conditions. Lengthy structured area called cribs or unroofed terrace or threshing floors are used for postharvest drying. Unroofed terrace drying causes quality issues and loss due to rodents and birds and their droppings. The unbroken bean is highly preferred; damage to bean seed coat like shattering, cracking, and splitting may happen during mechanical processing, so proper care has to be taken to avoid damages. Apart from on-farm storage system, commercial distribution system with monitoring and controlling devices is used for drying and storage (Grizzell 1961).

Threshing

After enough postharvest drying, harvests are taken for threshing to remove pods. Threshed beans are heaped and stored. If too damp beans are heaped immediately after threshing, it results in microbial attack, affecting their quality. Harvested, dried, and threshed beans are transported in trucks from field to storage area. Air aspiration is a primary cleaning process, which is done to remove stems, leaves, and pods; thus storage quality can be increased. Before additional cleaning, dried beans are stored in large silos or steel bins for short term or long term. Large bins have an elevating system, which lifts and deposits beans into the bins. Bins have an in-built bean ladder, which enables the bean to slide down into the container, thus reducing seed coat damage. Quality control checkpoints are present to inspect the quality and color of beans and detect foreign material. Bean moisture content and seed coat retainment are important concerns following harvest (Kincade 1985).

Cleaning and Grading

Hand-harvested beans are carefully supervised to ensure uniform cleaning, grading, and packaging. In small-scale growers, belt conveyors are attached near the packaging area to eliminate debris and scraps in small-scale industries. Field heat has to be removed before packaging and shipping. This is achieved by spreading beans on belt or flat surface. Grading line for beans includes off-loading belt, which is used for unloading freshly harvested beans from the truck onto a belt conveyor. Gravity separator is used to remove soil, heavy field trash, and rocks based on gravity. Trash eliminator removes light field trash like leaves and stem using air blast. Pin-bean eliminator, which consists of a slotted rotating drum tumbler, eliminates immature pods by size separation. Broken-bean eliminator, which consists of a rotating tumbler with shallow depression, removes broken pods. Vibrating tables are used for segregating good pods from any remaining field trash. Vibrating washers are used for wet cleaning of pods to remove soil particles and field heat using clean water. Grading tables are used for manually removing decayed, defective, blemished, and overmature pod. Automatic box filler which is used for filling cleaned graded beans into pre-weighed crates and cartons and unloaded onto an automatic sealing machine. Containers are filled and then cooled by hydrocooling or forced-air cooling and stored in cold storage or shipped directly.

Postharvest Cooling

Harvested beans produce respiration heat, which in turn affects their quality. Thorough postharvest cooling will eliminate this heat and preserve quality, increasing its shelf life. Postharvest cooling also reduces the effect of dehydration and damage by organisms. In case of immediate refrigeration unavailability after harvesting, keeping the produce in shade, drenching with cold well water, and harvesting during the coolest part of the day are alternatives. For local sale, properly cleaned, cooled, field-packed beans are used. Wetting of beans with proper air circulation results in evaporative cooling. Air circulation should be continued until the beans are packed in cartons, wetted, and properly refrigerated. Green beans intended for distant market are cooled immediately after harvest. Warm beans from field are kept in cooling chamber. Cooling the room by means of natural conductive and convective heat transfer is a time-consuming process. More than 16 h is required for sufficient cooling of bean-filled bulk containers. Beans have to be loosely stacked for air circulation inside containers to prevent generation of respiration heat.

Forced-Air Cooling

Forced-air cooling system increases the cooling rate. Circulating fans are fitted inside cooling room to pull air through bean containers. In forced-air cooling, rate of cooling is eight times higher than still-air room cooling.

Hydrocooling

In hydrocooling large quantity of chilled water is placed in contact with bean cartons and is preferred for cooling green beans. In large-scale industries, hydrocooling is done for quick cooling and delivery to distant markets since water is a good heat transfer medium than air. Bean is wetted with hydrocooler, which rains water onto containers, or immersed in tank of chilled water.

In flume hydrocooling, cleaned and graded beans are dropped directly into a long flume system with chilled chlorinated water from 34 to 38 °F. Quick and uniform cooling can be achieved through this method, which lowers the temperature of beans from 85 °F to about 45 °F in 6 min. Brown-end discoloration can be avoided by rapid cooling. Drawback of flume hydrocooling is beans are moistened, which results in postharvest diseases if rewarming of beans or improper chlorination happens. Postharvest diseases occurring in beans are caused by *Rhizopus* species, *Botrytis cinerea*, and *Sclerotinia* species. Sufficient refrigeration is required for continuous cooling and storage. Abrasion, cut, and scar during handling of bean result in infection, so proper washing in water with chlorine concentration of about 55–70 ppm at neutral pH is recommended (Boyette et al. 1994).

Packaging

Beans are packed in containers like crates, hampers with a capacity of 26–31 lb each, and fiberboard cartons with a capacity of 25–30 lb each. Cutting and shelling greatly increases rate of respiration and the vulnerability to disease organisms. LDPE bags with oxygen transmission rate around 10,000–15,000 $cm^3/m^2/day$ at 73 °F create modified atmospheric packaging with 3–5% oxygen and 20–30%

carbon dioxide. This will reduce the brown discoloration, sliminess, and spotting in freshly cut beans. Polyethylene bags (1–5 lb) are used for dry bean packaging, in which dry beans are packaged in polyethylene film using a form-fill seal machine. In this system, polyethylene film is used in making sachets which are filled and sealed automatically (Uebersax et al. 1996). Polypropylene bags, which have a capacity of 100 lb with a dimension of 21 × 37 in., allow stackability of the bags, thus avoiding slippage and shifting during shipment. Polyethylene totes (2000 lb) are also used for packaging. Large-capacity containers are made of polyethylene with a capacity of 1 ton. In the case of overseas shipment packages, 100-pound jute sacks are directly included within cargo container of dimension 8 × 8 × 20 ft (Thompson 1962).

Storage

Effective and long-term storage can be achieved through facilities, hygiene, and monitoring. Less than 18% moisture content is the limit to assure storage stability for beans. High moisture content leads to intrinsic fungal spoilage. Warehouse, bins, and granaries with temperature control are used as bean closed storage systems. Storage condition of green beans is 95% relative humidity and 37–45 °F. Temperatures lower than that lead to chilling injury. Adverse storage conditions result in "bin burn," "hard-shell" and "hard-to-cook" phenomena, which adversely affect bean quality and economic value (Pirhayati et al. 2011; Paredes-López et al. 1989). Some fruits like apples, bananas, tomatoes, and cantaloupes will produce ethylene gas, which changes the quality of green beans if stored and transported with them. Beans readily adsorb the odor of pepper, cantaloupes, and onions, so it is not stored and shipped with these items.

Dry Bean Storage

Dry beans can be stored in different storage conditions. Concrete silos are used for storing dry beans because of their capacity and structural strength which resist rapid temperature change. Flat storages are reinforced concrete floors which can be used for storing dry beans. Easy handling and cost-effectiveness are advantages of flat storage systems. Steel bins of different sizes can be used for storing dry beans; they are simple and installed in site. Tote boxes are made of polypropylene, wood, and cardboard material and can be used for storing dry beans. Capacity can be up to 1 ton; flexible handling and high bean movement are advantages of this system.

Postharvest storage quality of dry beans is maintained using different food safety standards of dry bean handling, which are ISO 9000, SQF, HACCP, GMPs, and GAPs.

Product Development

Processed Products

The impact of pressure cooking (PC) and traditional cooking (TC) of dehulled and undehulled of "hepo" (*Lablab purpureus L.*) on nutritional and mineral compositions was studied. Proximate compositions showed deviation in carbohydrate, fat, ash, fiber, and energy value, but protein content remained the same after processing. The analysis showed that the methods of processing caused a significant ($p < 0.05$) difference in all minerals, i.e., P, Ca, and Zn, except iron (Fe) (Mosisa and Tura 2017).

The effect of soaking (6, 12, and 18 h), time for germination (40, 60 h), cooking (unsoaked and soaked samples), autoclaving (unsoaked and soaked samples), and roasting on antinutritional factors and nutritional content was studied for *Dolichos lablab* (HA-4). Germination of bean increased the protein content and moisture and reduced crude lipid, carbohydrate, and ash contents. Processing significantly reduced phytic acid content but increased tannins. Development of food products from germinated bean can be beneficial for people suffering from cardiovascular disease, diabetes mellitus, and hypercholesterolemia as it contains less carbohydrates and lipid [D'souza 2013].

The effects of untreated (raw), dry-heated, and pressure-cooked samples on antioxidant activity and total phenolic content in *Dolichos lablab* L. were determined. The results show decrease in total phenolic content in processed bean than raw. The pressure-cooked samples possess low antioxidant activity than untreated (raw) and dry-heated samples of *D. Lablab* [Maheshu et al. 2013].

Isolates, Coagulants, and Additives

Hyacinth bean (*Lablab purpureus (L.)* sweet) seeds cultivated in Indonesia were used as protein source to prepare protein isolate by isoelectric method. Protein isolate was further used to characterize physiological and functional characteristics. Hyacinth bean (seeds) contains 17 ± 1.5% of protein and 1.1 ± 0.1 mg/100 g of HCN. Pretreatments are required to decrease the antinutritional factors of seed, i.e., phytate (18.9 ± 0.2 mg/g) and trypsin inhibitor (0.15 ± 0.02 TIU/mg). The isoelectric method gave low yield of protein isolate (approximately 7.38 g per 100 g of seeds), but other qualitative properties like color and odor were satisfactory with high amount of protein, i.e., 89.8 ± 0.82%. The protein isolate had good solubility, foaming capacity, and emulsifying activity. Foaming and emulsifying stabilities of protein isolate were low (Subagio 2006).

Protein isolate extracted from seeds of hyacinth bean (*Lablab purpureus L.* sweet) was used in cake to improve baking properties. Cake made with addition of protein isolate (up to 1%) from hyacinth bean seed showed improvement in volume development, specific volume, and softness. Cake prepared by adding 1% protein isolate showed development in volume (206%) and specific volume (2.63 mL/g), whereas the controls were 160% volume improvement and specific volume 2.17 mL/g. The addition of protein isolate more than 1% decreased the quality, i.e., low staling rate and undesirable color compared with that at 1% [Subagio and Morita 2008].

Hyacinth bean peels were also used as natural coagulants. The dosage of hyacinth bean peel was optimized to 20 mg/L for treatment of Kukkarahalli Lake (Karnataka, India) water. This treatment reduced bacterial count by 11.52%, and turbidity removal efficiency was found to be 77.10% [Shilpaa et al. 2012].

White hyacinth bean was used for extraction of water-soluble polysaccharides (WPs) by two methods, i.e., water extraction and alcohol precipitation method. The extracted WPs had scavenging activity on three different radicals (hydroxyl > DPPH > superoxide). The yield of WPs was 1.15 ± 0.07% under the following optimum extraction parameters: water-to-material ratio 50, 2 h for extraction, and extraction at temperature 95 °C. The research indicated that "The WPs from white hyacinth bean could promote the growth of *L. acidophilus LA5, B. bifidum BB01* and *L. bulgaricus LB6*" [Ni Lei et al. 2016].

Flours and Gels

The nutritional analysis of the Kenyan lablab seed flour proved that it is a good source of calories and protein. Carbohydrate content of the *Dolichos lablab* seed flour contains 34.96 ± 0.06% carbohydrate, 24.1.61 ± 1.61% protein, and 21.56 ± 0.44% dietary fibers. The other constituents of *Dolichos lablab* seed flour were moisture (10.62%), fat (5.74%), and ash (2.91%) approximately [Habib et al. 2017].

The study evaluated the thermal, rheological, and pasting properties of *Dolichos* bean (*Dolichos lablab* L.) and legume flour and gels. The comparative study of *Dolichos lablab* L. with pigeon pea and jack bean showed high viscosities and jack bean had high pasting temperature. The solid gel was prepared by heating it to 95 °C with estimated minimum flour concentration of 6–8%. Differential scanning calorimetry was used to form two endothermic peaks in all flours at 80–89 and 96–100 °C. Hardening kinetics of *Dolichos lablab* L. proved that flour is useful for self-supporting gels. This flour can be used to prepare gel-like products [Acevedo et al. 2013].

Future Trends in Processing and Product Development

The hyacinth bean seeds contain high amount of protein, which can be used as vegan protein source in products. Research is required to develop different methods for reduction of trypsin inhibitor and phytic acid content before using it for human consumption. Different methods can be used to extract protein isolate from hyacinth bean to increase the yield and quality of protein. The protein isolate has scope in nutraceuticals and functional foods as additive. Germination and other processing methods have potential to improve the nutritive and sensory values of hyacinth bean. The hyacinth bean can be used to prepare functional food products for iron deficiency as it contains high amount of iron and stability during processing.

Hyacinth bean is a versatile, variable, and adaptable crop, and research is required to avoid its loss. Research is required to increase the use of this underutilized but versatile plant in food products.

References

Acevedo, B. A., Avanza, M. V., Cháves, M. G., & Ronda, F. (2013). Gelation, thermal and pasting properties of pigeon pea (Cajanus cajan L.), dolichos bean (Dolichos lablab L.) and jack bean (Canavalia ensiformis) flours. *Journal of Food Engineering, 119*(1), 65–71.

Akpapunam, M. (1996). Hyacinth bean (Lablab purpureus (L.) sweet). In *Food and feed from legumes and oilseeds* (pp. 103–108). Boston: Springer.

Al-Snafi, A. E. (2017). The pharmacology and medical importance of Dolichos lablab (Lablab purpureus)-A review. *IOSR Journal of Pharmacy, 7*(2), 22–30.

Boyette, M. D., Schulteis, J. R., Estes, E. A., Hurst, W. C., & Sumner, P. E. (1994). Postharvest cooling and handling of green beans and field peas. *AG (North Carolina Agricultural Extension Service)(USA)*.

D'souza, M. R. (2013). Effect of traditional processing methods on nutritional quality of field bean. *Advances in Bioresearch, 4*(3), 29.

Duke, J. A. (1983) *Handbook of Legumes of World Economic Importance*. Plenum Press, New York, pp 102–106, 275–278, 293–296.

Grizzell, W. G. (1961). *Handling pea beans in country elevators* (No. 479). US Department of Agriculture, Agricultural Marketing Service, Transportation and Facilities Research Division and Market Quality Research Division.

Habib, H. M., Theuri, S. W., Kheadr, E. E., & Mohamed, F. E. (2017). Functional, bioactive, biochemical, and physicochemical properties of the Dolichos lablab bean. *Food & Function, 8*(2), 872–880.

Kala, B. K., Soris, P. T., Mohan, V. R., & Vadivel, V. (2010). Nutrient and chemical evaluation of raw seeds of five varieties of Lablab purpureus (L.) sweet. *Advances in Biology Research, 1*(1), 44–53.

Kincade, K. P. (1985). The A-to-Z of dry bean handling. In *Proc Technical Conference on Dry Beans Research*.

Lei, N., Shu, G., Cao, J., Wan, H., & Feng, J. (2016). Extraction and antioxidant activities assay of polysaccharides from white hyacinth bean and promoting growth to probiotics. *Annals of the University Dunarea de Jos of Galati Fascicle VI--Food Technology, 40*(2), 51–62.

Maass, B. L., Knox, M. R., Venkatesha, S. C., Angessa, T. T., Ramme, S., & Pengelly, B. C. (2010). Lablab purpureus—A crop lost for Africa? *Tropical Plant Biology, 3*(3), 123–135.

Maheshu, V., Priyadarsini, D. T., & Sasikumar, J. M. (2013). Effects of processing conditions on the stability of polyphenolic contents and antioxidant capacity of Dolichos lablab L. *Journal of Food Science and Technology, 50*(4), 731–738.

Mosisa, M. T., & Tura, D. C. (2017). Effect of processing on proximate and mineral composition of hepho, a black climbing bean (Lablab purpureus L.) flour. *Journal of Food and Nutrition Sciences, 5*, 16–22.

Paredes-López, O., Maza-Calviño, E. C., & González-Castañeda, J. (1989). Effect of the hardening phenomenon on some physicochemical properties of common bean. *Food Chemistry, 31*(3), 225–236.

Pirhayati, M., Soltanizadeh, N., & Kadivar, M. (2011). Chemical and microstructural evaluation of 'hard-to-cook' phenomenon in legumes (pinto bean and small-type lentil). *International Journal of Food Science & Technology, 46*(9), 1884–1890.

Sheahan, C. M. (2012). Plant guide for Lablab (Lablab purpureus), USDA-Natural Resources Conservation Service, Cape May Plant Materials Center. Cape May, NJ, 8210.

Shilpaa, B., Akankshaa, K., & Girish, P. (2012). Evaluation of cactus and hyacinth bean peels as natural coagulants. *International Journal of Chemical and Environmental Engineering, 3*(3), 187.

Subagio, A. (2006). Characterization of hyacinth bean (Lablab purpureus (L.) sweet) seeds from Indonesia and their protein isolate. *Food Chemistry, 95*(1), 65–70.

Subagio, A., & Morita, N. (2008). Effects of protein isolate from hyacinth beans (Lablab purpureus (L.) sweet) seeds on cake characteristics. *Food Science and Technology Research, 14*(1), 12–17.

Thompson, J. A. (1962). *Maintaining quality of pea beans during shipment overseas* (No. 519). US Department of Agriculture, Agricultural Marketing Service, Market Quality Research Division.

Uebersax, M. A., Kim, J. N., & Chung, Y. S. (1996). Packaging and handling systems for dry edible beans. *Michigan Dry Bean Digest, 20*(2), 5–13.

Uebersax, M. A., & Siddiq, M. (2013). Postharvest storage quality, packaging and distribution of dry beans. In M. Siddiq & M. A. Uebersax (Eds.), *Dry beans and pulses: Production, processing and nutrition* (pp. 75–100). Ames: Wiley.

Valenzuela, H., & Smith, J. (2002). *Sustainable agriculture green manure crops*. SA-GM-7. Cooperative Extension Service, College of Tropical Agriculture and Human Resources, University of Hawaii at Manoa.

Lentils

V. Chelladurai and C. Erkinbaev

Introduction

Lentil is a type of edible pulse that ranks fifth in the total worldwide production after chickpea, pigeon pea, bean, and mung bean. The average yield of lentils is 850–1100 kg/ha, and the annual worldwide production is 4.5 million tonnes (FAOSTAT 2014). Lentils are produced throughout the world, even in areas experiencing drought. The major lentil production countries are Canada, the United States, Turkey, Australia, and India, with these five countries encompassing around 75% of the world's total lentil production (Alexander 2015). Due to the increase in the consumption of lentils throughout the world, lentil production has shown a positive trend in the last two to three decades. In 1997, world lentil production was 2.76 million tonnes, and in 2007 production was increased to 3.29 million tonnes. Lentil production was 7.59 million tonnes, which was more than double the total yield when compared to 2007 (FAOSTAT 2019). Canada, India, and Turkey are the top three lentil-producing countries, with Canada ranking number one in lentil exports, as well as contributing to more than 50% of the annual worldwide lentil production (3.73 million tonnes in 2017). Western Canadian provinces like Saskatchewan, Alberta, and Manitoba produce most of the pulses in Canada, with Saskatchewan being the major lentil-producing province. Almost 70% of Canadian lentil production takes place in Saskatchewan, with most of Canada's lentil production being exported. Every year Canada exports nearly 60,000 tonnes of lentils to European countries, 63,000 tonnes to Mediterranean countries, 92,000 tonnes to South America, and 42,000 tonnes to Arab and African countries (Pulse Canada 2019).

V. Chelladurai (✉)
Department of Agricultural Engineering, Bannari Amman Institute of Technology, Sathyamangalam, India
e-mail: chelladurai@bitsathy.ac.in

C. Erkinbaev
Department of Biosystems Engineering, University of Manitoba, Winnipeg, MB, Canada

A. Manickavasagan, P. Thirunathan (eds.), *Pulses*,
https://doi.org/10.1007/978-3-030-41376-7_8

Fig. 1 Types of lentils: (**a**) green, (**b**) red, (**c**) small brown, (**d**) French green, (**e**) black lentils. (Reproduced with permission from Pulse Canada)

Table 1 Nutritional facts of red (split) and green (whole) lentils per serving (per 100 g, 1/2 cup, or 125 mL cooked) (adapted from www.lentils.org 2019)

S. no	Nutrition	Split red lentil (cooked)	Whole green lentil (cooked)
1	Calories	120 cal	112 cal
2	Total fat	0.4 g	0.4 g
3	Total carbohydrate	20 g	18.4 g
4	Protein	9.6 g	9.6 g
5	Dietary fiber	3.2 g	7.2 g
6	Calcium	9.6 mg	20 mg
7	Iron	2.4 mg	1.6 mg
8	Potassium	218 mg	202 mg
9	Folate	44 µg DFE	31 µg DFE

There are five major varieties of lentils that are produced throughout the world: red lentils, green lentils, French green lentils, brown lentils, and black lentils (Fig. 1). Among these five types, red and green lentils are the most common types, and red lentils are the major cultivars produced throughout the world. Red lentils play a major role in daily diet in South Asian countries like India, Pakistan, Sri Lanka, and Bangladesh.

Similar to all other major pulses, lentils have high amount of protein (20.6 to 31.4%) and lower fat (0.7 to 4.3%) content (Urbano et al. 2007). Lentils are also excellent sources of minerals and complex hydrocarbons (Gujral et al. 2011). Various studies conducted throughout the world have shown that consumption of legumes like lentils reduces the risk of type 2 diabetes, coronary heart disease, and obesity. Higher amount of legume intake also lowers the low-density lipoprotein (LDL) cholesterol level and increases the high-density lipoprotein (HDL) cholesterol level (Rizkalla et al. 2002; Pistollato et al. 2015). Lentils have a softer seed coat, reducing their cooking time which aids in reducing nutrient loss in pulses due to cooking (Satya et al. 2010). Total calories and carbohydrate values of the cooked split red lentil are higher than the cooked whole green lentil, but the cooked whole green lentil has higher dietary fiber and calcium. Table 1 summarizes the nutritional facts of red and green lentils, which are the major types of lentils produced and

Table 2 Glycemic index (GI) of various food products (Source: www.lentils.org)

S. no	Food item	Glycemic index (GI) value
1	Split red lentils	21
2	Green lentils	22
3	Kidney beans	23
4	Chickpeas	33
5	Oatmeal, rolled oats	58
6	White bread	76
7	Polished rice	89
8	Skinless potato	98

consumed. One cup of cooked red lentils provides almost 62% of the daily dietary fiber requirement for adults (15.6 g), which is the highest among all high-fiber foods such as chickpeas, kidney beans, and whole wheat bread. Split red lentil also has the lowest glycemic index (GI) when compared with rice and white bread, which makes the lentil an ideal food for people with diabetes (Lentils.org 2019) (Table 2).

Postharvest Processing

Lentil crops are generally sown in the fall and are harvested at the end of a very hot and dry summer. Lentils need to be harvested on time, with a delay in harvesting resulting in both quantity and quality losses due to shattering, lodging, loss of pod, and diseases. Quantity and quality losses also occur due to mechanical damages during harvesting and threshing, seed staining, weathering, and the high or low harvesting moisture content of the lentils. Harvesting the lentils outside the recommended moisture content affects the quality of the lentils during storage, and delays in lentil harvesting cause pod splitting as well as pods to drop from the plants. In Western Australia, it was estimated that delays in harvesting of 1–3 weeks caused an economic loss of 150–450 AUD per hectare due to pod splitting, pod dropping, shattering, and weathering (GRDC 2017). In general, early harvesting of lentils at around 14% moisture content has many advantages such as a decrease in grain losses due to pod dropping and pod splitting, a reduction in the percentage of cracked lentils, and a decrease in lentil discoloration. Early harvesting also decreases the snail contamination in lentils.

Ideally lentils should be threshed at 16–18% moisture content; therefore lentils need to be dried for safe, long-term storage. General harvesting time of lentils in Canada is in the middle of August when the weather is warm and dry, which aids for natural near-ambient air-drying of pulses. In Canada, the normally natural air-drying (near-ambient drying) technique is used to reduce the moisture content of the harvested lentils using ambient air.

Harvesting

Major postharvest processing applications involving lentils are threshing, drying, storage, and splitting (red lentils). The recommended harvesting moisture content for green lentils is 16–18% in order to reduce the threshing losses and maximize the benefits for producers. The recommended storage moisture content for green lentils is 14%. Red lentils need to be harvested and threshed at 14–16% moisture content and dried to less than 13% in order to store the red lentils for a long duration. Storage and handling guidelines for green and red lentils are almost the same other than the storage moisture content of red lentils (13%) being slightly less than that of green lentils (14%) (CGC 2009). Since most of the red lentils are used in split form, the harvesting and storage moisture of red lentils is lesser than that of green lentils to reduce losses during splitting (Agblor 2006).

Threshing

After harvesting, red lentils are threshed, and the inert materials are removed, after which the lentils are cleaned before being stored at proper temperature. As red lentils are decupled before consumption, the suitability of red lentils for secondary processing such as dehulling and splitting is important. Due to the softer seed coat, lentils are prone to have more mechanical damage than the cereal grains during harvesting and handling operations. Lower moisture content of lentils during harvesting results in the detachment of seed coats and chip on edges of seed coats, lowering the market price for lentils. Preferable moisture content to reduce mechanical damage and other damages to seed coat is 14%, but the milling quality is poor at this moisture content. Vandenberg (2009) found that swathing of red lentils with higher moisture and higher humidity weather conditions reduce shattering of red lentils during harvesting and threshing. Wind damage is the major cause of loss of lentil swaths during initial period right after swathing. So, threshing of lentils at a moisture content of 14–16% and then drying them to 13% using aeration are recommended for lentils to avoid handling and storage losses.

Storage

After threshing, lentils are stored in silos or bags at 13% moisture content until further processing. Threshed lentils have around 14% moisture, and for longer storage, lentils are dried to less than 13% using near-ambient drying or hot air-drying methods. Lower than 13% moisture also increases the mechanical losses during milling process (SaskPulse 2019). Sravanthi et al. (2013) studied the changes in quality parameters of red lentils during storage at various temperature conditions

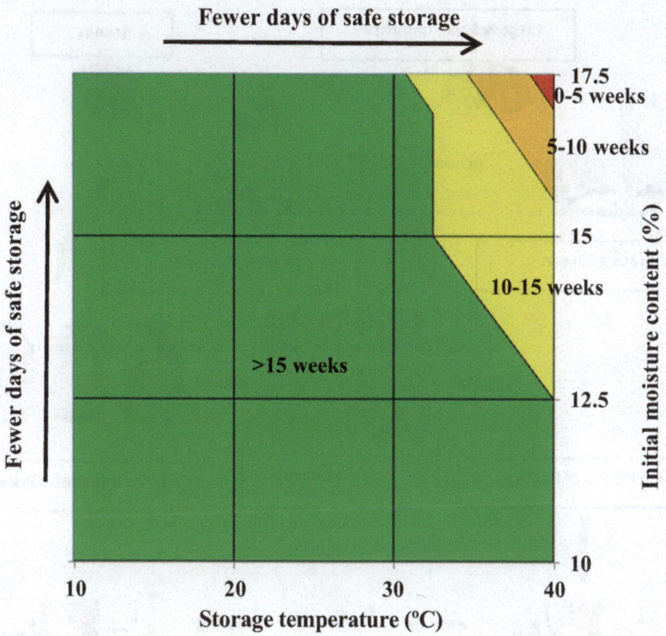

Fig. 2 Safe storage condition of red lentils under various temperatures. (Reproduced with permission from Sravanthi et al. (2013))

and found that the seed germination was more than 90% when they are stored below 13% moisture content; they have noticed visible molds from third week onward along with reduction in germination (66.7%) in the 17.5% moisture lentils stored at 40 °C. If the storage moisture of red lentils is below 13%, then lentils can be stored for more than 15 weeks at regular room temperature (<30 °C) (Sravanthi et al. 2013) (Fig. 2).

Milling

The whole red lentils are milled to get split using large carborundum emery rollers for de-husking and burr grinders for splitting. During this process normally 1.5% of grain is lost due to fines. China Win Tone Machinery (Lushan, China) developed a complete set of machinery for lentil processing, starting from cleaning to packaging (Fig. 3). The cleaning section contains a vibrating screener with two screens, a gravity destoner and a magnetic separator to remove impurities, stones, and magnetic materials from the lentils which arrive at the processing plant from farm storage or directly after combining. The peeling and splitting sections contain machinery for peeling and splitting lentils, with the peeler having a brusher assembly to remove micro bran from the lentils. Graders separate broken split lentils from the intact

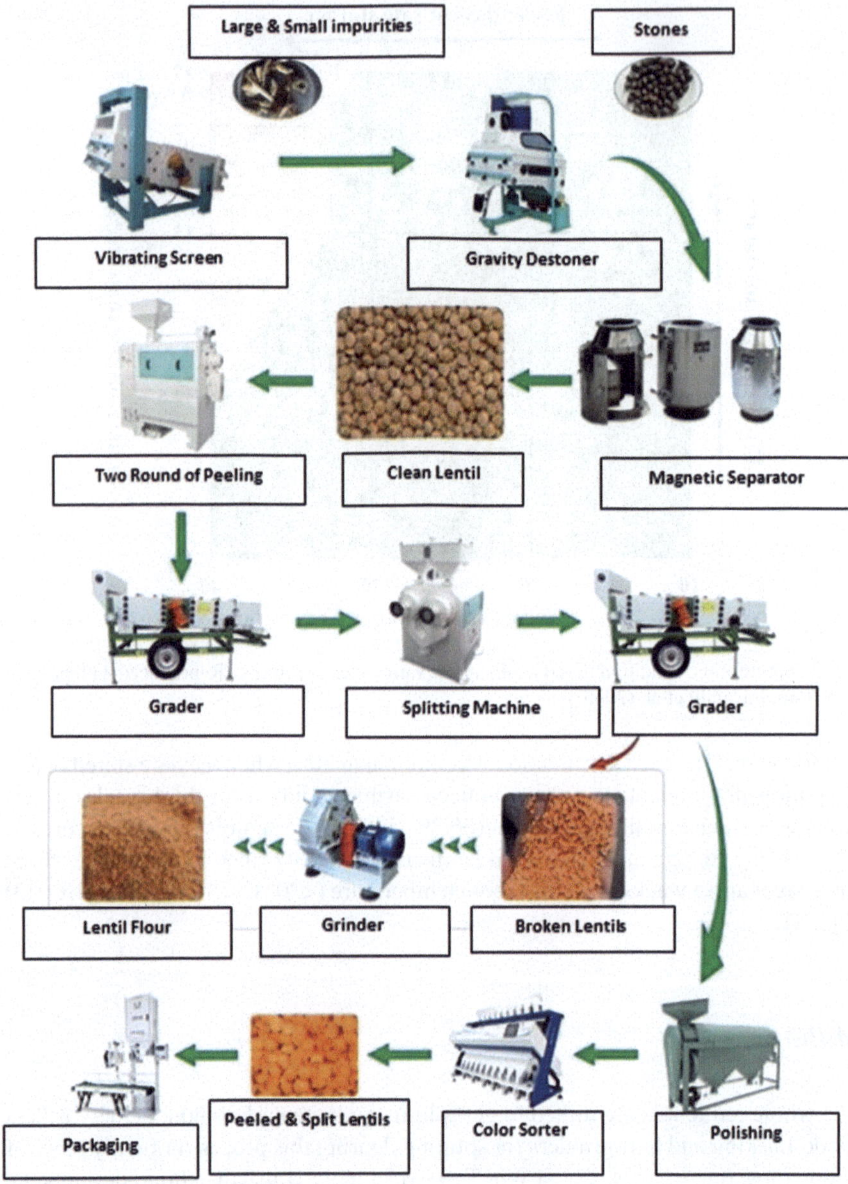

Fig. 3 Postharvest processing machineries for lentils. (Adapted from China Win Tone Machinery company)

ones, where the broken lentils can then be grinded to make lentil flour. The good split lentils can then be polished using a polisher (scourer), where the split lentils move between the screen and a rotor with leather, and the fine dust made by scouring action of the rotor is removed through the screen. The lentils can then be

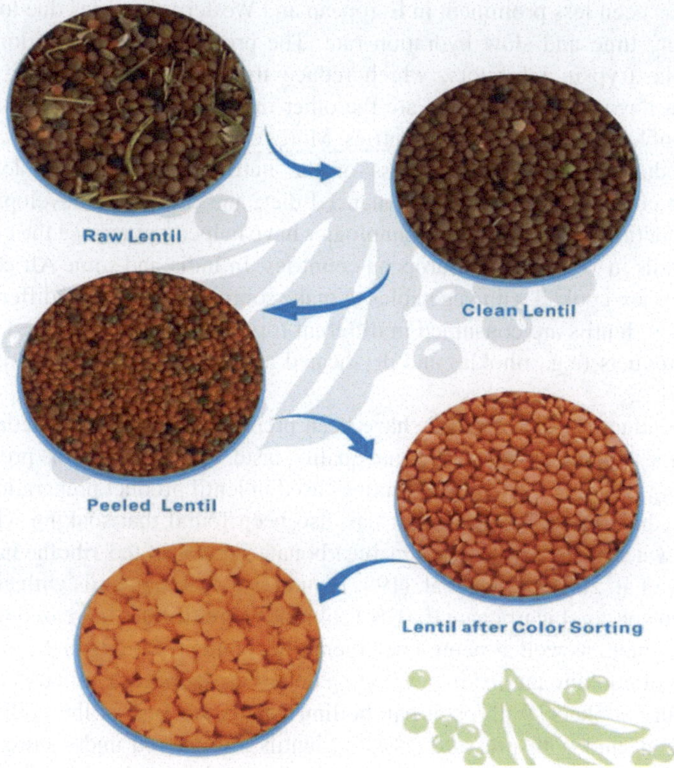

Fig. 4 Various products obtained during lentil postharvest processing line. (Adapted from China Win Tone Machinery company)

packaged using the packing machine after sorting using a color sorter to remove foreign materials and moldy lentils. Various products obtained at different stages of this processing line are shown in Fig. 4.

Product Development

Traditional Asian Products

Due to higher dietary protein and complex carbohydrate content, lentils have been considered as a staple plant protein source in many developing countries. Split pulses (dhal) and the flour made from different pulses have been used to make different products like "Phutna" (roasted grains) and "roti" (chickpea flour mixed with wheat flour) in South Asian countries like India, Pakistan, and Bangladesh for many centuries (Singh and Jambunathan 1990). Even though the consumption of lentils and other pulses is high in developing countries, the usage of lentils as a food product has been less prominent in European and Western countries due to the longer cooking time and slow hydration rate. The presence of antinutritional compounds like trypsin inhibitors, which reduce the protein digestibility, and the unpleasant flavor during cooking are the other major reasons for the lack of consumption of lentils in developed countries. More recently, formulation of new lentil-based products, increased awareness of the nutritional benefits of lentils, an increasing consumer demand of a balanced diet, and the recent developments in lentil production and processing technologies have helped to increase the consumption of lentils in Western and European countries. In India and some African countries, lentils are cooked with vegetables to make soups and curries. In different parts of the world, lentils are consumed in different forms like boiled, sprouted, and fermented products (e.g., dhokla) and dry-heated and fried products (Cokkizgin and Shtaya 2013).

Food products based on lentils have been prepared using several methods, and based on these methods, the nutritional quality, taste, and texture of the product differ. Soaking is one of the major techniques used in lentil product preparation which helps to reduce the cooking time. It has also been found that soaking with clean water or water treated with sodium bicarbonate increases the riboflavin content (Prodanov et al. 2004). Vidal et al. (1992) found that soaking lentils with citric acid (0.1%) and sodium bicarbonate (0.07%) solutions significantly increased the hemicellulose content as well as neutral detergent fiber content. Boiling is the most common form of cooking lentils. In developing countries, lentils are boiled in open pan and pressure cookers. With open pan boiling, lentils are kept in the boiling water until cooked, and with pressure cooking, lentils are cooked under pressure over 100 °C in quick time. With the boiling method, use of excessive water leached out nearly 30–70% of water-soluble nutrients during cooking (Raghuvanshi and Singh 2009). Pratibha et al. (1999) tested the changes in fat, carbohydrate, and protein content during open pan and pressure boiling of lentils and found the loss in carbohydrate and protein were similar in both methods, but the protein retention was higher in pressure cooking. Naveeda and Jamuna (2006) also obtained similar results with the retention of protein and found higher protein digestibility in pressure-cooked lentils. Although the boiling method caused losses in some nutrients, it also increased the digestibility of the lentils. Boiling also helped to reduce the antinutrient content in lentils. Porres et al. (2003) found that pressure-cooking lentils for 30 min (at 1 atm pressure and 120 °C) reduced the antinutrient compounds phytate content, tannin content, and trypsin inhibitor activity by 8%, 12%, and 76%, respectively.

Roasting of lentils also reduced the trypsin inhibitor activity, as well as the starch content of the lentils (Urbano et al. 1995). There was no significant change in nutritional quality of the lentils cooked with microwave heating compared to pressure cooking, but microwave cooking time was low when compared with pressure cooking (Naveeda and Jamuna 2006). The frying process helped to make various lentil-based products like cutlets, dosa, pancakes, and papad in South Asian countries. Lentils are also used to make salads and soups, and in South Asia (especially in India), lentils are boiled with or without vegetables and seasoning agents to make

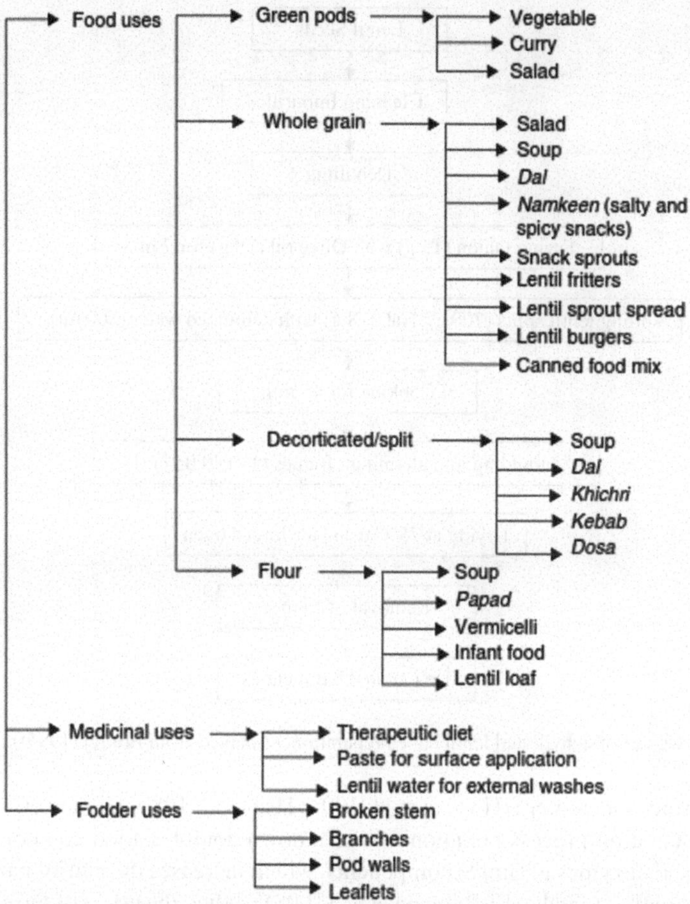

Fig. 5 Application of lentils for food, fodder, and medicinal purposes. (Adapted from Ereifej (1995))

side dishes for rice or roti like sambar, curry, rasam, and dal. Various products obtained from lentils for food, fodder, and medicinal purposes are given in Fig. 5.

Sprouted Products

Lentils are simple to sprout, which is beneficial as it was noted that germination/sprouting of lentils eliminated the α-galactosidase and reduced the antinutrient factors like trypsin inhibitor activity as well as reducing the total starch content (Urbano et al. 1995). Sprouting is commonly done by soaking the lentil seeds overnight, followed by drying off the excess water and keeping the seeds in a warm, dark place for 1 day. Sprouting of lentils also increased the thiamin, niacin, riboflavin, biotin,

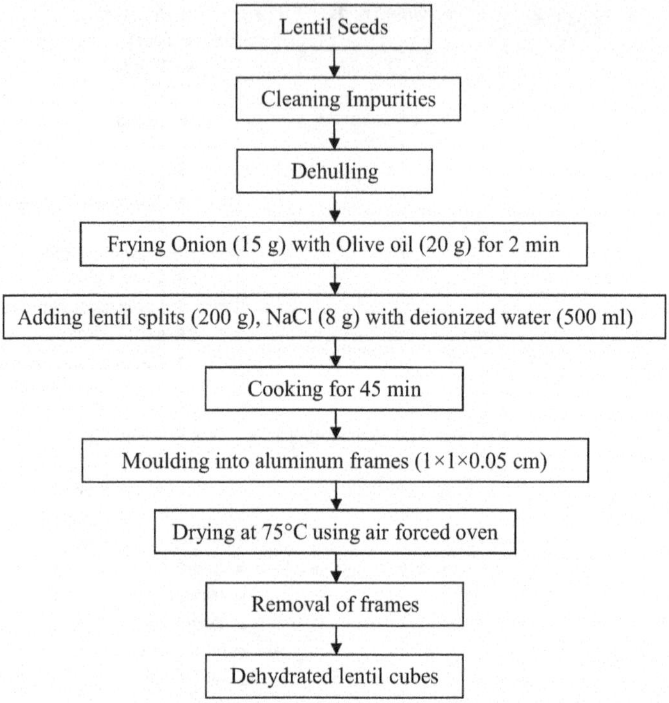

Fig. 6 Flowchart of dehydrated lentil cube preparation. (Adapted from Ereifej (1995))

and ascorbic acid content (Hozova et al. 1994; Hsu et al. 1980; Urbano et al. 1995). The fermentation process commonly breaks down complex food components via bacteria and enzymes to simple components, which increases the nutritional content as well as nutrient intake. El-Rahman et al. (1998) studied the effect of fermentation of lentils on nutrient and antinutrient properties and found that the fermentation of lentils for 2 days decreased the trypsin inhibitor activity and total galactosidase by 50% and 26%, respectively. El-Rahman et al. (1998) also studied the effect of fermentation of lentils on extruded product production and concluded that 24-hour fermentation increased the rheological properties (thixotropy values, yield stress, and consistently coefficient) of lentil slurry during the extrusion cooking process. Dry heating methods like roasting, microwave heating, and baking of lentils reduced the moisture content of lentils, which resulted in an increased storage life.

Lentil Soup

Lentils have been commonly used for soup preparation in Europe and North America with meat, vegetables, and spices. To increase the shelf life of lentils, dehydrated lentil cubes were developed in Jordan (Ereifej 1995) (Fig. 6). There were no significant differences in nutritional and sensory properties between the soup prepared with the dehydrated lentil cube and regular split lentils. In South Asian countries, lentils named namkeen had nearly 16.6% of protein and 1.95% crude fiber (Verma and Raghuvanshi 2002). Snack sprouts made by sprouting the lentils (soaked for 12 h and then allowed to sprout for 5 days) and baking had crispy texture and were stored for long periods in airtight conditions. Traditional Indian food products like papads were made by mixing different types of lentil flours along with the regular black gram flour, and the taste, color, aroma, and texture of the papads made with lentil flour, mung bean, and black gram flour (15:25:60 ratio) were similar to regular papads made with black gram flour (Saxena et al. 1989).

Baked Products

Baked products incorporated with pulse flour showed higher protein, fiber, and ash content than the baked products made from regular wheat flour (Faheid and Hegazi 1991; Sironi et al. 2005). Researchers throughout the world have developed baked products like cakes and breads with various combinations of pulse flour and wheat flour, and they have found that baked products fortified with pulse flour had higher nutritional content than regular baked products (Dalgetty and Baik 2006; Sathe et al. 1981; Wang et al. 2002). Cookies made by incorporating up to 15% pulse flour showed an elevation in protein and dietary content when compared to the regular wheat-based cookies (Faheid and Hegazi 1991). Functional and rheological properties of bakery products began to show negative effects when the lentil and small white bean flour were mixed with wheat flour at a mass that was more than 10% by weight (Kohajdová et al. 2012). Stability of the dough prepared with lentil and small white bean flours was reduced from 6.67 to 2.30 min and increased the dough development time (3.50–5.50 min). Incorporation of more than 10% lentil and bean flour also affected the crust color, hardness, and shape of the baked rolls. The sensory evaluation of the baked rolls made with the 10% incorporation of lentil and bean flour with wheat flour had the highest (92.70–93.70%) consumer acceptability when compared with other combinations (Kohajdová et al. 2012).

Extruded Products

Pasta is commonly made from durum wheat flour, but the increase in awareness about the health benefits of complex carbohydrates and dietary fiber has made pasta producers fortify their pasta with a certain percentage of pulse flour. Spaghetti prepared with pinto bean, navy bean, and lentil flour has higher trypsin inhibitor activity (TIA) than regular spaghetti, and the sensory tests indicated that the incorporation of legume flours of up to 10% did not alter the sensory properties (Bahnassey and Khan 1986). Gluten-free pasta made with faba bean, lentil, or black gram flour was

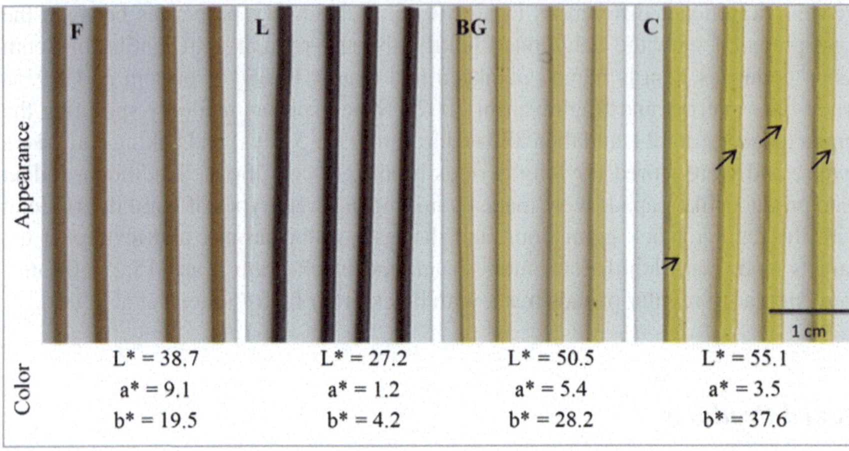

Fig. 7 Hunter color $L*$, $a*$, $b*$ values of (F) faba bean flour, (L) lentil flour, (BG) black gram flour, and (C) commercial pasta. (Adapted from Laleg et al. (2016))

tested for its structural, rheological, and cooking properties along with regular cereal-based gluten-free pasta by Laleg et al. (2016). The tests found that all legume pasta had higher protein and fiber contents, and the TIA activity was reduced up to 82% during the pasta making process. Cooking losses were less in all legume pastas when compared with regular pastas, but the springiness and cohesiveness of legume pastas were lower than commercial pastas. The color of the pasta developed from lentil flour was darker than other legume pastas as well as commercial pasta (Fig. 7) (Laleg et al. 2016). Zhao et al. (2006) prepared spaghetti with lentil, green pea, and chickpea flour along with wheat flour and found that an increase in lentil flour content increased the cooking losses and firmness of the spaghetti. Wang et al. (2014) tested the characteristics of the noodles prepared from pea and lentil flour and found that the noodles made from lentil starch had higher cooking loss than noodles made from cooked pea starch. The textural profile analysis showed that the texture of the cooked lentil noodles and pea noodles were superior to the regular commercial noodles.

Future Trends in Processing and Product Development

At present, the lentil is consumed through the products made by either whole grain or split form. In Asian countries, lentils have been used as a main ingredient in preparation of curries and dal. In developed countries, lentils are consumed through lentil-based products like salad, soup, and baked products. Lentil-based bakery items, rotis, and extruded products like pasta and noodles are getting consumers' attention throughout the world due to their higher nutritional contents. The recent advancements in product development and processing techniques opened new ways to use lentil protein and starch in snacks, extruded products, and baby foods. Since lentil starch has lower glycemic index (GI) value and higher functional properties than legumes and cereal starches, lentil starch can be used as an excellent alternative for conventional cereal starches. In recent times, the consumption of lentils is increasing throughout the world due to the increased awareness about more balanced diets. Consumer preference is also shifting toward plant-based protein sources like lentil and other sources. Recent studies also proved that the preprocessing and processing techniques like soaking, sprouting, and cooking reduce the antinutritional components present in the lentils, which helps to remove the major obstacles in the consumer's mind about including lentil-based products in their diet (Hardeep et al. 2011). Consumer's eagerness toward healthy and tasty snack and baked products drives the food industry to incorporate lentils with regular cereal-based flours for production of baked and snack products. Increasing trend of including lentils in soups, dried food products, and extruded products like pasta, noodles, and sauce dips in recent times was identified by Lakkakula et al. (2017). The IMARC survey clearly indicates the growth in lentil production and their market share in recent times, and IMARC forecasts the compound annual growth rate for market volume of lentils will be 4.8% between 2019 and 2024 (IMARC 2019). Development of lentil-based innovative products, increased health consciousness, demand on organically produced lentil-based products, and awareness on healthy diet among the consumers are the major driving forces for the increase in lentil production and application. Improved functionality of the lentil protein and lentil starch extracts opens up new arena for food processors to produce novel products with lentil protein and lentil starch extracts. More research on new lentil-based product and process development and assessment of improved physiochemical and nutritional properties of lentil-based products will help expanding the application of lentils. Development of new preprocessing technologies to reduce cooking time and antinutritional components as well as processing and preprocessing techniques to improve digestibility and nutrition intake will help processors to explore new avenues of application and also help the producers to increase their benefits.

References

Agblor, K. (2006). The cropportunity strategy for red lentil. *Pulse Point, 6*, 26–27.

Alexander, W. (2015). *Lentil trading and marketing: Australian grain exports*. Retrieved from https://grdc.com.au/resources-and-publications/grdc-update-papers/tab-content/grdc-update-papers/2015/08/lentil-trading-and-marketing

Bahnassey, Y., & Khan, K. (1986). Fortification of spaghetti with edible legumes. II. Rheological, processing, and quality evaluation studies. *Cereal Chemistry, 63*(3), 216–219. Retrieved from https://www.aaccnet.org/publications/cc/backissues/1986/Documents/CC1986a52.html

CGC. (2009). *Canadian grain commissions*. Retrieved from https://www.grainscanada.gc.ca/en/

Cokkizgin, A., & Shtaya, M. J. (2013). Lentil: Origin, cultivation techniques, utilization and advances in transformation. *Agricultural Science, 1*(1), 55–62.

Dalgetty, D. D., & Baik, B. K. (2006). Fortification of bread with hulls and cotyledon fibers isolated from peas, lentils, and chickpeas. *Cereal Chemistry, 83*(3), 269–274. https://doi.org/10.1094/CC-83-0269.

El-Rahman, H. A. A., Heikal, Y. A., Rasmy, N. M., & Idris, A. M. (1998). Influence of lactic acid fermentation on selected physiochemical characteristics of some legumes. Proceedings of Seventh Conference of Agricultural Development Research, Cairo, Egypt, 15–17 December 1998. Volume 3. *Annals of Agricultural Science, Cairo Special Issue, 3*, 771–792.

Ereifej, K. I. (1995). Preparation of acceptable traditional dehydrated lentil soup at small scale. *Journal of Food Science and Technology, 32,* 486–488. Retrieved from https://www.researchgate.net/publication/294749909_Preparation_of_acceptable_traditional_dehydrated_lentil_soup_at_small_scale

Faheid, S. M., & Hegazi, N. A. (1991). Effect of adding some legume flours on the nutritive value of cookies. *Egyptian Journal of Food Science (Egypt), 19*(1–2), 147–159.

FAOSTAT. (2014). *FAOSTAT statistics database—Agriculture*. Retrieved from http://faostat3.fao.org/browse/Q/QC/E

FAOSTAT. (2019). *FAOSTAT statistics database—Agriculture*. Retrieved from July 15, 2019, from http://faostat3.fao.org/browse/Q/QC/E

GRDC. (2017). *Annual report 2017-2018*. Retrieved from https://grdc.com.au/data/assets/pdf_file/0022/370372/GRDC-Annual-Report-2017-18.pdf

Gujral, H. S., Angurala, M., Sharma, P., & Singh, J. (2011). Phenolic content and antioxidant activity of germinated and cooked pulses. *International Journal of Food Properties, 14*(6), 1366–1374.

Hardeep, S. G., Mamta, A., Paras, S., & Singh, J. (2011). Phenolic content and antioxidant activity of germinated and cooked pulses. *International Journal of Food Properties, 14*(6), 1366–1374. https://doi.org/10.1080/10942911003672167.

Hozova, B., Danisova, C., & Holotnakova, H. (1994). Group B vitamins is macrobiotic food. *Potravinarske Vedy, 12*, 197–201.

Hsu, D., Leung, H. K., Finney, P. L., & Morad, M. (1980). Effect of germination on nutritive value and baking properties of dry peas, lentils and faba beans. *Journal of Food Science, 45*, 87–92.

IMARC. (2019). *Lentil market: Global industry trends, share, size, growth, opportunity and forecast 2019-2024*. Retrieved from https://www.imarcgroup.com/prefeasibility-report-lentil-processing-plant

Kohajdová, Z., Karovičová, J., & Magala, M. (2012). Effect of lentil and bean flours on rheological and baking properties of wheat dough. *Chemical Papers – Slovak Academy of Sciences, 67*(4), 398–407. https://doi.org/10.2478/s11696-012-0295-3.

Lakkakula, P., Olson, F., & Ripplinger, D. (2017). *Pea and lentil market analysis*. Final research report. Fargo: North Dakota State University.

Laleg, K., Cassan, D., Barron, C., Prabhasankar, P., & Micard, V. (2016). Structural, culinary, nutritional and anti-nutritional properties of high protein, gluten free, 100% legume pasta. *PLoS One, 11*(9), e0160721. https://doi.org/10.1371/journal.pone.0160721.

Lentils.org. (2019). *Canadian lentils: A staple in a diabetic diet*. Retrieved from https://www.lentils.org/health-nutrition/dietary-specific/

Naveeda, K., & Jamuna, P. (2006). Nutritive value and sensory profile of microwave and pressure cooked decorticated legumes (*dals*). *Journal of Food Processing and Preservation, 30*, 299–313.

Pistollato, F., Cano, S. S., Elio, I., Vergara, M. M., Giampieri, F., & Battino, M. (2015). Plant-based and plant-rich diet patterns during gestation: Beneficial effects and possible shortcomings. *Advances in Nutrition, 6*(5), 581–591. https://doi.org/10.3945/an.115.009126.

Porres, J. M., Aranda, P., López-Jurado, M., & Urbano, G. (2003). Effect of natural and controlled fermentation on chemical composition and nutrient dialyzability from beans (Phaseolus vulgaris L.). *Journal of Agricultural and Food Chemistry, 51*(17), 5144–5149.

Pratibha, P., Mishra, A., Gupta, O. P., Rajput, L. P. S., Singh, A., Parihar, P., & Singh, A. (1999). Effect of cooking processes on nutritional quality of some common pulses. *Advances in Plant Sciences, 12*, 15–20.

Prodanov, M., Sierra, I., & Vidal-Valverde, C. (2004). Influence of soaking and cooking on the thiamin, riboflavin and niacin contents of legumes. *Food Chemistry, 84*(2), 271–277.
Pulse Canada. (2019). *Pulses and protein quality*. Retrieved June 20, 2019, from http://www.pulsecanada.com/ resources/
Raghuvanshi, R. S., & Singh, D. P. (2009). 25 food preparation and use. The Lentil, 408.
Rizkalla, S. W., Bellisle, F., & Slama, G. (2002). Health benefits of low glycaemic indcx foods, such as pulses, in diabetic patients and healthy individuals. *British Journal of Nutrition, 88*(Suppl 3), S255–S262. https://doi.org/10.1079/BJN2002715.
SaskPulse. (2019). *Lentils: Harvest*. Retrieved June 20, 2019, from https://saskpulse.com/growing-pulses/lentils/harvest/
Sathe, S. K., Ponte, J. G., Rangnekar, P. D., & Salunkhe, D. K. (1981). Effects of addition of great northern bean flour and protein concentrates on rheological properties of dough and baking quality of bread. *Cereal Chemistry, 58*(2), 97–100. Retrieved from https://www.aaccnet.org/publications/cc/backissues/1981/Documents/CC1981a25.html
Satya, S., Kaushik, G., & Naik, S. N. (2010). Processing of food legumes: A boon to human nutrition. *Mediterranean Journal of Nutrition and Metabolism, 3*(3), 183–195. https://doi.org/10.1007/s12349-010-0017-8.
Saxena, A.K., Kulkarni, S.G., Manan, J.K. & Berry, S.K. (1989). Studies on the blends of different pulses (Bengal gram, green gram, lentils and arhar) in the preparation of north Indian spiced papads. *Journal of Food Science and Technology, 26*(3), 133–136. Retrieved from http://ir.cftri.com/5712/
Singh, U., & Jambunathan, R. (1990). *Pigeonpea: Postharvest technology* (pp. 435–455). Wallingford: CAB International.
Sironi, E., Sessa, F., & Duranti, M. (2005). A simple procedure of lupin seed protein fractionation for selective food applications. *European Food Research and Technology, 221*(1–2), 145–150. https://doi.org/10.1007/s00217-005-1151-2.
Sravanthi, B., Jayas, D. S., Alagusundaram, K., Chelladurai, V., & White, N. D. G. (2013). Effect of storage conditions on red lentils. *Journal of Stored Products Research, 53*, 48–53.
Urbano, G., Lopez, J. M., Hernandez, J., Fernandez, M., Moreu, M. C., Frias, J., Diaz, P. C., Prodanov, M., & Vidal, V. C. (1995). Nutritional assessment of raw, heated and germinated lentils. *Journal of Agricultural and Food Chemistry, 43*, 1871–1877.
Urbano, G., Porres, J. M., Frías, J., & Vidal-Valverde, C. (2007). Nutritional value (Chapter 5). In S. S. Yadav, D. McNeil, & P. C. Stevenson (Eds.), *Lentil: An ancient crop for modern times* (Vol. 3, pp. 47–93). Berlin: Springer.
Vandenberg, A. (2009). Postharvest processing and value addition. In W. Erskine & F. J. Muehlbauer (Eds.), *The lentil: Botany, production and uses* (pp. 391–424). Wallingford: CAB International. https://doi.org/10.1079/9781845. 934873.0391.
Verma, T., & Raghuvanshi, R. S. (2002). Nutriguide: Manual for calculation of dietary adequacy using nutrient composition of Indian recipes. In *All India Co-ordinated Research Project (AICRP) (Home Science)*. New Delhi: Indian Council of Agricultural Research.
Vidal, V. C., Frias, J., & Valverde, S. (1992). Effect of processing on the soluble carbohydrates content of lentils. *Journal of Food Protection, 55*, 301–303.
Wang, J., Rosell, C. M., & de Barber, C. B. (2002). Effect of the addition of different fibres on wheat dough performance and bread quality. *Food Chemistry, 79*(2), 221–226. https://doi.org/10.1016/S0308.8146(02)00135-8.
Wang, N., Warkentin, T. D., Vandenberg, B., & Bing, D. J. (2014). Physicochemical properties of starches from various pea and lentil varieties, and characteristics of their noodles prepared by high temperature extrusion. *Food Research International, 55*, 119–127. https://doi.org/10.1016/j.foodres.2013.10.043.
Zhao, Y. H., Manthey, F. A., Chang, S. K., Hou, H. J., & Yuan, S. H. (2006). Quality characteristics of spaghetti as affected by green and yellow pea, lentil, and chickpea flours. *Journal of Food Science, 70*(6), s371–s376. https://doi.org/10.1111/j.1365-2621.2005.tb11458x.

Lima Bean

M. Sandoval-Peraza, G. Peraza-Mercado, D. Betancur-Ancona, A. Castellanos-Ruelas, and L. Chel-Guerrero

Introduction

According to the taxonomy, the bean belongs to the genus *Phaseolus*, which includes approximately 35 species of which 4 are cultivated: *P. vulgaris* L.; *P. lunatus* L.; *P. coccineus* L., and *P. acutifolius* L. (Arias-Restrepo et al. 2007). *Phaseolus lunatus* L. belongs to the Fabaceae family, and there are two domesticated genetic stocks from two different wild forms with two seed morphologies, small and large (Debouk 2019). The small seeds are known as ib., patashete and futuna (Yucatan, Chiapas, and Jalapa, Mexico, respectively), caballero bean (Cuba), ixtapacal (Guatemala), chilipuca (El Salvador), haba (Puerto Rico and Panama), sieva and comba (Colombia), and guaracaro (Venezuela), among others. The large seeds are known as lima, layo and pallar (Peru), torta (Colombia), palato (Bolivia), and manteotto (Argentina) (Debouk 2019).

It is proposed that *P. lunatus* could have originated in the Neotropical region of America, ranging from Mexico to Chile, passing through the Andean region of Peru. It is believed that its origin is found in Guatemala since in this area the wild progenitor of this species was found; on the other hand, molecular studies propose that its origin is found in the Andean zone and that its distribution throughout the Americas was given by domestication (FAO 2018).

M. Sandoval-Peraza
Universidad del Valle de México, Escuela de Ciencias de la Salud, Mérida, Yucatán, Mexico

Facultad de Ingeniería química—Campus de Ciencias Exactas e Ingenierías, Universidad Autónoma de Yucatán, Mérida, Yucatán, Mexico

G. Peraza-Mercado
Instituto de Ciencias Biomédicas (Departamento de Ciencias Químico-Biológicas), Universidad Autónoma de Ciudad Juárez, Ciudad Juárez, Chihuahua, Mexico

D. Betancur-Ancona · A. Castellanos-Ruelas · L. Chel-Guerrero (✉)
Facultad de Ingeniería química—Campus de Ciencias Exactas e Ingenierías, Universidad Autónoma de Yucatán, Mérida, Yucatán, Mexico

A. Manickavasagan, P. Thirunathan (eds.), *Pulses*,
https://doi.org/10.1007/978-3-030-41376-7_9

Table 1 Nutrient content in *Phaseolus lunatus* L. (g/100 g)

Components	Values
Moisture	12.0
Protein	10.1–20.7
Fat	0.1–1.2
Total carbohydrates	30.7–62.4
Crude fiber	2.1–4.9
Ash	1.7–3.7
Calcium (mg)	40–113
Phosphorus (mg)	124–330
Iron (mg)	3.5–4.8
Thiamine (mg)	0.27–0.34
Riboflavin (mg)	0.12–0.21
Niacin (mg)	1.66–2.2
Ascorbic acid (mg)	40

The nutrient content according to FAO (2020) for *Phaseolus lunatus* L. is (Table 1):

Other researches have shown its seeds have high protein (21–26%) and carbohydrate (55–64%) contents; high levels of minerals such as K, Zn, Ca, and Fe; and low levels of Na and P (Chel-Guerrero et al. 2012).

In Mexico, *Phaseolus lunatus* L. is widely grown in the tropic and is known as ib (Mayan dialect). It is common to find this legume in backyard home gardens; in Yucatan it is traditionally planted in the system rub-overthrow-burn (nomad agriculture) with corn and wild pumpkins (Debouk 2019). Betancur-Ancona et al. (2009) mentions its seeds are an underexploited protein source and could be a potential ingredient in industrial food systems.

Chronic Noncommunicable Diseases (CNCDs)

These are currently a major public health problem, since these diseases are one of the main causes of death and disability, both in developed countries and in those that are in development. These diseases include cardiovascular diseases, cancer, diabetes mellitus, and chronic respiratory diseases, among others (Gómez et al. 2010). CNCDs share a set of risk factors, including smoking, hypertension, high serum cholesterol levels, obesity, physical inactivity, and diabetes. A strategy to reduce the occurrence of these cases is primary prevention through community programs that promote and modify positively individual and group life habits (PAHO 2002).

Exercise and physical activity are important, due to the positive effects they cause on the improvement of health, increasing functional capacity and improving the quality of life of people (Gómez et al. 2010). Other aspects to take into account are changes in habits, specifically in the consumption of tobacco, alcohol (immoderate), and food. In Mexico, CNCDs are one of the greatest challenges which the health system is facing; due to the large number of cases that occur as they contribute to an increasing general mortality, they are also the most frequent cause of premature disability, together with the complexity and high cost for its treatment (Córdova-Villalobos et al. 2008).

Therefore, in Mexico the government has implemented the program Chécate, Mídete, & Muévete (2019) (Check Yourself, Measure Yourself, and Move), which contains tools and information to create healthy habits; tips are given regarding diets and exercise routines; in this portal people are encouraged to improve their health through the following steps: (1) visit to the doctor at the corresponding health clinic to keep track of the weight and measurements of the waist circumference; (2) BMI calculation and explanation of the relationship between food consumption and daily activity; and (3) calculation of the necessary physical activity based on sex, age, and activity level.

Postharvest Processing

Production

Nowadays, the cultivation of *Phaseolus lunatus* is practiced in different countries of Latin America and the Caribbean. The sieve ecotype is cultivated from Mexico to Argentina, while the Lima is present in a smaller area, comprised of the western zone of the Andes, mainly in Peru (FAO 2018); 7000 ha are sown in this country, and 11,000 tons are harvested annually (López-Alcocer et al. 2016). During the harvest season, harvesting can be done manually; however, it is important not to delay it since the pods can be opened, and most of the grains would be lost. The yield in terms of dry grains is in the order of 800–2000 kg/ha and can reach 3000 kg/ha in some regions depending on the variety (FAO 2018).

Storage Conditions

The first step after harvesting is to ensure that the pods reach a humidity of 12–14%, to subsequently be able to perform the extraction of the seed, always avoiding unnecessary breakage of the grains. Afterward, the storage of the seeds must be done in metal silos, avoiding the proliferation of insects and the increase of humidity (FAO 2018).

Poor handling and storage after the harvest induces the hardening process in the grain. According to the method dictated in the applicable Mexican regulations NMX-FF-038-SCFI-2002, hardened grain is that for which cooking time has increased significantly in relation to cooking freshly harvested grain. The hardening process in the grain is caused by the effects of aging or grain storage under conditions of high relative humidity in combination with high temperatures (temperature >25 °C and relative humidity >65%). Beans are considered hard when their cooking time is over 55 min. The aforementioned is very important because the temperature and relative humidity in the state of Yucatan are the same that induce the hardening effect (temperature >25 °C and relative humidity >65%, respectively), so storage becomes more important to avoid grain losses.

Toxicity

Legumes contain different antinutritional compounds, such as flatulence factors, saponins, protease inhibitors, tannins, phytic acid, lectins, etc. (De Dios et al. 2009); in the case of wild lima bean, plants contain cyanogenic glycosides that are known to defend the plant against leaf herbivores (Cuny et al. 2019); these seeds contain phaseolunatin, a glycoside which unfolds under the influence of an enzyme in glucose, hydrocyanic acid, and acetone. This compound is abundant in wild varieties making them poisonous (FAO 2020). Betancur-Ancona et al. (2004a) mention that all legumes contain antinutritional components and in the case of *P. lunatus* L. contain cyanogenic glycosides in quantities of 0.0369 g/kg, and these can limit its direct consumption in food and feed. During the soaking and cooking time of the grains, the glucoside is not completely eliminated, so the remnant portion of this compound continues to have its toxic properties (FAO 2020); the authors abovementioned proposed a wet-fractionation process for the detoxification of *P. lunatus* L.

In recent research, the lethal dose of hydrolysates and peptide fractions from *P. lunatus* L. have been evaluated (Nuñez-Aragón et al. 2019), founding that there was no mortality due to the administration of such materials at different concentration doses (10, 100, 1000, 1600, 2900, and 5000 mg/kg). The period of test in male ICR mice was for 14 days, and no behavioral alterations, signs of toxicity, or weight loss was detected during the observation period; the estimated LD_{50} was above 5000 mg/kg. Therefore, the wet-fractionation process proposed by Betancur-Ancona et al. (2004a) allows the detoxification of peptide fractions from *P. lunatus* L.

Nutritional Content of P. lunatus *L.*

Proximate Composition of Flour

In our work team, the *P. lunatus* L. legume has been extensively explored since 2000 until today. All lima bean seeds of our researches were purchased in the local market of the city of Merida; specifically, from the major distributor of the entity (supplier of the farmer, downtown, Merida, Yucatan). The harvest of the beans was done in the communities of the neighboring state of Campeche (Mexico).

Table 2 Proximate composition of flours from *P. lunatus* L. (g/100 g) harvested in Mexico

Components	2002	2004	2009	2015	2018
Moisture	14.88	10.24	14.80	11.02	9.30
Protein	24.07	25.50	23.70	23.82	20.86
Crude fiber	5.10	5.90	5.10	5.83	4.60
Fat	3.77	0.75	3.70	1.19	1.42
Ash	3.40	4.90	3.30	4.01	3.83
NFE	63.66	62.90	63.40	65.15	69.29

Chel-Guerrero et al. (2002), Betancur-Ancona et al. (2004a, 2009), Franco-Miranda (2015), Arias-Trinidad (2018)
NFE nitrogen-free extract

Table 2 shows the proximate composition of flours from *P. lunatus* in different years of harvest. It can be seen that despite the different harvest times, the proximal composition of grain has not been modified, and the results are similar to the report by FAO (2020, Table 1), positioning this legume as a stable source of macromolecules such as proteins and carbohydrates.

Giami (2001) reports the proximal composition of three new improved lines of Nigerian lima beans, observing similar values in the protein content (23.8, 24.4, and 27.3%), fat (1.5, 1.8, and 2.1%), ash (3.4, 3.5, and 3.6%), and crude fiber (2.4, 2.5, and 2.7%); the only differences between materials were in the NFE (53.4, 56.8, and 57.3%); these may be due to the genetic variations in the *P. lunatus* analyzed by the aforementioned authors.

Amino Acid Composition

Table 3 shows the amino acid composition of lima bean flour at different years of harvest, observing that the obtained flours from lima bean (ungerminated and germinated) supply the recommended amino acid scoring patterns of histidine, threonine, tryptophan, tyrosine, valine, isoleucine, leucine, phenylalanine, and lysine; but these flours are deficient in methionine and cysteine (FAO 2011). It is important to highlight that the obtention of germinated lima bean flour was made; however, this process does not improve sulfur amino acid deficiencies. An alternative to improve these deficiencies could be improved with a wet-fractionation process as mentioned by Betancur-Ancona et al. (2004a) with the objective to obtain protein isolates.

Wet-Fractionation Process

As mentioned, previously this process would allow obtaining protein isolates, starches, and fibrous residues; Fig. 1 shows the general scheme for this process. Betancur-Ancona et al. (2004a) report different flour/water ratio, pH and time of agitation to extract protein, and starch fractions from *P. lunatus* L. The

Table 3 Amino acid composition (g/100 g) of flours of *Phaseolus lunatus* L. harvested in Mexico

	(2004)	(2009)	(2009) germinated	(2015)	(2018)	FAO (2011)
Essentials						
Histidine	3.20	3.08	3.13	3.00	3.00	2.0
Threonine	4.87	4.46	4.40	4.10	4.36	3.1
Tryptophan	1.32	0.97	1.07	1.66	1.63	0.85
Tyrosine	10.67[a]	3.77	3.72	3.45	3.45	5.2[a]
Valine	5.12	4.40	4.41	5.90	4.54	4.3
Methionine	2.05[b]	0.81	0.63	0.81	1.00	2.7[b]
Cysteine	[b]	0.55	0.53	0.29	1.00	[b]
Isoleucine	4.29	3.86	3.92	5.71	5.28	3.2
Leucine	8.54	8.65	8.49	9.35	7.36	6.6
Phenylalanine	[a]	6.22	6.05	6.18	5.84	[a]
Lysine	7.97	7.10	6.85	5.69	5.94	5.7
Nonessentials						
Asp + Asn	14.43	12.98	13.86	13.16	10.98	
Glu + Gln	14.85	16.70	16.63	13.81	12.89	
Serine	7.99	7.97	7.93	5.44	5.99	
Glycine	4.87	4.46	4.34	2.48	2.74	
Arginine	7.01	6.37	6.42	10.13	12.25	
Alanine	5.01	4.82	4.83	3.31	3.19	
Proline	NR	2.84	2.80	5.43	8.56	

¥FAO (2011), recommended amino acid scoring patterns for child (6 months to older child), teenage and adult group
Betancur-Ancona et al. (2004a), Domínguez-Magaña (2009), Sandoval-Peraza (2015), Arias-Trinidad (2018). *NR* not reported
[a]Tyrosine + phenylalanine
[b]Methionine + cysteine

experimental design used in that study proved advantageous for choosing conditions to obtain starch and protein yield, depending on which of the products is required. For example if the starch is the priority product, the best treatment was with a 1:10 ratio (flour/water), pH 9, and 3 h of agitation; on the other hand, if the protein is the priority product, the best treatment was with a 1:6 ratio (flour/water), pH 11, and 1 h of agitation.

Protein Isolates

Table 4 shows the proximate composition of protein isolates (PI) from *P. lunatus* in different years of harvest. All PI were obtained by wet-fractionation process with the following conditions: 1:6 flour/water ratio, pH 11, and 1 h of stirring. It should be noted that the seeds of all studies were acquired from the same distributor, so the

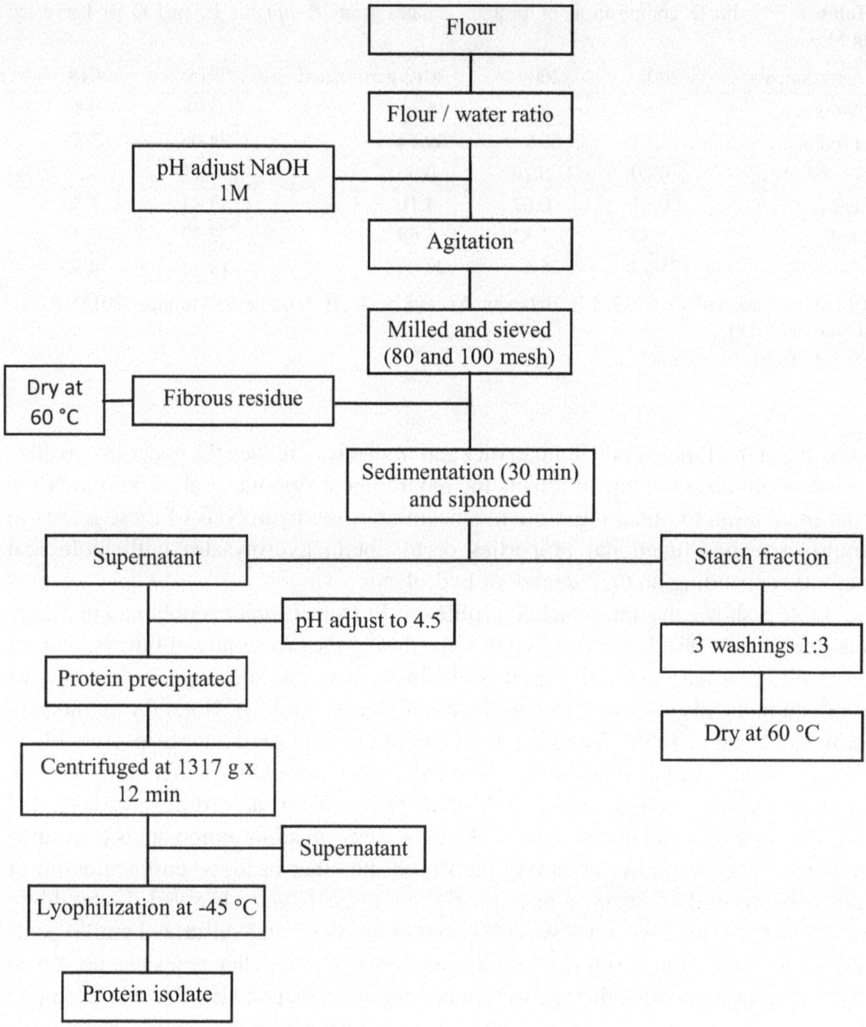

Fig. 1 Wet-fractionation process for integral use of *P. lunatus* L. flours

grains correspond to the same species and area of cultivation, but in different years of harvest, observing small differences in the protein content in the isolates (5%); therefore, these differences may be due to the protein extraction process.

Habitually the consumption of proteins in the human diet comes from animal sources such as meat, milk, eggs, etc.; however, a viable alternative could be the one mentioned by Linnemann and Dijkstra (2002), where proteins of vegetable origin can be used for the production of protein-rich foods, allowing to replace the

Table 4 Proximate composition of protein isolates from *P. lunatus* L. (g/100 g) harvested in Mexico

Components	2002	2009	2012 germinated	2015	2018
Moisture	7.87	8.6	5.4	3.38	8.8
Protein	71.13	69.9	69.14	74.06	62.37
Crude fiber	0.20	0.20	0.96	0.11	1.5
Fat	0.67	0.67	4.16	3.84	3.30
Ash	2.82	2.82	3.68	2.99	4.01
NFE	25.12	26.4	22.06	19	28.82

Chel-Guerrero et al. (2002, 2012), Betancur-Ancona et al. (2009), Franco-Miranda (2015), Arias-Trinidad (2018)
NFE nitrogen-free extract

consumption of meat in the human diet and in this way reduce the excessive exploitation of animals and how it affects the environment. Vioque et al. (2006) mention that in addition to obtaining protein concentrates, the hydrolysis of these serves to improve techno-functional properties or to obtain hydrolysates with biological activity, depending on their degree of hydrolysis.

Table 5 shows the amino acidic profiles of PI from *P. lunatus* obtained in different years of harvest. It can be observed that during the processing of lima bean flour from different harvests until the PI is obtained, there was a change in the amounts of all amino acids in comparison to the initial flours (Table 3). Note: the amino acid profiles shown in Tables 3 and 5 correspond to the flours and their respective PI.

The amino acid composition quantified in the materials of this study can be approached in the compliance with the daily requirements according to FAO (2011) for infants from 6 months to adults. The deficiency in sulfur amino acids continued in some cases even after obtaining the PI; on the other hand, when the amount of sulfur is covered, a decrease in aromatic amino acids was observed; it should be noted that the same behavior was observed in aromatic and sulfurized amino acids for PI obtained from germinated lima bean flours. These deficiencies can be solved by combining materials that are rich in the amino acids that lack IPs. For example, the proteins of corn and beans complement each other, providing significant amounts of each respective limited amino acids (Treviño-Mejía et al. 2016).

The BV calculated for the PI obtained by Sandoval-Peraza (2015) was 43.42 (Table 5) which is higher in comparison to that reported for beach pea (36.5) according to Chavan et al. (2001); also the value of this study is within the range reported by Pastor-Cavada et al. (2011) for 28 different species of beans with values between 18.7 and 67. The PER showed a value of 2.98, being higher than that reported by Betancur-Ancona et al. (2004a) for PI obtained from *Phaseolus lunatus* (2.5); these authors mentioned that the value of PER obtained classifies the PI as a good-quality protein. With the information obtained through the use of these formulas, it can be observed that the PI of *Phaseolus lunatus* can be an alternative source in the formulation of products enriched with protein.

Table 5 Amino acid composition (g/100 g) of protein isolates of *Phaseolus lunatus* L. harvested in Mexico

	(2004)	(2012)	(2012) germinated	(2015)	(2018)	FAO (2011)
Essentials						
Histidine	3.70	3.24	3.17	3.00	3.09	2.0
Threonine	4.71	4.40	4.41	3.95	3.95	3.1
Tryptophan	1.20	0.98	0.79	1.37	2.01	0.85
Tyrosine	11.33[a]	3.54	3.09	3.63	3.63	5.2[a]
Valine	5.80	4.79	4.88	4.65	4.65	4.3
Methionine	2.98[b]	0.35	0.50	0.57	1.60	2.7[b]
Cysteine	[b]	3.87	3.90	0.82	0.82	[b]
Isoleucine	4.78	4.30	4.11	4.35	4.95	3.2
Leucine	9.28	9.19	9.01	8.52	8.52	6.6
Phenylalanine	[a]	0.52	0.65	5.62	5.62	[a]
Lysine	7.55	5.99	5.94	7.07	7.32	5.7
Nonessentials						
Asp + Asn	12.79	12.80	12.47	12.05	10.43	
Glu + Gln	15.21	15.30	16.10	13.62	12.62	
Serine	7.38	7.39	7.26	5.25	5.25	
Glycine	4.70	4.67	4.84	4.55	5.12	
Arginine	6.13	4.96	5.02	9.50	9.50	
Alanine	5.17	6.08	6.19	2.81	2.81	
Proline	NR	7.62	7.68	8.59	8.59	
BV	–	–	–	43.2	–	
c-PER	2.5	–	–	2.98	–	

¥FAO (2011) recommended amino acid scoring patterns for child (6 months to older child), teenage, and adult group
Betancur-Ancona et al. (2004a), Chel-Guerrero et al. (2012), Sandoval-Peraza (2015), Arias-Trinidad (2018)
NR no reported
[a]Tyrosine + phenylalanine
[b]Methionine + cysteine

Starch Isolates

One of the products resulting from wet-fractionation is starch; approximately 10–40% of starch flour can be obtained for every 10 kg of lime beans (Betancur-Ancona et al. 2001, 2004a); Miranda-Villa et al. (2013) reported a 15% recovery for *Phaseolus lunatus*; however, the aforementioned authors did not solubilize the protein by changing the pH, observing that this process during the extraction plays an important role in the starch extraction yield.

Table 6 shows the proximal composition of starch isolates from lima bean noticing the lowest values of protein (under 0.2%); this property allows the use of these starches in the high-glucose syrup industry. 0.35% is the maximum allowed by the FDA in corn syrup to avoid the formation of undesirable dark syrups resulting from Maillard reactions during processing.

Table 6 Proximate composition of starches isolates from *P. lunatus* L. (g/100 g) harvested in Mexico

Components	2003	2005	2008
Moisture	0	11.93	10.9
Protein	0.12	0.10	0.14
Crude fiber	0.67	NR	0.10
Fat	0.54	0.12	0.12
Ash	0.14	0.04	0.65
NFE	98.53	98.49	98.9

Betancur-Ancona et al. (2003), Novelo-Cen and Betancur Ancona (2005), Segura-Campos et al. (2008). *NFE* nitrogen-free extract

It has been reported that the isolated starch of *P. lunatus* grown in Mexico has a composition of 60–70% amylopectin and 30–35% amylose, in addition to a total dietary fiber content of 1.25% (Betancur-Ancona et al. 2003; Novelo-Cen and Betancur-Ancona 2005; Segura-Campos et al. 2008); these authors mention that the values of amylose are similar to the reported in other legume starches like pinto bean, navy bean, and field pea (32–34%). The starch composition extracted from lime beans harvested in Colombia showed amylopectin and amylose values of 78.19 and 21.81%, respectively, presenting a slight difference (Miranda-Villa et al. 2013).

Fibrous Residue

After wet-fractionation it is common to have about a 30% yield of the fibrous fraction, Betancur-Ancona et al. (2004b) report a 35.43% of yield for *P. lunatus*.

Table 7 shows the proximal composition of fibrous residues from lima bean noticing the highest values of NFE and the difference content of crude fiber between samples; this occurs because NFE represents part of the cellulose, hemicellulose, pectin, and other carbohydrates not included in the crude fiber content as a result of the limitations of the method used; therefore a more adequate determination for this estimate would be total dietary fiber combined with the Van Soest fractions. Betancur-Ancona et al. (2004b) report the dietary fiber composition in the *P. lunatus* fibrous residues (29.4% total dietary fiber) observing that the insoluble fraction has the highest value (28.6%) followed by the soluble fraction (0.77%).

Functional Properties

This section shows the functional properties reported in the flour, protein isolates, starch, and fibrous residue obtained from *P. lunatus* L.

Table 7 Proximate composition of fibrous residues from *P. lunatus* L. (g/100 g) harvested in Mexico

Components	2004b	2004a
Moisture	4.72	10.54
Protein	11.1	6.30
Crude fiber	12.88	32.84
Fat	0.66	7.9
Ash	1.77	3.34
NFE	68.9	56.73

Betancur-Ancona et al. (2004a and b)
NFE nitrogen-free extract

Table 8 Water-holding capacity in *P. lunatus* flour and isolated fractions

Material	(g/g sample)	Reference
P. lunatus flour	2.65	Chel-Guerrero et al. (2002)
P. lunatus protein isolates (PI)	3.50	
Total globulin from *P. lunatus* at pHs 5, 7, and 9	1 1.05 1	Chel-Guerrero et al. (2011)
7S globulin fraction from *P. lunatus* at pHs 5, 7, and 9	3.4 1.5 3.3	
11S globulin fraction from *P. lunatus* at pHs 5, 7, and 9	3.3 2.9 3.2	

Water-Holding Capacity (WHC)

Table 8 shows the values reported for WHC in flour and different fractions obtained from *P. lunatus* L.

It is a fact that the best value of WHC was from PI compared to the seed meal; the authors of this research mention that a factor that can affect WHC is the concentration of carbohydrates present in the flour. Regarding PI, WHC may be due to the different protein fractions of this compound (e.g., albumins, globulins, etc.). The previous information was reported by Chel-Guerrero et al. (2011) where the 7S and 11S fractions of globulin had a higher WHC at different pHs compared to the total globulin; these authors demonstrate that extraction conditions and raw material composition play an important role in the resulting WHC properties, coinciding with the reports that at a higher pH, there is a greater interaction with water by proteins.

Oil-Holding Capacity (OHC)

Table 9 shows the values reported for OHC in flour and different fractions obtained from *P. lunatus* L.

Table 9 Oil-holding capacity in *P. lunatus* flour and isolated fractions

Material	(g/g sample)	Reference
P. lunatus flour	1.83	Chel-Guerrero et al. (2002)
P. lunatus protein isolates (PI)	4.59	
Total globulin from *P. lunatus* at pHs 5, 7, and 9	4	Chel-Guerrero et al. (2011)
7S globulin fraction from *P. lunatus* at pHs 5, 7, and 9	2.3	
11S globulin fraction from *P. lunatus* at pHs 5, 7, and 9	4.2	

PI from *P. lunatus* had the highest value of OHC in comparison with their respective flour; the authors of this study attribute the OHC to the high levels of nonpolar residues in the PI protein molecules. This functional capacity makes the PI from lima bean potentially useful in structural interactions in food, especially in flavor retention, improvement of palatability, and extension of shelf life in meat products through reduction of moisture and fat loss.

In accordance with OHC, it can be observed that the total globulin and its 11S fraction had similar values; this may be due to the fact that this fraction conforms approximately 60% of the total globulin. The authors mention that the differences in OHC between the 7S and 11S fractions were probably due to their conformational characteristics, which can influence a proteins capacity to entrap oil.

WHC and OHC of the fibrous fraction of *P. lunatus* obtained after wet-fractionation have also been reported, reporting values of 26.5% and 18%, respectively, observing that the fiber also has functional properties that can be used in the production of rich foods in fiber (Betancur-Ancona et al. 2004b).

Emulsifying Activity (EA) and Emulsion Stability (ES)

Chel-Guerrero et al. (2002) report the EA of lime bean flour in a pH range of 2–10, observing an EA comprised between 48 and 52%, presenting a stability decrease zone (2%) at a pH range of 6–7, increasing again to 50% from pH 8 and ending at 52% at pH 10. According to PI, a higher EA was observed at pH 2 (56%); however, at a pH range from 2 to 5, there was a notable decay of EA (42%), increasing again at pH 6 (52%) and ending at 50% at pH 10. The behavior of decrease and subsequent increase in each percentage presented by PI may possibly be due to the protein composition of the isolate as well as the amino acid content. Regarding ES, it was observed that both emulsions formed with lime bean flour and PI show a greater stability (80–100%) from pH 6. The authors mention that these results indicate that both (flour and PI) materials are effective emulsifiers, making them useful in applications such as the production of sausages, mayonnaise, and seasonings, especially in products that require heating.

A way to improve the emulsifying activity can be by making use of the enzymatic hydrolysis of the PI. The above was reported by Betancur-Ancona et al. (2009) for lime bean PI hydrolysates with the enzyme flavourzyme at 5 and 15 min of reaction. Noting that the highest values of EA were hydrolysates with flavourzyme at 15 min, at pHs 2 and 6 (80% at both pHs) followed by hydrolysates with flavourzyme at 5 min at pHs 2, 8 and 10 (65 and 60%, respectively). It should be noted that both hydrolysates showed an activity decrease at pH 4 (40%). These authors mention that the highest emulsifying capacity observed in the flavourzyme hydrolysates is probably because of the limited hydrolysis attained with this enzymatic complex, which provides a better hydrophobic/hydrophilic balance.

Another aspect that can be observed is that the hydrolysis of PI improves the functional properties; the aforementioned authors reported hydrolysis degrees of 2.8% and 6.3% for lime bean hydrolysates with flavourzyme at 5 and 15 min, respectively. Vioque et al. (2006) mention that obtaining a hydrolysis degree (DH) below 10% improves the functional properties (solubility, physicochemical and sensory properties) of a protein concentrate; values of DH above 10% allow the production of bioactive peptides; these could be used as food supplements or in medical diets.

Polanco-Lugo et al. (2014) report the EA for a limited hydrolysate of a *P. lunatus* PI with an enzymatic sequential system (pepsin-pancreatin), noting that the hydrolysate with a degree of hydrolysis of 1.7% at different pH values (2–10) had higher values of EA (99.3–261.11 m^2/g) compared to those of their respective PI (67.61–180.2 m^2/g). These authors mention that the increase in EA is probably due to suitable solubility of the limited hydrolysate in the system, as well as the increased hydrophobic surface due to limited hydrolysis of the globulin structure. On the other hand, the PI which had less interaction between nonpolar side chains of the proteins and lipidic chains might be attributed to the high extent of protein aggregation in PI caused by the extraction process.

Regarding the fibrous residue obtained from the wet-fractionation of *P. lunatus*, it has been found that it also has EA (49.3%) and ES (28.25%) (Betancur-Ancona et al. 2004b). The aforementioned authors indicate that the use of this fibrous fraction into food will depend on the type of product, for example, in products in which long shelf life is required, and thus higher emulsifying stability and the fibrous fraction from *P. lunatus* might not be so appropriate.

Foaming Capacity (FC)

The values reported to FC in flour, PI, and hydrolysates from *P. lunatus* L. are shown in Table 10.

Starch

In the case of starch from *P. lunatus* L., a high gelatinization temperature (80.16 °C) has been reported; this high gelatinization temperature of lima bean starch may allow use of higher temperatures than those used for other common starches during thermal processing food, in order to achieve complete gelatinization and to assure

Table 10 Foaming capacity in flour, PI, and hydrolysates from *P. lunatus* L.

Material	pH	Values (%)	Reference
Flour from *P. lunatus* (120 min)	2	23	Chel-Guerrero et al. (2002)
	4	18	
	6	16	
	8	15	
	10	15	
PI from *P. lunatus* (120 min)	2	30	
	4	10	
	6	5	
	8	15	
	10	10	
Limited hydrolysates from *P. lunatus* (flavourzyme at 5 and 15 min separately; hydrolysis degree (DH) 2.8 and 6.3, respectively; 120 min)	2	142 and 160	Betancur-Ancona et al. (2009)
	4	138 and 140	
	6	141 and 142	
	8	148 and 159	
	10	150 both	

its thickening effect. The solubility and the swelling power were correlated in a direct way with the temperature; the swelling pattern for *P. lunatus* starch shows swelling resistance at lower temperatures than 75 °C; in the range of 70–90 °C, the granules swell gradually when the temperature increases (Betancur-Ancona et al. 2001).

Campechano-Carrera et al. (2007) report the effect of the pyrodextrinization of starch extracted from *P. lunatus* indicating that this process dramatically reduces viscosity compared to native starch (13.2 and 752 cP, pyrodextrinized and native starch from lima bean, respectively). These authors indicated that the pyroconversion preferentially affects amorphous areas mainly formed of the ramified chains causing a reduction in the molecular weight and conversion to more highly branched molecules that are responsible for imparting viscosity. Another fact which may cause a drastic reduction in viscosity was the addition of HCL during the pyrodextrinization process, because this reagent caused hydrolysis and rearrangement of the starch molecules.

The functional properties that have been reported for flours, protein concentrates, hydrolysates, starches, and fibrous residues can provide a basis for choosing the best terms of use and application of these materials for the production of functional foods or for the improvement of those already present in the market.

Table 11 Bioactive properties for hydrolysates from *P. lunatus* L.

Hydrolysate conditions	Values	Activity	Reference
Sequential hydrolysis (60 min with pepsin-pancreatin), hydrolysis degree (DH) 15.97%	IC_{50} 0.321 mg/mL TEAC 13.2 mM/protein	I-ACE Antioxidant	Polanco-Lugo et al. (2014)
Hydrolysis with alcalase (90 min), DH 32%	IC_{50} 0.056 mg/mL TEAC 9.89 mM/protein	I-ACE Antioxidant	Torruco-Uco et al. (2009)
Hydrolysis with flavourzyme (90 min), DH 22%	IC_{50} 0.0069 mg/mL 8.42 mM/protein		
Hydrolysis of germinated seeds with alcalase (30 min), DH 30.34%	IC_{50} 0.61 mg/mL	I-ACE	Domínguez-Magaña et al. (2015)
Hydrolysis with alcalase (120 min), DH 51.28%	IC_{50} 0.56 mg/mL		
Sequential hydrolysis (60 min with pepsin-pancreatin), DH 32.16	IC_{50} 0.25 mg/mL		
Sequential hydrolysis of germinated seeds (60 min with pepsin-pancreatin), DH 32.16	IC_{50} 0.28 mg/mL		

Bioactive Properties

This section shows the bioactive proprieties reported from *P. lunatus* L. hydrolysates; Table 11 shows these activities.

Within our working group, extensive hydrolysis conditions have been analyzed extensively to obtain hydrolysates with biological activity, among which I-ACE and antioxidant stand out. Within the different bioactivities shown in Table 11, it seems that the best I-ACE values are obtained at 60–120 min of enzymatic action, either the individual or sequential enzyme, flavourzyme being the one that produces hydrolysates with the highest I-ACE activity.

An important aspect is mentioned by Torruco-Uco et al. (2009) that there is no correlation between I-ACE and antioxidant activity. This suggests that the antioxidant activity of the peptides may depend on the specific proteases used to produce them, the degree of hydrolysis attained, the nature of the released peptides (molecular weight, composition, and amino acid sequence), as well as the combined effects of their properties, including their ability to locate free radicals, acting as chelating agents of metal ions or as a hydrogen donor (Tang et al. 2009).

An alternative to improve the biological activity of the hydrolysates generated can be by obtaining peptide fractions of different molecular weight, by ultrafiltration (Cho et al. 2004). In this regard, Ciau-Solís et al. (2017) report the I-ACE activity of a molecular weight fraction >3 kDa from a lime bean hydrolysate obtained by sequential hydrolysis with pepsin-pancreatin at 90 min with a DH of 32.33%. This fraction showed an IC_{50} value of 0.172 mg/mL which is higher for hydrolysates obtained with the same enzymatic sequential system (IC_{50} of 0.321, 0.25, and 0.28 mg/mL Table 11). This shows that obtaining peptide fractions can be an alternative to improve the biological properties that the hydrolysate can present.

Cordova-Lizama et al. (2013) report the antithrombotic and anticariogenic activity in lime bean hydrolysates obtained with the enzyme pepsin, finding an 88% inhibition of platelet aggregation at a hydrolysate concentration of 4.5 mg/mL. Regarding the anticariogenic activity, a reduction in the demineralization of calcium and phosphorus was found in a hydroxypatite matrix by 50 and 55.8%, respectively.

Bojorquez-Balam et al. (2013) report the antimicrobial activity of hydrolysates and peptide fractions (>10 and <10 kDa) of lima beans. Finding that both the hydrolysate and the peptide fractions showed antimicrobial activity against *S. aureus* and *S. flexneri*, the <10 kDa fraction presented the highest inhibitory activity with 392.04 μg/mL for *S. aureus* and 993.17 μg/mL for *S. flexneri*.

Renin-inhibitory activity is another bioactive property that has been found in peptide fractions of *P. lunatus*, with inhibition values of 31.73 and 30.05% being observed for >3 kDa molecular weight fractions obtained with sequential alcalase-flavourzyme and pepsin-pancreatin systems, respectively (Ciau-Solís et al. 2017).

Another fraction that has shown biological activity is the fibrous residue obtained from the wet-fractionation processing of lima beans; it has been found that this portion has antioxidant activity (35.5%) which is slightly lower than those presented by hydrolysates obtained from the same seed (51.43–61.34%) (Torruco-Uco et al. 2009).

The biological properties that can be obtained from protein hydrolysates and fibrous residues from *P. lunatus* could be implemented in the preparation of functional products that provide some benefit to the organism beyond its nutritional value or elaborate nutraceutical products that could be used both in the prevention and treatment of chronic degenerative diseases.

Product Development

Functional Foods

This term was first used in Japan (1984) as a result of a study on the relationships between nutrition, sensory satisfaction, fortification, and modulation of physiological systems in order to define those food products fortified with special constituents that possess advantageous physiological effects (Bigliardi and Galati 2013). Otherwise this term is considered a marketing term only, and there is no consistent definition recognized globally by regulatory institutions. Crowe and Francis (2013) mention that the Academy of Nutrition and Dietetics define functional foods as:

Whole foods along with fortified, enriched, or enhanced foods that have a potentially beneficial effect on health when consumed as part of a varied diet on a regular basis at effective levels.

In Mexico the term functional food is widely used in the scientific field; but to date there are no laws that specifically regulate the use and production of these foods; however, their consumption is aimed at obtaining a health benefit. According to Hartmann and Meisel (2007) in the market, there is a wide variety of functional products added with bioactive peptides, for example, Calpis (Calpis Co., Japan), Evolus (Valio, Finland), BioZate (Davisco, USA), C12 Peption (DMV, the Netherlands), peptide soup (Nippon, Japan), ProDiet F200 (Ingredia, France), Capolac (Arla Foods, Denmark), etc. In Mexico, companies such as LALA® (2019) have de-lactose with fiber within their milk line, which provides more than 10% of the fiber required per day per serving. Bimbo® (2019) commercializes double fiber bread, which provides 25% per serving of the daily fiber requirement. The previously mentioned shows the need of the population to acquire food containing any compound that could provide some benefit to the organism during its consumption.

When talking about functional foods, these are commonly confused with nutraceuticals; however, they differ from each other, since a nutraceutical product is a substance of natural origin that, when ingested, in the organism behaves as a medicine, providing a beneficial effect for health beyond its nutritional value (Cortés et al. 2015). Recently the definition was modified by Health Canada, defining nutraceutical as a product isolated or purified from foods, and generally sold in medicinal forms (pills, capsules, powders, etc.) not usually associated with food and demonstrated to have a physiological benefit or provide protection against chronic disease (Monge 2008).

Foods Added with **Phaseolus lunatus** *L.*

This section shows the different foods that have been prepared with the addition of hydrolysates, starch, and fibrous residue extracted from *P. lunatus*.

Functional Foods

(a) *Flour*

Pérez-Navarrete et al. (2006) report the preparation of extrudates prepared with mixtures of cornmeal and lima beans in relation to 75:25, 50:50, and 25:75, respectively. It was observed that the property of expansion decreases as the content of bean flour increases; according to the density and maximum breaking force by compression, an increase of these parameters was observed as the proportion of bean flour increased. The 50:50 mixture was considered potentially nutritious because it met the requirements of essential amino acids such as lysine and tryptophan in 100%.

(b) *Protein Isolates*

Davalos-Cervera (2003) reports the elaboration of Frankfurt-type sausages added with protein isolates of *P. lunatus* with addition levels of 3.5 and 7%. An increase in the protein content of 46.49 and 51.66%, respectively, was observed, being higher than the control (39.71%); sensory evaluation showed acceptance for sausages with a 3.5% addition and rejection for sausages made with 7%. For both products, minimal concentrations of cyanogenic glycosides (0.188 and 0.559 mg/100 g) were found without jeopardizing their consumption.

(c) *Hydrolysates*

Franco-Miranda et al. (2017) reports the incorporation of protein hydrolysates from lima bean (1 and 3%) in the production of concha-type Mexican sweet bread, observing that the addition of *P. lunatus* hydrolysates to bread dough affected its rheological properties, decreased tenacity, and slightly increased elasticity compared to the control. The I-ACE activity of bread with a 3% hydrolysate addition showed the highest activity (83.10%) compared to that added with 1% and the control (69.61 and 11.27%, respectively); regarding the antioxidant activity, bread with a 3% addition also showed the highest value (TEAC 17.19 μmol trolox/g) compared to that added with 1% and the control (TEAC 15.59 and 5.42 μmol trolox/g, respectively). The bread added with 3% lima beans had a low acceptance due to a slight acidic taste.

(d) *Starch*

Chim-Rodríguez (2000) reports the preparation of cookies added with 50% of lima bean starch; the cookies had a total dietary fiber content of 3.67%, 2.42% insoluble dietary fiber and 1.24% soluble dietary fiber. The sensory test showed that cookies added with *P. lunatus* starch were accepted even over the control cookies.

(e) *Fibrous Residue*

Peraza-Mercado (2000) reports the addition of fibrous residue of *P. lunatus* in an addition percentage of 2.6%. The total, insoluble, and soluble dietary fiber content was 9.23, 7.14 and 2.08%, respectively; the sensory evaluation reported a low acceptance for these cookies, presenting softness compared to the control cookies.

Perspectives

With the information presented in this section, it is shown that both the flour and each of the fractions that can be obtained from *P. lunatus* may well be used in the production of functional foods with sensory acceptance. The aforementioned will depend on choosing the product to be prepared in conjunction with the functional properties of the different fractions of lima beans that were discussed in "Functional Properties" section.

Nutraceuticals

Encapsulation is a process to entrap one substance defined as an active agent within another substance defined as wall material. In the food industry, an encapsulation process can be applied for different reasons, for example, improve delivery of bioactive molecules (e.g., antioxidants, minerals, vitamins, phytosterols, fatty acids, etc.) and living cells (e.g., probiotics) into foods. Encapsulation technology refers to a technique in which the bioactive component is completely enveloped, covered, and protected by a physical barrier, without any protrusion of the bioactive component. Also, microcapsules release their contents at controlled rates over prolonged periods and under specific conditions (Nedovic et al. 2011).

To guarantee the activity of these bioactive peptides, it must remain active and intact during the gastrointestinal digestion and absorption in order to achieve their physiological effects. But once it is in the organism, all peptides go through different barriers that can inactivate them and consequently lose their efficiency (Segura-Campos et al. 2011). In addition to improve the beneficial effect of bioactive peptides, the microencapsulation could be a way to protect the peptides against the environment to which they are exposed. Microencapsulation protects bioactive peptides and ensures the release of an appropriate dosage at a gastric or intestinal pH. In this way the encapsulation of hydrolysates and peptide fractions with I-ACE and antioxidant activity from *P. lunates* were evaluated.

Ruiz-Ruiz et al. (2013) reports an IC_{50} of residual I-ACE in intestinal media of 2.9 mg/mL for encapsulated hydrolysates and released by in vitro digestion; the encapsulation conditions were pH 4, 1 mM $CaCl_2$, and a 50:50 ratio of alginate sodium and carboxymethylated flamboyant gum. It was observed that the conditions selected for the encapsulation of this hydrolysate allowed an efficiency of 71.7%.

Sandoval-Peraza et al. (2014) report the encapsulation of a peptide fraction with a <10 kDa weight obtained from protein hydrolysates of *P. lunatus* by ultrafiltration, by means of the ionic gelation technique using alginate and carboxymethylated flamboyant gum. The treatment with the highest encapsulation efficiency achieved a 33.43% entrapment of the peptide fraction; the encapsulation conditions for this treatment were 50:50 gum ratio (alginate/flamboyan), 0.1 M $CaCl_2$, and 25 min of hardening time. After an in vitro digestion, the antioxidant activity and remnant I-ACE in the intestinal environment were 615.10 mTEAC/mg protein and 0.035 mg/mL of IC_{50}; it should be noted that the IC_{50} value and antioxidant activity of the fraction released were higher compared to the one that was initially encapsulated (0.37 mg/mL of IC_{50} and 26.94 TEAC mM/mg protein, respectively).

Notably hydrolysates as well as their fractions can be used in the preparation of nutraceutical products, and it has been demonstrated that these fractions maintain their biological activity even after being exposed to an in vitro digestion.

Future Trends in Processing and Product Development

Currently, Mexico and the United States occupy the first places of worldwide prevalence of obesity in the adult population (30%), which is ten times higher than that found in countries such as Japan and Korea. Regarding the child population, Mexico ranks fourth in the worldwide prevalence of obesity, approximately 28.1% in boys and 29% in girls, surpassed only by Greece, the United States, and Italy (Dávila-Torres et al. 2015).

In our country, according to ENSANUT (2012), the state of Yucatan is among the first places in obesity nationwide. This is alarming, since the development of obesity is the first step to develop chronic degenerative diseases. Therefore, it is important to implement national strategies such as those mentioned before (Check yourself, Measure yourself, and Move), along with the search for functional and nutraceutical products that can help in the treatment and prevention of chronic degenerative diseases.

This review intends to explore the seeds of *Phaseolus lunatus* (underexploited species) grown and harvested in the region (Yucatan, Mexico) as a possibility to obtain compounds with functional and biological activity and their subsequent implementation in the production of functional foods and nutraceuticals. It should be noted that the data presented here can be used as a starting point and improvement for research involving the use of *Phaseolus lunatus*, both nationally and internationally, since the cultivation, consumption, and use of this legume extend from Mexico to South America (Chile).

References

Arias-Restrepo, J., Rengifo-Martínez, T., Jaramillo-Carmona, M. (2007). *Manual Técnico, Buenas prácticas agrícolas (BPA) en la producción de frijol voluble* (1st ed.). Medellin: FAO. ISBN 978-92-5-305827-3. Retrieved May 15, 2019, from http://www.fao.org/3/a-a1359s.pdf

Arias-Trinidad, A. (2018). *Evaluacion de la goma nativa de Guazuma ulmifolia, en la encapsulación de fracciones peptídicas con actividad antioxidantes de Phaseolus lunatus*. PhD Thesis. Autonomous University of Yucatan, Merida, Yucatan, Mexico.

Betancur-Ancona, D., Chel-Guerrero, L., Camelo-Matos, R., & Dávila-Ortiz, G. (2001). Physicochemical and functional characterization of baby lima bean (*Phaseolus lunatus*) starch. *Starch/Stärke, 53*, 219–226. https://doi.org/10.1002/1521-379X(200105)53:5<219::AID-STAR219>3.0.CO;2-R.

Betancur-Ancona, D., López-Luna, J., & Chel-Guerrero, L. (2003). Comparison of the chemical composition and functional properties of *Phaseolus lunatus* prime and tailing starches. *Food Chemistry, 82*, 217–225. https://doi.org/10.1016/S0308-8146(02)00515-0.

Betancur-Ancona, D., Gallegos-Tintoré, S., & Chel-Guerrero, L. (2004a). Wet-fraction of *Phaseolus lunatus* seeds: Partial characterization of starch and protein. *Journal of the Science of Food and Agriculture, 84*, 1193–1201. https://doi.org/10.1002/jsfa.1804.

Betancur-Ancona, D., Peraza-Mercado, G., Moguel-Ordoñez, Y., & Fuertes-Blanco, S. (2004b). Physicochemical characterization of lima bean (*Phaseolus lunatus*) and jack bean (*Canavalia ensiformis*) fibrous residues. *Food Chemistry, 84*, 287–295. https://doi.org/10.1016/S0308-8146(03)00213-9.

Betancur-Ancona, D., Martínez-Rosado, R., Corona-Cruz, A., Castellanos-Ruelas, A., Jaramillo-Flores, E., & Chel-Guerrero, L. (2009). Functional properties of hydrolysates from *Phaseolus lunatus* seeds. *Journal of Food Science and Technology, 44*, 128–137. https://doi.org/10.1111/j.1365-2621.2007.01690.x.

Bigliardi, B., & Galati, F. (2013). Innovation trends in the food industry: The case of functional foods. *Trends in Food Science and Technology, 31*, 118–129. https://doi.org/10.1016/j.tifs.2013.03.006.

BIMBO. (2019). *Pan doble fibra*. Retrieved May 17, 2019, from https://www.bimbo.com.mx/es/productos-pan/doble-fibra?product_category=16817

Bojorquez-Balam, E., Ruiz-Ruiz, J., Segura-Campos, M., Betancur-Ancona, D., & Chel-Guerrero, L. (2013). Evaluacion de la capacidad antimicrobiana de fracciones peptídicas de hidrolizados proteínicos de frijol lima (*Phaseolus lunatus*). In *Bioactividad de péptidos derivados de proteínas alimentarias* (pp. 139–154). Mexico: Lilu Enterprises.

Campechano-Carrera, E., Corona-Cruz, A., Chel-Guerrero, L., & Betancur-Ancona, D. (2007). Effect of pyrodextrinization on available starch content of lima bean (*Phaseolus lunatus*) and cowpea (*Vigna unguiculata*) starches. *Food Hydrocolloids, 21*, 472–479. https://doi.org/10.1016/j.foodhyd.2006.06.006.

Chavan, U., McKenzie, D., & Shahidi, F. (2001). Functional properties of protein isolates from beach pea (*Lathyrus maritimus* L.). *Food Chemistry, 74*, 177–187. https://doi.org/10.1016/S0308-8146(01)00123-6.

Chécate, Mídete, & Muévete. (2019). *(Check yourself, measure yourself and move)*. Retrieved May 17, 2019, from http://checatemidetemuevete.gob.mx/

Chel-Guerrero, L., Pérez-Flores, V., Betancur-Ancona, D., & Dávila-Ortiz, G. (2002). Functional properties of flours and protein isolates from *Phaseolus lunatus* and *Canavalia ensiformis* seeds. *Journal of Agricultural and Food Chemistry, 50*, 584–591. https://doi.org/10.1021/jf010778j.

Chel-Guerrero, L., Gallegos-Tintoré, S., Martínez-Ayala, A., Castellanos-Ruelas, A., & Betancur-Ancona, D. (2011). Functional properties of proteins from lima bean (*Phaseolus lunatus* L.) seeds. *Food Science and Technology International, 17*(2), 119–126. https://doi.org/10.1177/1082013210381433.

Chel-Guerrero, L., Domínguez-Magaña, M., Martínez-Ayala, A., Dávila-Ortiz, G., & Betancur-Ancona, D. (2012). Lima bean (*Phaseolus lunatus*) protein hydrolysates with ACE-I inhibitory activity. *Food and Nutrition Sciences, 3*, 511–521. https://doi.org/10.4236/fns.2012.34072.

Chim-Rodríguez, A. (2000). *Incorporación de almidones primarios y secundarios de Canavalia ensiformis L. y Phaseolus lunatus L. en galletas*. Bachelor Thesis, Autonomous University of Yucatan, Merida, Yucatan, Mexico.

Cho, M., Unklesbay, N., Hsieh, F., & Clarke, A. (2004). Hydrophobicity of bitter peptides from soy protein hydrolysates. *Journal of Agricultural and Food Chemistry, 52*(19), 5895–5901. https://doi.org/10.1021/jf0495035.

Ciau-Solís, N., Acevedo-Fernández, J., & Betancur-Ancona, D. (2017). In vitro renin—Angiotensin system inhibition and in vivo antihypertensive activity of peptide fractions from lima bean (*Phaseolus lunatus* L.). *Journal of the Science of Food and Agriculture, 98*, 781–786. https://doi.org/10.1002/jsfa.8543.

Córdova-Lizama, A., Ruiz-Ruiz, J., Segura-Campos, M., Betancur-Ancona, D., & Chel-Guerrero, L. (2013). Actividad antitrombótica y anticariogénica de hidrolizados proteínicos de frijol lima (*Phaseolus lunatus*). In *Bioactividad de péptidos derivados de proteínas alimentarias* (pp. 123–137). Mexico: Lilu Enterprises.

Córdova-Villalobos, J., Barriguete-Meléndez, J., Lara-Esqueda, A., Barquera, S., Rosas-Peralta, M., Hernández-Ávila, M., León-May, M., & Aguilar-Salinas, C. (2008). Las enfermedades crónicas no transmisibles en México: sinopsis epidemiológica y prevención integral. *Salud Pública de México, 50*(5):419–427. Retrieved May 20, 2019, from http://www.scielo.org.mx/pdf/spm/v50n5/a15v50n5.pdf

Cortés, G., Prieto, G., & Rozo, W. (2015). Bromatological and physicochemical characterization of *Physalis peruviana* L., and its potential as a nutraceutic food. *Revista Ciencia en Desarrollo, 6*(1):87–97. Retrieved May 20, 2019, from http://www.scielo.org.co/pdf/cide/v6n1/v6n1a11.pdf

Crowe, K. M., & Francis, C. (2013). Position of the academy of nutrition and dietetics: Functional foods. *Journal of the Academy of Nutrition and Dietetics, 113*(8), 1096–1103. https://doi.org/10.1016/j.jand.2013.06.002.

Cuny, M., La Forgia, D., Desurmont, G., Glauser, G., & Benrey, B. (2019). Role of cyanogenic glycosides in the seeds of wild lima bean, *Phaseolus lunatus*: Defense, plant nutrition or both? *Planta, 250*, 1281–1292. https://doi.org/10.1007/s00425-019-03221-3.

Davalos-Cervera, J. (2003). *Elaboración de salchichas tipo Frakfurt adicionadas con concentrado de proteína de Phaseolus lunatus*. Thesis, Autonomous University of Yucatan, Merida, Yucatan, Mexico.

Dávila-Torres, J., González-Izquierdo, J., Barrera-Cruz, A. (2015). Panorama de la obesidad en México. *Revista medica del Instituto Mexicano del* Seguro *Social, 3*(2):240–249. Retrieved May 15, 2019, from http://revistamedica.imss.gob.mx/editorial/index.php/revista_medica/article/viewFile/21/54

De Dios, A., Porrilla, Y., & Chaparro, D. (2009). Antinutritional factors in eatable seeds. *Facultad de Ciencias Agropecuarias, 7*(1), 45–54.

Debouk, D. G. (2019). La agricultura en Mesoamerica. Frijoles (*Phaseolus* spp.). In *Cultivos Andinos FAO*. Retrieved May 20, 2019, http://www.fao.org/tempref/GI/Reserved/FTP_FaoRlc/old/prior/segalim/prodalim/prodveg/cdrom/contenido/libro09/Cap2_2.htm#auto

Domínguez-Magaña, M. (2009). *Aislamiento de biopeptidos con actividad inhibitoria de la enzima convertidora de angiotensina-I a partir de hidrolizados de P. lunatus*. PhD Thesis, Autonomous University of Yucatan, Merida, Yucatan, Mexico.

Domínguez-Magaña, M., Segura-Campos, M., Dávila-Ortiz, G., Betancur-Ancona, D., & Chel-Guerrero, L. (2015) ACE-I inhibitory properties of hydrolysates from germinated and ungerminated *Phaseolus lunatus* proteins. *Food Science and Technology (Campinas)*. https://doi.org/10.1590/1678-457X.6551

ENSANUT. (2012). *Resultados Nacionales*. Retrieved May 10, 2019, from https://ensanut.insp.mx/informes/ENSANUT2012ResultadosNacionales.pdf

FAO. (2011). *Dietary protein quality evaluation in human nutrition*. Report of an FAO Expert Consultation. FAO Food and Nutrition Paper 92. ISSN 0254-4725. Retrieved May 20, 2019, from https://apps.who.int/iris/bitstream/handle/10665/41042/WHO_TRS_522_eng.pdf?sequence=1&isAllowed=y

FAO. (2018). *Nuestras legumbres. Pequeñas semillas, grandes soluciones*. FAO, Ciudad de Panamá. ISBN 978-92-5-131129-5. Retrieved May 20, 2019, from http://www.fao.org/3/ca2597es/CA2597ES.pdf

FAO. (2020). *Cultivos Andinos. Cultivos Autoctonos Subexplotados de Mesoamerica*. Retrieved Apr 03, 2020, from http://www.fao.org/tempref/GI/Reserved/FTP_FaoRlc/old/prior/segalim/prodalim/prodveg/cdrom/contenido/libro11/home11.htm

Franco-Miranda, H. (2015). *Incorporacion de hidrolizados proteicos de Phaseolus lunatus y Vigna unguiculata en productos de panaderia y extrudidos*. Master's Thesis, Autonomous University of Yucatan, Merida, Yucatan, Mexico.

Franco-Miranda, H., Chel-Guerrero, L., Gallegos-Tintoré, S., Castellanos-Ruelas, A., & Betancur-Ancona, D. (2017). Physicochemical, rheological, bioactive and consumer acceptance analyses of concha-type Mexican sweet bread containing lima bean or cowpea hydrolysates. *Food Science and Technology, 80*, 250–256. https://doi.org/10.1016/j.lwt.2017.02.034.

Giami, S. (2001). Quality attributes of three new improved lines of Nigerian lima beans (*Phaseolus lunatus* L. Walp.). *Plant Food for Human Nutrition*, *56*:325–333. Retrieved May 20, 2019, from https://www.ncbi.nlm.nih.gov/pubmed/11678438

Gómez, R., Monteiro, H., Cossio-Bolaños, A., Fama-Cortez, D., & Zanesco, A. (2010). Physical exercise and its prescription in patients with chronic degenerative diseases. *Revista Peruana de*

Medicina Experimental y Salud Pública, *27*(3):379–386. Retrieved May 20, 2019, from http://www.scielo.org.pe/pdf/rins/v27n3/a11v27n3.pdf

Hartmann, R., & Meisel, H. (2007). Food-derived peptides with biological activity: From research to food applications. *Current Opinion in Biotechnology, 18*, 163–169. https://doi.org/10.1016/j.copbio.2007.01.013.

LALA. (2019). *Leche LALA fibra semideslactosada*. Retrieved May 25, 2019, from https://www.lala.com.mx/producto/lala-fibra-semi/

Linnemann, A., & Dijkstra, D. (2002). Toward sustainable production of protein-rich foods: Appraisal of eight crops for Western Europe. Part I. Analysis of the primary links of the production chain. *Critical Reviews in Food Science and Nutrition, 42*(2), 377–401. https://doi.org/10.1080/20024091054193.

López-Alcocer, J., Lépiz-Idelfonso, R., González-Eguiarte, D., Rodríguez-Macías, R., Lípez-Alcocer, E. (2016). Morphological variability of wild *Phaseolus lunatus* L. from the western region of Mexico. *Revista Fitotecnia Mexicana*, *39*(1):49–58. Retrieved May 20, 2019 from http://www.scielo.org.mx/scielo.php?script=sci_arttext&pid=S0187-73802016000100009

Miranda-Villa, P., Marrugo-Ligardo, Y., & Montero-Castillo, P. (2013). Functional characterization of bean Zaragoza starch (*Phaseolus lunatus* L.) and quantification of the resistant starch. *TecnoLógicas, 30*, 17–32.

Monge, A. (2008). Functional foods. Reflexions of a scientist regarding a market in expansion. *Revista de la Sociedad Quimica del Peru, 74*(2):138–147. Retrieved May 20, 2019, from http://www.scielo.org.pe/pdf/rsqp/v74n2/a06v74n2.pdf

Nedovic, V., Kalusevic, A., Manojlovic, V., Levic, S., & Bugarski, B. (2011). An overview of encapsulation technologies for food applications. *Procedia Food Science, 1*, 1806–1815. https://doi.org/10.1016/j.profoo.2011.09.266.

NMX-FF-038-SCFI-2002. (n.d.) *Productos alimenticios no industrializados para consumo humano, fabáceas, frijol (Phaseolus vulgaris) especificaciones y métodos de prueba*. México: Secretaría de Economía. Retrieved May 20, 2019, from http://www.sagarpa.gob.mx/agronegocios/Lists/Instrumentos%20Tcnicos%20Normalizacin%20y%20Marcas%20Colecti/Attachments/91/NMX_FRIJOL.pdf

Novelo-Cen, L., Betancur-Ancona, D. (2005). Chemical and Functional Properties of Phaseolus lunatus and Manihot esculenta Starch Blends. *Starch - Stärke 57*(9):431–441.

Nuñez-Aragón, P., Segura-Campos, M., Negrete-León, E., Acevedo-Fernández, J., Betancur-Ancona, D., Chel-Guerrero, L., & Castañeda-Corral, G. (2019). Protein hydrolysates and ultra-filtered <1 kDa fractions from *Phaseolus lunatus, Phaseolus vulgaris* and *Mucuna pruriens* exhibit antihyperglycemic activity, intestinal glucose absorption and α-glucosidase inhibition with no acute toxicity in rodents. *Journal of the Science of Food and Agriculture, 99*, 587–595. https://doi.org/10.1002/jsfa.9219.

PAHO. (2002). *Health in the Americas* (PAHO Scientific and Technical Publication No. 587) (2002 ed). Washington, DC. ISBN 92-75-31587-6. Retrieved May 20, 2019, from http://iris.paho.org/xmlui/bitstream/handle/123456789/746/9275115877_vol_I_EN.pdf;jsessionid=D16467E7BD31C373D43D17080617375E?sequence=7.

Pastor-Cavada, E., Juan, R., Alaiz, M., & Vioque, J. (2011). Nutritional characteristics of seed proteins in 28 *Vicia* species (*Fabaceae*) from southern Spain. *Journal of Food Science, 76*(8), C1118–C1124. https://doi.org/10.1111/j.1750-3841.2011.02336.x.

Peraza-Mercado, G. (2000) *Caracterización de los residuos fibrosos de Canavalia ensiformis L. y Phaseolus lunatus L. y su incorporación a un producto alimenticio*. Master's Thesis. Autonomous University of Yucatan. Merida, Yucatan, Mexico.

Pérez-Navarrete, C., Cruz-Estrada, R., Chel-Guerrero, L., & Betancur-Ancona, D. (2006). Physical characterization of extrudates prepared with blends of QPM maize (*Zea maiz* L.) and lima bean (*Phaseolus lunatus* L.). *Revista de Ingenieria Quimica, 5*, 145–155.

Polanco-Lugo, E., Dávila-Ortiz, G., Betancur-Ancona, D., & Chel-Guerrero, L. (2014). Effects of sequential enzymatic hydrolysis on structural, bioactive and functional properties of *Phaseolus lunatus* protein isolate. Food Sci. *Technol (Campinas), 34*(3), 441–448. https://doi.org/10.1590/1678-457x.6349.

Ruiz-Ruiz, J., Segura-Campos, M., Betancur-Ancona, D. A., Chel-Guerrero, L. (2013). Encapsulation of *Phaseolus lunatus* protein hydrolysate with angiotensin-converting enzyme inhibitory activity. *ISRN Biotechnology, 2013*: 1–6. https://doi.org/10.5402/2013/341974. Retrieved May 20, 2019, from http://www.hindawi.com/isrn/biotechnology/2013/341974/

Sandoval-Peraza, V. (2015). *Microencapsulacion de hidrolizados proteicos de Phaseolus lunatus L. con gomas de flamboyán (Delonix regia bojer Raf.) y chia (Salvia hispánica)*. PhD Thesis. Autonomous University of Yucatan. Merida, Yucatan, Mexico.

Sandoval-Peraza, M., Betancur-Ancona, D., Gallegos-Tintoré, S., & Chel-Guerrero, L. (2014). Evaluation of some residual bioactivities of microencapsulated *Phaseolus lunatus* protein fraction with carboxymethylated flamboyant (*Delonix regia*) gum/sodium alginate. *Food Science and Technology (Campinas), 34*(4), 680–687. https://doi.org/10.1590/1678-457X.6425.

Segura-Campos, M., Chel-Guerrero, L., & Betancur-Ancona, D. (2008). Synthesis and partial characterization of octenylsuccinic starch from *Phaseolus lunatus*. *Food Hydrocolloids, 22*, 1467–1474. https://doi.org/10.1016/j.foodhyd.2007.09.009.

Segura-Campos, M., Chel-Guerrero, L., Betancur-Ancona, D., & Hernández-Escalante, V. (2011). Bioavailability of bioactive peptides. *Food Review International, 27*(3), 213–226. https://doi.org/10.1080/87559129.2011.563395.

Tang, C., Peng, J., Zhen, D., & Chen, Z. (2009). Physicochemical and antioxidant properties of buckwheat (*Fagopyrum esculentum* Moench) protein hydrolysates. *Food Chemistry, 115*, 672–678. https://doi.org/10.1016/j.foodchem.2008.12.068.

Torruco-Uco, J., Chel-Guerrero, L., Martínez-Ayala, A., Dávila-Ortíz, G., & Betancur-Ancona, D. (2009). Angiotensin-I converting enzyme inhibitory and antioxidant activities of protein hydrolysates from *Phaseolus lunatus* and *Phaseolus vulgaris* seeds. *Food Science and Technology, 42*, 1597–1604. https://doi.org/10.1016/j.lwt.2009.06.006.

Treviño-Mejía, D., Luna-Vital, D., Gaytán-Martínez, M., Mendoza, S., & Loarca-Piña, G. (2016). Fortification of commercial nixtamalized maize (*Zea mays* L.) with common bean (*Phaseolus vulgaris* L.) increased the nutritional and nutraceutical content of tortillas without modifying sensory properties. *Journal of Food Quality, 39*, 569–579. https://doi.org/10.1111/jfq.12251.

Vioque, J., Pedroche, J., Yust, M., Lqari, H., Megías, C., Girón-Calle, J., Alaiz, M., & Millán, F. (2006). Bioactive peptides in storage plant proteins. *Brazilian Journal of Food Technology, Especial Edition III International Day of Proteins and Food Colloids*, 99–102.

Lupin

Rizliya Visvanathan, Terrence Madhujith, Ashoka Gamage, and Na Zhang

Introduction

Lupin belonging to the genus *Lupinus* and family Leguminosae/Fabaceae is a diverse species with both annual and perennial crops (van de Noort 2017). It is a protein-rich legume with high-quality protein and acceptable amino acid content with virtually no starch (Pastor-Cavada et al. 2009). At present, there are more than 400 species reported globally (Sanz et al. 2010; Kinder and Knecht 2011; Confortin et al. 2018), of which only 164 are accepted and recorded as *Lupinus* species in the integrated taxonomic information system (Australian Government 2013).

The role of lupin in agriculture dates back to more than 4000 years. From ancient times, lupins were mainly used as animal feed, green manures and soil-improving agents, especially as nitrogen fixers (Sweetingham and Kingwell 2008). However, there are reports on the use of lupins as food in the Mediterranean and South American Andes regions from ancient times (Bader et al. 2011). Although lupins have been consumed as a human food for centuries, the extent of the health benefits and functionality of lupins and its compounds as a food ingredient was much less

R. Visvanathan
Department of Nutrition, Dietetics and Food, Monash University, Notting Hill, VIC, Australia

T. Madhujith (✉)
Department of Food Science and Technology, Faculty of Agriculture, University of Peradeniya, Peradeniya, Sri Lanka

A. Gamage
Department of Chemical and Process Engineering, Faculty of Engineering, University of Peradeniya, Peradeniya, Sri Lanka

N. Zhang
College of Food Engineering, Harbin University of Commerce, Harbin, China

A. Manickavasagan, P. Thirunathan (eds.), *Pulses*,
https://doi.org/10.1007/978-3-030-41376-7_10

Table 1 Nutritional composition (%) of lupin

Component	*L. albus*	*L. angustifolius*	*L. luteus*	*L. mutabilis*
Moisture	11	12	12	10
Protein	44	41	52	52
Fat	11	7	7	17
Ash	4	3	4	4
Crude fibre	2	9	2	10
Lignin	<1	<1	<1	<1
NSP	21	29	11	10
Oligosaccharides	8	6	12	6
Starch	–	–	–	–

Source: Adapted from Australian Government (2013)

understood and only recently acknowledged. Over the past century, there has been a drastic change in the role of lupins in the agricultural sector. Following the revelation of their reported functional properties against the development of several metabolic diseases, the trend in the use of lupin in human food has taken a favourable turn. In modern societies, consumers are increasingly concerned about the role of diet in health and wellbeing. In vivo studies on lupin report several health beneficial properties including antidiabetic, antihypertensive, antihyperlipidaemic and anticancer properties and antioxidant activity in both animal (Magni et al. 2004; Bertoglio et al. 2011; Kapravelou et al. 2013) and human clinical studies (Lee et al. 2006, 2009 Bouchoucha Bouchoucha et al. 2016a, b). The bioactive properties exerted by lupin are attributed to the presence of several classes of compounds such as dietary fibre (Kapravelou et al. 2013), starch, oligosaccharides (Duranti et al. 2008), phytochemicals (Baldeón et al. 2012; Mosaad et al. 2014) and proteins and peptides (Duranti et al. 2008; Kapravelou et al. 2013).

The unique chemical composition and functional properties of lupin seeds render it as a preferred alternative for dry beans and soybeans. At present, lupin has found its way into the bakery, confectionary, milk and meat processing industries where it is used in pasta, bread, cakes, pastries, milk, tofu, tempeh, miso, soy sauce and snack foods, among others (Petterson and Fairbrother 1996; Archer et al. 2004; Johnson et al. 2017; Clark and Johnson 2002; Jayasena et al. 2010a). Other than that, lupin hull is used as dietary fibre or fibre additive in bread (Clark and Johnson 2002; Villarino et al. 2016). Countries in Europe and North America and Australia are commercially involved in the production of lupin-based food products (Johnson et al. 2017). Apart from basic food applications, the use of lupins is also expanding in the functional food sectors. Thus, lupins now have received the global attention as a future food source that has the potential to be used as a food ingredient to enhance the nutritional and functional status of food products.

Chemical and Nutritional Composition

The overall nutrient composition of the four main domesticated lupin species is presented in Table 1. Similar to other grain legumes, lupin is a rich source of high-quality protein and dietary fibre with good-quality fats rich in unsaturated fatty acids, minerals and other nonnutritive bioactive compounds with virtually no starch.

Proteins

The major macronutrient present in lupin is protein. Protein content of lupins is approximately 34–44% on dry weight basis (dw) which is almost similar to soybean and is significantly higher than other legumes such as lentils, pea, chickpea and faba bean (Duranti et al. 2008; Bähr et al. 2014; Arnoldi et al. 2015; Johnson et al. 2017; van de Noort 2017). Among the four domesticated lupin species, *L. mutabilis* is reported to have the highest protein content of 44% followed by *L. luteus*, *L. albus* and *L. angustifolius* (Carvajal-Larenas et al. 2016; Johnson et al. 2017). In addition to species variation, the protein content of lupin is reported to differ within species based on variety, growing season, location and cultivation practices (Calabrò et al. 2015).

Apart from quantity, the quality of protein is mainly decided based on its amino acid composition and protein digestibility. Table 2 summarizes the essential amino acid content of the four main lupin species. Similar to other legume crops, lupin is a poor source of the essential sulphur-containing amino acids. According to the Food and Agricultural Organization (FAO) amino acid recommendations for adults, the major limiting essential amino acids in main lupin species are methionine, cysteine and valine (Belski 2012) (Table 2). However, the presence of other essential amino acids, i.e. lysine, isoleucine, leucine, tryptophan, phenylalanine and tyrosine, is within the FAO recommendations (Johnson et al. 2017). Pastor-Cavada et al. (2009) reported a wild lupin species, *L. cosentinii*, to satisfy the FAO recommendations for sulphur amino acids (2.5 g/100 g protein). According to Carvajal-Larenas et al. (2016), there was little variation in the essential amino acid profile of raw lupins within species (Table 2). However, higher cystine and leucine content was found in *L. luteus* species, while *L. albus* and *L. mutabilis* had high tyrosine and lysine content, respectively (Carvajal-Larenas et al. 2016). Compared to soybeans, lupins are a good source of arginine, lysine, leucine and phenylalanine (Australian Government 2013). Thus, consumption of lupin along with cereals will complement for the lack of lysine in cereals and provide higher-quality protein (Sathe 2016).

In lupin seeds, the globulins correspond to about 90% of the total protein content with albumins making up the remainder (Duranti et al. 2008; Czubinski and Feder 2019). Based on the electrophoretic mobility of lupin seed proteins, it is grouped into four groups: the two major storage proteins, α-conglutin (11S legumin-like lupin globulin) and β-conglutin (7S vicilin-like lupin globulin), and two minor

Table 2 Essential amino acid content of different lupins

	L. albus	*L. angustifolius*	*L. luteus*	*L. mutabilis*	FAO[a]
Amino acid	Content (g/100 g)				
Essential amino acids					
Histidine	2.0	2.6	3.1	3.5	1.6
Threonine	3.4	3.4	3.0	3.5	2.5
Valine	3.8	3.7	3.4	3.8	4.0
Methionine	0.7	0.7	0.6	0.8	2.3[b]
Isoleucine	4.1	4.0	3.6	4.2	3.0
Tryptophan	0.9	0.9	0.9	0.8	0.66
Leucine	6.8	6.9	7.8	7.0	6.1
Lysine	4.5	4.6	4.5	5.8	4.8
Phenylalanine	3.4	3.7	3.7	3.5	4.1[c]
Conditionally essential amino acids					
Cystine[d]	1.5	1.6	2.4	1.6	NA
Tyrosine	4.8	3.4	2.9	4.0	4.1[c]
Arginine	12.4	12.0	9.1	10.2	NA
Glutamine	NA	NA	NA	24.3	NA
Glycine	NA	NA	NA	3.8	NA
Proline	NA	NA	NA	3.8	NA

Source: Adapted from Carvajal-Larenas et al. (2016)
[a]Suggested pattern of amino acid requirements (FAO/WHO/UNU 2007) (Pastor-Cavada et al. 2009)
[b]Met + Cys
[c]Phe + Tyr
[d]Equivalent to 3.168 g/100 g cysteine

groups: γ-conglutin (7S globulin family) and δ-conglutin (2S sulphur-rich albumin family) (Duranti et al. 2008; Cabello-Hurtado et al. 2016). β-Conglutin is the most abundant lupin globulin representing nearly 44–45% of white lupin seed storage proteins followed by 35–37% of α-conglutin. The content of γ-conglutin in mature white lupin seeds accounts to about 4–5% of the total amount of proteins, while the least present is the δ-conglutin representing only around 3–4% of total conglutins in white lupin (Duranti et al. 2008).

Bioactive peptides released upon digestion of lupin seed proteins are reported to exert antiobesity, antidiabetic, antihypertensive and antihyperlipidaemic effect (Fontanari et al. 2012; Cabello-Hurtado et al. 2016; Lammi et al. 2016; Pavanello et al. 2017). Among these, β- and γ-conglutin have received more attention recently due to its hypocholesterolaemic and hypoglycaemic activities (Terruzzi et al. 2011; Rosa Lovati et al. 2012; Muñoz et al. 2018). As a result of the high homology of lupin β-conglutins with the α-subunit of β-conglycinin of soybean, β-conglutins are hypothesized to be a potential hypocholesterolaemic agent of white lupin proteins (Duranti et al. 2008; Pihlanto et al. 2017). So far, among the lupin components, γ-conglutin is the only protein known to elicit a significant glucose decrease response (Magni et al. 2004). The first report on the hypoglycaemic property of γ-conglutin on humans was published by Bertoglio et al. (2011). In a study by

Terruzzi et al. (2011), γ-conglutin was found to mimic the activity of insulin by modulating muscle glucose metabolism through activating the IRS-1/PI3/Akt/p70S6k signaling pathway. In addition, γ-conglutin have also been reported with hypocholesterolaemic, ACE-inhibitory, antioxidant and bile acid binding properties (Sirtori et al. 2004; Yoshie-Stark and Wäsche 2004).

Carbohydrates and Dietary Fibre

Unlike other legumes such as peas and chickpeas having nearly 50–70% of starch, lupins are relatively a poor source of available carbohydrate with most species having less than 1.5% in the seeds (Petterson 2015). The amount of polysaccharides in lupin seeds is dependent on the species, with high contents found in *L. albus* and *L. angustifolius* and the lowest in *L. mutabilis*, due to its thinner seed hull (Petterson 2015). Johnson et al. (2017) reported the carbohydrate content in six varieties of *L. angustifolius* to range between 7.3 and 9.4%. In another study, the carbohydrate content in *L. albus* and *L. luteus* was less than 3% (Martínez-Villaluenga et al. 2006). Raffinose family oligosaccharides (RFO) make up the majority of carbohydrates in *L. angustifolius* kernels, which are α-galactosyl derivatives of sucrose that cannot be metabolized by monogastrics and are well-known for their flatulence-causing effects (Johnson et al. 2017). The main galactosides present in lupin seeds are raffinose, stachyose and verbascose (Mohamed and Rayas-Duarte 1995; Trugo et al. 2016; Kaczmarska et al. 2017). RFOs are reported to range from 7 to 15% in raw lupin seeds (Huyghe 1997; Martínez-Villaluenga et al. 2006). Many studies have reported fermentation, germination, enzyme treatment, dehusking, soaking and cooking to reduce or remove α-galactosides present in lupins (Martínez-Villaluenga et al. 2006; Kaczmarska et al. 2017). In addition, removal of α-galactosides by these means also aids in extracting the α-galactosides separately, which can be then purified and used as prebiotic (Martínez-Villaluenga et al. 2006). According to most of the studies published so far, lupins are devoid of available carbohydrates, and the major carbohydrates in mature seeds are oligosaccharides and non-starch polysaccharides, primarily resulting from cell walls (Trugo et al. 2016; Johnson et al. 2017). Thus, the glycaemic index (GI) of lupin food ingredients is almost close to zero. In a study done by Hall et al. (2005b), addition of lupin flour into standard white bread reduced the GI significantly compared to white bread without compromising its palatability. The low glycaemic effect of lupin flour has encouraged the use of lupin flour in many flour-based food products as a source to reduce its GI value (Ruales et al. 1988; Johnson et al. 2003; Archer et al. 2004; Hall et al. 2005b).

Lupin is a rich source of dietary fibre, with a fibre content higher than that of other legumes including soybean (Martínez-Villaluenga et al. 2006; Johnson et al. 2017). In lupins, 30–40% of the kernel weight is represented by dietary fibre (Hall et al. 2005b; Martínez-Villaluenga et al. 2006). The fibre content in lupin varies widely among species and cultivars (Martínez-Villaluenga et al. 2006; Carvajal-Larenas et al. 2016; Musco et al. 2017). Based on reported literature, Carvajal-Larenas et al. (2016) reported whole raw *L. mutabilis* to have the lowest fibre content of 8.2% (dw) among the four domesticated species, while whole raw *L. angustifolius* had the highest value of 16% (dw) on average. The total dietary fibre in lupin seeds are generally above 30%, with >80% of it being insoluble and <10% soluble fibre (Mohamed and Rayas-Duarte 1995; Martínez-Villaluenga et al. 2006; Belski 2012; Musco et al. 2017).

The characteristics of lupin carbohydrates make it a useful material in producing special high-fibre food for humans (Turnbull et al. 2005; Johnson et al. 2017). Lupin kernel fibre extracted from Australian sweet lupin (*L. angustifolius*) is available as a novel fibre-rich food ingredient (Hall et al. 2005a), which has been shown to aid in the production of palatable fibre-enriched products such as baked goods and pasta (Johnson et al. 2017). Consumption of lupin kernel fibre-enriched foods has demonstrated several health beneficial properties (Hall et al. 2005a; Lee et al. 2009, 2018; Kapravelou et al. 2013). In a study by Hall et al. (2005a), consumption of lupin kernel fibre diet favourably modulated the serum lipid profile in healthy men compared to the control diet with no fibre.

Lupin Seed Oil

The fat content and fatty acid profile of lupin seed varies considerably among species (Erbaş et al. 2005; Trugo et al. 2016; Curti et al. 2018). The oil content of lupin varies from 4 to 21% (Erbaş et al. 2005; Chiofalo et al. 2012; Curti et al. 2018). In general, the oil level in lupin is much lower than that of soya (Dijkstra 2015; Liu 2015), however, is significantly higher than most other legumes (Sathe 2016). *L. luteus*, *L. angustifolius* and *L. albus* have relatively low oil contents, whereas *L. mutabilis* has a higher oil content which is almost similar to soybean in some varieties (Sweetingham and Kingwell 2008; Trugo et al. 2016). Thus, *L. mutabilis* is considered as a potential material in lupin oil production, and present genetic breeding programmes are focused on obtaining seeds with higher oil yields. As reviewed by Carvajal-Larenas et al. (2016), the lipid content in *L. luteus*, *L .angustifolius*, *L. albus* and *L. mutabilis* is reported to be 5.5, 6.3, 11.2 and 18.9 g/100 g (dw), respectively. There is no significant difference in the lipid content of *L. luteus* and *L. angustifolius*, and the difference between varieties including *L. albus* is also very marginal (Fraser et al. 2005; Chiofalo et al. 2012; Musco et al. 2017). However, the oil content of *L. mutabilis* varieties shows a significant variation ranging from 13.0 to 24.6 g/100 (dw) which was attributed partly to genetic and agronomical factors (Carvajal-Larenas et al. 2016).

Generally, lupin oil is low in saturated fatty acids (SFA) and is a good source of unsaturated fatty acids (USFA) (Erbaş et al. 2005; Calabrò et al. 2015; Trugo et al. 2016). The saturated fatty acid content in lupin oil is lower than many vegetable oils including soybean, sesame, olive and wheat oil (Erbaş et al. 2005). Among the unsaturated fatty acids, oleic (ω-9, 18:1) and linoleic (ω-6,18:2) are the predominant fatty acids in all four species followed by alpha linolenic acid (ω-3, 18:3) (Chiofalo et al. 2012; Carvajal-Larenas et al. 2016; Trugo et al. 2016). The individual fatty acid profiles differ depending on the species where *L. angustifolius* and *L. luteus* are particularly rich in linoleic acid (Chiofalo et al. 2012; Curti et al. 2018) and *L. albus* and *L. mutabilis* contain high levels of oleic acid (Carvajal-Larenas et al. 2016; Trugo et al. 2016). Particularly, *L. albus* and *L. luteus* contain very high level of linolenic acid (Trugo et al. 2016; Musco et al. 2017). Based on the fatty acid profile, the oil derived from *L. mutabilis* and *L. albus* is reported to show more similarity towards peanut oil (Erbaş et al. 2005), while *L. angustifolius* and *L. luteus* resemble corn oil (Trugo et al. 2016). A low SFA and high PUFA level in diet is considered beneficial to health; especially the ratio between ω-6 and ω-3 fatty acids is an important determinant in the prevention of several metabolic diseases including coronary heart diseases (Simopoulos 2002, 2016; Huerta-Yépez et al. 2016). van de Noort (2017) reported the ω-6 and ω-3 fatty acids content in lupin to have similar health benefits as in soy oil. In a study done by Chiofalo et al. (2012), *L. albus* had the lowest ω-6/ω-3 polyunsaturated fatty acid ratio (1.97) followed by *L. angustifolius* (3.75) and *L. luteus* (5.37).

Minerals and Vitamins

Information on minerals and vitamins in lupin seed is scarce. As reviewed by Carvajal-Larenas et al. (2016) the mineral (ash) content in lupin species fluctuates between 3.0 and 3.9 g/100 g dry matter with *L. mutabilis* having the highest and *L. angustifolius* having the lowest values. Typical mineral contents reported in lupin are (mg/100 g dw) calcium between 120 and 330; magnesium 90–330; phosphorus 210–900; sodium 3–11; potassium 280–1400; manganese 2.1–377; iron 2.1–15; zinc 2.2–8.2; and copper 0.25–1.2 (Petterson 2015; Carvajal-Larenas et al. 2016). In general, compared to cereals, lupins have a higher Ca and P content (Carvajal-Larenas et al. 2016). In terms of manganese content, a wide variation is observed in *L. albus* (2.3–377 mg/100 g dw) when compared to other major lupin species (2.1–8.6 mg/100 g dw) (Petterson 2015; Carvajal-Larenas et al. 2016). The essential mineral content in raw mature lupin seeds is almost similar to raw mature pinto beans except for the level of zinc being higher in lupin (United States Department of Agriculture (USDA) National Nutrient Database undated, cited from Johnson et al. (2017)).

Only few studies have reported the vitamin content in lupin seeds. Similar to mineral content, the vitamin content in lupin is also reported to be closely related to the level in pinto beans (USDA, National Nutrient Database undated, cited from Johnson et al. (2017)). Lupin seeds can be considered as a fairly good source of vitamin B. The consumption of 100 g of lupin seeds is capable of satisfying approximately 50% of thiamin (B1), 30% of niacin (B3) and 20% of riboflavin (B2) daily requirement for a diet of 2000 kcal/day (Erbaş et al. 2005). According to Erbaş et al. (2005), lupin had the highest niacin content and the lowest thiamin content than haricot bean, lentil, soybean and wheat. However, riboflavin content was higher than wheat and haricot bean but was lower than lentil and soybean. In terms of vitamin E, γ-tocopherol was the main isomer of vitamin E in lupins (Martínez-Villaluenga et al. 2006). *L. luteus* cv. 4486 and *L. albus* var. Marta had higher tocopherol content and vitamin E activity than *L. luteus* cv. 4492 and *L. albus* var. Multolupa (Martínez-Villaluenga et al. 2006). The vitamin E content in lupin oil is nearly similar to that of soybean, however, lower than that of sunflower and rapeseed oil (Johnson et al. 2017). The presence of vitamin C was only detected in *L. albus* var. Multolupa and *L. luteus* cv. 4486 (Martínez-Villaluenga et al. 2006). In terms of carotenoids, the presence of zeaxanthin and β-carotene in lupins has been reported with *L. mutabilis* and *L. luteus* having higher total carotenoids level than *L. angustifolius* and *L. albus* (Mohamed and Rayas-Duarte 1995). Another study reported the presence of lutein, zeaxanthin and α-carotene in some commercial varieties of *L. angustifolius* grown in Western Australia with *L. angustifolius* var. Mandelup having the highest total carotenoid content of 2.01 mg/100 g (dw) (Johnson et al. 2017).

Phytochemicals

The predominant phytochemical group present in lupins is the phenolic acids and flavonoids. Unlike other legumes, in lupins, flavonoids are the predominant group over phenolic acids (Siger et al. 2012; Johnson et al. 2017; Magalhães et al. 2017a, b). The presence of phenolic acids such as protocatechuic, *p*-hydroxybenzoic, chlorogenic, vanillic, *p*-coumaric and ferulic in lupin seeds is reported (Lampart-Szczapa et al. 2003; Siger et al. 2012). Siger et al. (2012) analysed three lupin species cultivars, *L. albus,* cultivars Butan and Boros; *L. luteus,* cv. Parys and Lord; and *L. angustifolius,* cv. Bojar and Zeus, and found Parys (*L. luteus*) (731.14 mg/100 g dw) to have the highest total phenolic content (TPC) while the lowest was found in Butan (*L. albus*) (491.51 mg/100 g dw). Protocatechuic acid was the predominant acid in *L. luteus* (up to 73.60 mg/kg dw), whereas in *L. angustifolius* it was *p*-hydroxybenzoic acid (about 43 mg/kg dw). In addition, gallic, caffeic and *p*-coumaric acids were also detected. Apigenin-6,8-di-*C*-ß-glucopyranoside and apigenin 7-*O*-ß-apiofuranosyl-6,8-di-*C*-ß-glucopyranoside were the two most abundant flavonoids in lupins with *L. luteus* having the highest content of the apigenin glycosides while *L. albus* the lowest (Siger et al. 2012). The abundant presence of apigenin-6,8-di-C-glucoside (rarely present in dietary plants) was also reported in wild and cultivated *L. albus* species (Karamać et al. 2018).

In contrast to other seed legumes, the testa of lupins contain lower level of polyphenols than the cotyledon (Lampart-Szczapa et al. 2003; Ranilla et al. 2009). The seed coats of legumes such as chickpea, faba bean, field pea, lentil and mung bean are reported to contain more phenols than the whole seeds (Zhong et al. 2018; 2019a, b). Ranilla et al. (2009) reported the total polyphenolic content in seed coats of *L. mutabilis*, *L. albus* and *L. angustifolius* (1.02–4.49 mg catechin equivalents (CE)/g dw) to be far more less than in cotyledons (7.38–12.42 mg CE/g dw). The seed coat of *L. angustifolius* was mainly comprised of three flavones (apigenin-7-*O*-β-apiofuranosyl-6,8-di-C-β-glucopyranoside, vicenin 2 and apigenin-7-*O*-β-glucopyranoside), one isoflavone (genistein) and one dihydroflavonol derivative (aromadendrin-6-C-β-D-glucopyranosyl-7-*O*-[β-D-apiofuranosyl-(1→2)]-*O*-β-D-glucopyranoside) and several hydroxybenzoic and hydroxycinnamic acid derivatives (Zhong et al. 2019a, b).

The phytochemical level in lupins changes during processing and digestion. Germination of lupins significantly changes the phenolic composition (Dueñas et al. 2009). Isoflavones (genistein), flavones (luteolin and diosmetin) and dihydroflavonols in free and conjugated forms were identified in raw and germinated *L. angustifolius* seeds, and germination favourably increased the flavonoid content. Among these, flavones were the largest subclass of phenolic compounds in *L. angustifolius* seeds (Dueñas et al. 2009). Digestion resulted in a lower level of total phenolic compounds compared to its undigested counterparts in *L. albus*, *L. luteus* and *L. angustifolius* (Czubinski et al. 2019). Czubinski et al. (2019) reported digestion to result in a modified phenolic profile and increased bioaccessibility (90%).

Antinutritional Compounds

Generally compared to other plant sources, legumes are a rich source of antinutritional compounds. Like most grain legumes, lupins also contain antinutritional factors such as phytic acid, oligosaccharides, trypsin inhibitors, lectins and saponins; however, the concentrations are significantly lower than those in other legume species (Pastor-Cavada et al. 2009; Carvajal-Larenas et al. 2016).

Alkaloids, especially belonging to the group of quinolizidine, are the main antinutritional substances found in lupin seeds (Musco et al. 2017). Quinolizidine alkaloids (QAs) contribute to the bitter taste of lupin and are toxic when consumed above the recommended level (Frick et al. 2017). The QAs reportedly found in lupin are lupinine, lupanine, sparteine, lupinidine, hydroxylupanine, anagirine, monolupine, termophsine, puziline, angustifoline and others (Sujak et al. 2006; Carvajal-Larenas et al. 2016; Frick et al. 2017). The alkaloid profile of lupin is dependent upon species where each lupin species is reported to have a characteristic alkaloid profile (Frick et al. 2017). For example, the principal alkaloid in *L. albus* and *L. angustifolius* is lupanine, and sparteine is found abundantly in *L. luteus* (Magalhães et al. 2017a). Lupanine and sparteine, the most common QAs in lupin, have proven to be the most toxic to both humans and laboratory animals (Resta et al. 2008; Frick et al. 2017).

As reported by Carvajal-Larenas et al. (2016), the total alkaloid content in lupins ranges between 0.186 and 2.8 g/100 g (dw) with *L. albus* having the lowest and *L. mutabilis* having the highest value. The presence of QA compounds has been a major concern in the utilization of lupin for human consumption (Kohajdová et al. 2011). The wild-type lupins contain more than 150 alkaloids representing the quinolizidine, piperidine and indole groups. However, through breeding programmes, the concentration in domesticated species are reduced to a safe level as low as 0.02% (Mohamed and Rayas-Duarte 1995). The maximum permitted level for total alkaloids in lupin flours and derived products by the health authorities of the UK, France, Australia and New Zealand is 0.2 g/kg dry matter (Resta et al. 2008; Magalhães et al. 2017a). The first alkaloid-free lupin was produced in the early 1930s by von Sengbusch (Huyghe 1997). In general, lupins having an alkaloid content below 0.05% are called sweet lupins (Calabrò et al. 2015; Magalhães et al. 2017a). The Australian sweet lupins are generally regarded to contain an alkaloid level usually below 0.02%, with a characteristic alkaloid profile comprising 42–59% lupanine, 24–45% 13-hydroxylupanine, 7–15% angustifoline and 1–1.5% α-isolupanine (Belski 2012).

In terms of other antinutritional factors, lupins contain very low levels of lectins, saponins, trypsin inhibitors and phytoestrogens which is of little concern (Ruiz-López et al. 2000; Carvajal-Larenas et al. 2016). The phytic acid and saponin content reported for lupin varies from 1.11 to 2.74 g/100 g (dw) and up to 0.05–1.7 g/100 g, respectively (Múzquiz et al. 1989; Ruiz-López et al. 2000; Karnpanit 2016). In terms of isoflavones, Ranilla et al. (2009) reported total isoflavones in *L. mutabilis* species to range from 9.8 to 87, 16.1 to 30.8 and 1.3 to 6.1 mg/100 g fresh weight in seed coat, cotyledon and hypocotyl fractions, respectively; whereas no isoflavones were detected in *L. angustifolius* and *L. albus*. In addition, the presence of minute amounts of cyanogenic compounds, hemagglutinins and trypsin inhibitor activity has been reported. However, it is not considered to be significant enough to cause any antinutritional effect (Múzquiz et al. 1989; Carvajal-Larenas et al. 2016).

Furthermore, lupins can trigger allergic reactions in humans. So far, worldwide at least 151 cases of lupin allergy as a result of immunoglobulin E (IgE)-mediated allergic reactions to lupin, either as a primary response or as a result of cross-reactivity with other legumes, have been reported (Sirtori et al. 2011; Jimenez-Lopez et al. 2018). In the EU, lupin has been declared as a source of allergens, and the European Food Safety Authority (EFSA) has made it mandatory to be mentioned on food labels (Sirtori et al. 2011; Bingemann et al. 2019). Lupin proteins are the main implicated agents in these allergenic reactions (Guillamón et al. 2010; Sanz et al. 2010; Jimenez-Lopez et al. 2018). Although not fully characterized yet, the major proteins responsible for the allergenic sensitization in lupin are Lup-1/β--conglutin and Lup-2/α-conglutin. The role of γ-conglutin and δ-conglutin is reported to be minor (Guillamón et al. 2010; Sanz et al. 2010).

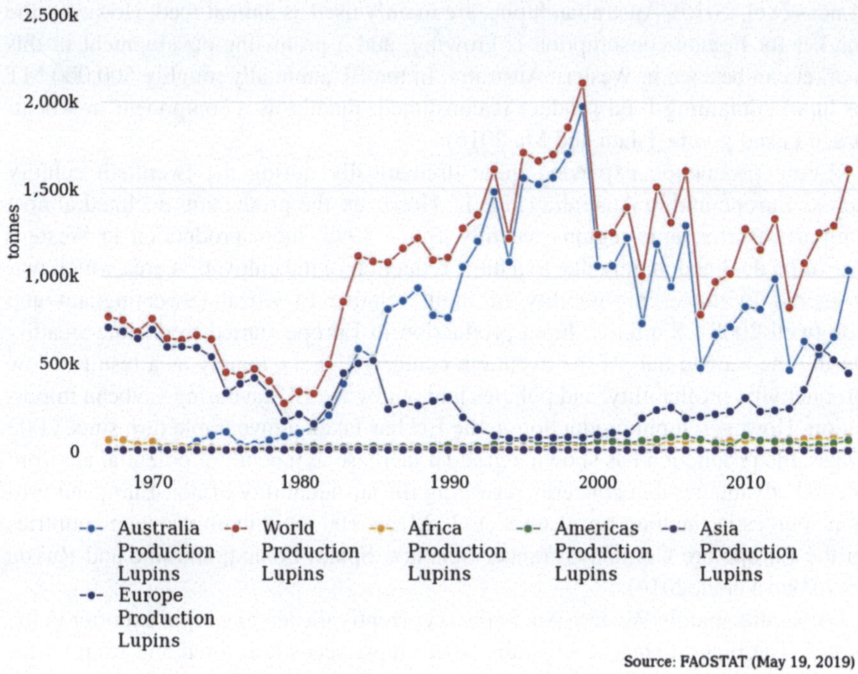

Fig. 1 Lupin production (MT) of major regions over the past 50 years. Extracted from FAO (2019)

Overview of Production, Export and Import

According to FAO (2019), more than 1,610,969 metric tons (MT) of lupin were produced in 2017. The top 10 lupin-producing countries in 2017 were Australia (about 103,1425 MT), followed by Poland (around 168,678 MT), the Russian Federation (161,680 MT), Morocco (64,355 MT), Germany (52,800 MT), Chile, Ukraine, France, Peru and South Africa (FAO 2019).

Over the past 30 years, Australia has been the leading producer of lupins accounting to about 80–85% of total lupin produced worldwide followed by the European region (Fig. 1). Until the early 1980s, the European region dominated the lupin market. From thereon, Australia took on, and the production of lupin in Australia has increased dramatically. The average national production of lupin in Australia in recent years is around 750,000 MT per year. The main lupin-producing areas in Australia are Western Australia, New South Wales, Victoria and South Australia. Approximately 70% of Australia's lupin production comes from Western Australia (Islam and Ma 2016). *L. angustifolus* and *L. albus* are the main varieties grown in Australia, the former representing more than 80% of the total area sown to lupin in Australia. The typical European and South American varieties are *L. luteus* and *L. albus*,respectively (Duranti et al. 2008; Islam and Ma 2016). At present, *L. mutabilis* (sweet) has been brought into cultivation in some parts of South America

(Lucas et al. 2015). Australian lupins are mainly used as animal feed. However, the market for human consumption is growing, and a promising development in this market can be seen in Western Australia. In the EU, annually roughly 500,000 MT of lupin-containing food product is consumed, mainly as a component in wheat-based baked goods (Islam and Ma 2016).

Lupin production expanded quite dramatically during the twentieth century across Europe and in Australia (Fig. 1). However, the production declined almost similarly in the same regions recently. Since 1999, lupin production in Western Australia declined mainly due to a huge reduction in the cultivation area which was triggered by lower profitability of lupin relative to wheat (Sweetingham and Kingwell 2008). Similarly, lupin production in Europe started to decline steadily during the second half of the twentieth century (Fig. 1), mainly as a result of low productivity, profitability, and policies laid out by the EU favouring soybean importation. However, lupin production in the EU has taken a favourable turn since 2003 where the production has shown a gradual increase as a result of potential environmental advantages and concerns regarding the sustainability of non-European protein sources in the long run (Lucas et al. 2015). The main lupin-growing countries in the Europe are Germany, France, Benelux, Spain, Poland, Ukraine and Russia (Sedláková et al. 2016).

Australia, mainly Western Australia, is currently the leading lupin exporter in the world. The biggest silos in Australia where lupin seeds are stored and from where they are distributed to all over the world are located in Western Australia (Sedláková et al. 2016). Approximately 90–95% of the lupin produced in Australia are exported all over the world. From 2000 to 2005, Australia exported approximately 41% of its annual lupin production accounting to about 2% of the total volume of Australian exports of grains and oilseeds. The main export destinations of Australian lupins have been the EU, Japan and Korea. The top five export destinations of Australian lupin in 2010–2011 were South Korea, Japan, the Netherland, Malaysia and Germany (Australian Government 2013).

Postharvest Processing

Compared to other legumes, generally, postharvest processing of lupins is time-consuming and labour-intensive. Commonly, harvesting of lupins is done soon after maturity. Delayed harvesting results in lodging, pod shattering and pod drop (Azeze et al. 2016). Generally, harvesting is done as soon as the moisture content reaches 14%. However, in Europe, due to slow maturity of lupins, at times it is harvested with moisture level as high as 30% (French 2015). After harvest, the grains are stored in bags or silos and are sold immediately or else stored for future sales. Based on the final use, lupins are subjected to drying, dehulling, milling and fractionation. Debittering is performed in bitter varieties to specially reduce the alkaloids content to a safe level. Debittering can be performed through biological processes (fermentation, germination), acidic or alkaline treatments and by soaking the seeds in abundant water (Carvajal-Larenas et al. 2016; Magalhães et al. 2017a).

Drying

Drying of lupins takes place just after harvesting and before threshing, storage and primary processing. Drying is an important aspect in lupin and other legume processing to avoid postharvest losses incurred by insect and mould infestation and to preserve grain quality. Furthermore, drying also facilitates further processing. The prime objective of drying is to remove excess moisture. Sun-drying is the most common drying method practiced worldwide in tropical and subtropical countries. Normally lupins are dried in the field just after harvesting by spreading the grains in the field in a thin layer to ensure adequate exposure to sun and wind for drying. Drying is carried out until the final moisture content of lupin seed is below 13%.

Dehulling

Before dehulling, the seeds are first sorted, graded and made free of any extraneous materials using a vibrating screen or metal detector. Compared to most other grains, the relatively high thickness and hardness of the lupin testa necessitates the dehulling process in lupins before further processing and also makes the dehulling process a bit harder. The hulls are removed from whole seeds by passing it through a dehuller. The simplest means of dehulling lupin is by using a ripple flow mill or by using a tangentially abrasive dehulling device. Finally, the removed hulls are separated from the cotyledons by air classification. During this process, the germ fraction of the seed is lost along with the hulls (Petterson 2015; Villarino et al. 2016; Johnson et al. 2017).

Milling

Milling of the dehulled cotyledons results in the production of lupin flour with different particle size. Initially, whole lupins are milled with the hull resulting in flour that is too coarse. Normally, the dehulled seeds are hammer and/or roller milled to produce meal or flour for animal feed. The milling process is of utmost importance in producing flour of optimal quality for food use. For food applications, a refined milling process is used to produce high-quality lupin flour with minimal heat damage. Processing of lupins cannot be done using conventional mills, such as those used to produce wheat flour, as the high temperature generated during milling tends to burn the oil in lupin seeds resulting in a rancid flour (http://grdc.com.au). After milling, the flour is sieved to particle sizes ranging from <150 to >600 μm resulting in the production of chips, flakes, grits, meal and flour (Villarino 2014; Johnson et al. 2017). Based on the requirement, these products are used in various food applications. The particle size of milled lupin plays an important role in determining the quality of the final product. For example, an increase in the particle size of lupin flour significantly increased the amount of lupin flour incorporation in bread before affecting its quality (Villarino et al. 2015).

Fractionation

Lupin flour can be further fractionated into valuable products such as protein isolates, purified dietary fibre and oil- and water-soluble by-products (e.g. whey proteins and oligosaccharides). Fractionation generally involves wet grinding of the cotyledons followed by separation of fractions depending on solubility at different pH or in different solvent systems (Petterson 2015). At present, food manufacturers have paid attention in using lupin fractions as ingredients in food applications.

Lupin as a Food Ingredient

Lupin Kernels, Flour, Flakes, Grits and Meals

Lupin flour is commercially produced in Australia, Austria and Germany and is used in bread, pasta, snacks and bakery, confectionary and meat products. In Australia, Irwin Valley Pty Ltd., Coorow Seeds and Mirfak Pty Ltd. are involved in commercial lupin flour production (http://www.lupins.org/products/#). Flour mixes such as enzyme-active flour and fibre-rich flour are being sold in the Netherlands. The CBH Group Board-owned plant in Australia manufactures a number of lupin food products including lupin flour, flakes, hull, kibble and splits (https://www.cbh.com.au/customers/commodities-and-products). Compared to lupin flour, the particle size of flakes, grits, crumbs and meal is higher. Besides Australia, lupin grits are also produced in Austria, Chile, the Netherlands and France, while flakes are produced in Australia and France (Johnson et al. 2017). There are no standards defining the particle size for these products. However, according to US standards for soy, the flour should pass through a mesh with an opening of 149 μm (#100 mesh), and grits should pass through mesh openings with 2000 to 177 μm (#10 to #80 mesh) (Johnson et al. 2017). Lupin flakes are manufactured in a different way to have a flake-like geometry with a particle size of 1.5–3 mm. Lupin grits and flakes are used in some food products such as bread, snack and cereal products as a texture, taste and shelf life improver (Johnson et al. 2017). Moreover, chips from lupins have been proposed as a new product in the Food Ingredients Europe, which can be used as a nutritional ingredient in salads and soups (Kohajdová et al. 2011).

Lupin Protein Concentrates and Isolates

In addition to being used as flour directly, the fractions of lupin including protein isolates and protein concentrates are used in various food applications including bread, pasta, snacks, drinks, yoghurt, sauces, infant formula and meat products (Petterson and Fairbrother 1996; Islam and Ma 2016; Johnson et al. 2017). Lupin protein isolates can be extracted in multiple ways. The most common ways used to isolate lupin proteins from lupin flour are via isoelectric precipitation (Sironi et al. 2005; Wong et al. 2013) or ultrafiltration (Wong et al. 2013). Protein is conventionally extracted by alkaline solubilization (e.g. pH 9) of protein from previously defatted or non-defatted lupin flour/wet milled kernels, removal of the insoluble dietary fibre fraction by centrifugation, and followed by acid precipitation of the main classes of globulins (α- and β-conglutins) at pH 4.5. Next the acid-precipitated proteins are recovered from the acid-soluble whey fraction by centrifugation. Finally, both the isolated acid-precipitated protein and the fibre fractions are made into powdered ingredients by drying (Villarino et al. 2016; Johnson et al. 2017). The acid-soluble whey fraction, once considered a waste, is also presently dried to be used as food ingredients, that is, as foaming agents, bioactive peptides and prebiotic oligosaccharides (Sironi et al. 2005; Wong et al. 2013; Villarino et al. 2016). Moreover, the acid-soluble protein fraction of lupin has been reported to be rich in γ-conglutin (Wong et al. 2013), a protein fraction that has showed promising antidiabetic activity in vitro and in vivo (Magni et al. 2004; Bertoglio et al. 2011; Terruzzi et al. 2011).

Commercially, lupin protein fractions can be classified as protein isolates, having >90 g protein/100 g (dw), protein concentrates with 60–90 g protein/100 g (dw) and protein-meal (<60 g protein/100 g). Lupin protein isolates are reported to show good emulsifying, solubility and foam- and gel-forming properties, which can beneficially influence the structure, texture and sensory properties of food and can be used as milk, egg and meat substitutes (Sironi et al. 2005; Wong et al. 2013). Burgos-Díaz et al. (2019) have reported the potential applications of the protein isolate AluProt-CGNA* extracted from *L. luteus*, cv., as a food-grade Pickering emulsion stabilizer. Furthermore, the particular protein isolate is also being used as a functional ingredient for preparing different food prototypes (Burgos-Díaz et al. 2019). In addition, commercial protein concentrates are reported to show improved emulsifying properties and modify the crispness and stickiness of cake and batters (Johnson et al. 2017). The lupin protein fractions, α-, β- and δ-conglutin, have been categorized as E fraction due to its excellent emulsifying properties, which can be successfully used in fat-reduced spreads, spread cheese, emulsified meat products, convenience foods and salad dressings. The γ-conglutin-rich fraction of lupin protein has promising foaming properties (F fraction) and can be used as an replacement for egg yolk in a number of food products, such as confectionery, marshmallows, egg glazes and ice creams (Sironi et al. 2005). Additionally, it should be noted that the concern regarding bitter-tasting alkaloids in lupins is minimal when protein isolates or concentrates are used in food manufacturing since majority of the alkaloids will be removed during the preparation of protein isolates (Islam and Ma 2016).

Lupin Fibre

Fibre derived from lupin hull is commercially available in Australia, the Netherlands, Chile and Germany (Johnson et al. 2017). Avelup® in Chile make a dietary fibre product from the hulls of *L. albus* which is used in bakery, cereal bars and snacks and is also marketed as a fat replacer and a dietary fibre source in meat products (https://avelup.cl/productos.html). In addition, milled hulls are also marketed as bran, which in addition to increasing the dietary fibre content in breads is also involved in improving the shelf life (Johnson et al. 2017). Several bakers in Australia have included hulls of *L. angustifolius* in bread for many years to produce fibre-rich breads (http://www.lupins.org/products/#). Lupin kernel fibre derived from the endosperm of *L. angustifolius* is used as an ingredient in foods such as baked goods in Australia (Johnson et al. 2003). This fibre has an 'invisible' nature, which can be considered valuable in the development of palatable high-fibre food products. Furthermore, compared to other insoluble fibres, lupin kernel fibres have also been reported to show unusually high water-binding capacity and viscosity (Johnson et al. 2003; Turnbull et al. 2005).

Lupin Oil

Lupin oil can be extracted from the grain by enzyme-assisted aqueous extraction with extrusion pretreatments (Jung 2009) or solvent extraction (Sbihi et al. 2013). According to Johnson et al. (2017), lupin oil extracted from the seeds of *L. angustifolius* is reportedly sold by a German company, which is recommended for use in bakery goods, meat products and pasta manufacture. However, due to the low level of oil in most of the lupin varieties and due to its unusually favourable fatty acid composition, lupin oil may remain a high-cost niche product, and its mass scale production in the near future seems unrealistic.

Product Development

Historically, lupins have been used as food in both the Mediterranean and the Andean cultures (Dervas et al. 1999). At present, in some part of the world, lupin seeds are still consumed after boiling or after washing with abundant water to remove the antinutritional seed alkaloids. For example, lupin seeds are consumed raw as a snack (salted or unsalted) in the Middle East after soaking in water, and in some European countries, lupin seed pickle is produced (Tizazu and Emire 2010; Awad et al. 2014). In Ethiopia, white lupin is consumed as a snack and is used in local soup and alcohol preparations (Azeze et al. 2016). In Portugal, Spain, France, Italy, Greece, Egypt and Algeria, debittered lupins preserved in brine (called *lupini*) are consumed as a snack food. At present, *lupini* preserved in a salt solution is available packed in jars (http://www.lupins.org/products/). Though lupin has a unique nutrient composition that will benefit humans, until recent times, the main use of lupin has been in the animal husbandry sector.

There is increasing interest in the use of lupins in food industry as an alternative or substitute to other grain legumes especially soybean (Doxastakis 2000), which can be attributed to the almost similar nutrient composition of these two legumes. One of the main advantages lupins have over other grain legumes is its low content of antinutritional compounds. Lupin flour, protein concentrates and isolates can be applied as a substance for enriching different kinds of food systems such as bakery products, pasta, noodles, fermented foods, milk and meat products. Lupin foods are commercially manufactured in Europe, North America and Australia (http://www.lupins.org/products/). Lupin ingredients found its way into the European diet since the early 2000s mainly as an ingredient in wheat-based baked goods. As of 2015, among the total bakery products produced with lupin flour, 22% came from Germany followed by France holding a share of 21% and Italy 11% (Johnson et al. 2017). Though the suitability of lupin for human consumption was endorsed by Food Standards Australia New Zealand in the late 1980s, to date, the use of lupin as human food in Australia has been limited compared to Europe where food producers have used it much more widely as an ingredient primarily in cereal-based foods (Johnson et al. 2017). Some examples of lupin-containing food products commercially available in Europe are lupine snacks, pasta, bread, cookies, lupine coffee and some vegetarian instant meals (Teshome et al. 2012).

Lupin and its fractions are commercially used as ingredients in many products including bread, crisp breads, cracker, cones, cookies, muffins, waffles, breakfast cereals, snacks, dips and mayonnaise, meat analogs, desserts and confectionary and in ready-to-eat meals, especially in Europe and Australia (Johnson et al. 2017). Commercially, a gluten-free lupin pan bread (10% lupin flour) and a sourdough rye bread (40% lupin flour) are available in Australia marketed as a niche healthy product (Villarino 2014). In addition, currently, a number of lupin-incorporated pasta products are also commercially available (Dervas et al. 1999). The water and oil absorption capacity, emulsifying and foaming properties and gelation capacity of lupin ingredients are considered valuable to the food industry in developing novel food products (Doxastakis 2000; Pollard et al. 2002). This part of the chapter will cover the use of lupin ingredients in food processing and will discuss its techno-functionality in food processing and consumer acceptability.

Bakery and Wheat-Based Value Added Products

Wheat flour is the main raw material used in baked and other starch-based food products owing to its desirable techno-functional properties. Gluten-forming proteins present in wheat flour provide industrially desired rheological and sensory properties for the end products (Dervas et al. 1999). However, recently health and nutritional issues associated with high wheat consumption, for example, in patients suffering from celiac disease (gluten intolerance) and diabetes have encouraged the substitution of gluten-free low glycaemic flours in food formulations (Hall et al. 2005b). Lupin is free of gluten (Villarino et al. 2016) and has shown promising effect against diabetes/hyperglycaemia (Fornasini and Baldeon 2012; Bouchoucha et al. 2016b; Tapadia et al. 2019). As a result, lupin flour and lupin proteins have found its way into bakery and wheat-based products aimed at producing gluten-free low glycaemic foods (Johnson et al. 2003; Hall et al. 2005b; Villarino et al. 2016). Studies have demonstrated the application of lupin flour and proteins in bakery (Dervas et al. 1999; Paraskevopoulou et al. 2010; Klupsaite et al. 2017) as well as other wheat-based value added food formulations (Doxastakis et al. 2007; Mahmoud et al. 2012). Lupin is used in the production of a range of products including bread, biscuits, cookies, cake, noodles, spaghetti, pasta, muffins and crisps (Villarino et al. 2016; Johnson et al. 2017).

In terms of food applications, the main use of lupin has been in the bakery sector. Several studies have reported the use of lupin flour, protein and fibre in bread making and its effect on quality (Dervas et al. 1999; Doxastakis et al. 2002; Johnson et al. 2003; Turnbull et al. 2005; Villarino et al. 2015). Incorporation of 5–10% lupin flour (concentrated lupin flour and defatted concentrated lupin flour) into medium strength wheat flour resulted in bread having almost similar quality parameters as the control with acceptable bread weight, volume, crumb structure and colour (Dervas et al. 1999). The same was observed in bread incorporated with lupin kernel fibre (total dietary fibre content 6.6 and 8.3%) where the breads were rated highly in terms of individual sensory attributes and the acceptability was similar to that of the control white bread having a total dietary fibre content of 3.5% (Johnson et al. 2003). However, in general, increasing level of lupin incorporation was associated with reduced overall acceptability and also markedly decreased the stability and the tolerance index of the dough and reduced the bread volume (Dervas et al. 1999; Doxastakis et al. 2002; Villarino et al. 2016), which was attributed to the disruption of the gluten matrix due to non-elastic nature of lupin proteins and high water-binding capacity of dietary fibre in lupin flour (Turnbull et al. 2005; Paraskevopoulou et al. 2010; Villarino et al. 2015, 2016). Paraskevopoulou et al. (2010) studied the impact of addition of lupin protein isolates (LPI) on wheat flour dough and bread characteristics. LPI addition increased dough development time, stability, resistance to deformation and the extensibility of the dough and impaired bread quality in terms of volume, internal structure and texture. Addition of gluten into the blends improved bread quality characteristics indicating the weakening effect of lupin proteins on wheat flour doughs due to dilution of the gluten structure by the added protein (Doxastakis et al. 2002; Paraskevopoulou et al. 2010). In general, studies published so far recommend maximum 10% substitution of lupin in wheat bread to increase dough and loaf quality and inhibit staling without affecting consumer acceptability (Dervas et al. 1999; Doxastakis et al. 2002; Villarino et al. 2016; Johnson et al. 2017).

Moreover, lupin has also been used in other baked goods such as muffins, cookies and brownies (Clark and Johnson 2002; Hall and Johnson 2006; Nasar-Abbas and Jayasena 2012; Rumiyati et al. 2015), cakes (Levent and Bilgiçli 2011; R. Ahmed 2014) and biscuits (Jayasena and Nasar-Abbas 2011) and other wheat-based foods including noodles (Jayasena et al. 2010b; Mahmoud et al. 2012) and pasta (Doxastakis et al. 2007; Martínez-Villaluenga et al. 2010; Jayasena and Nasar-Abbas 2012). Hall and Johnson (2006) studied the sensory acceptability of Australian sweet lupin (*L. angustifolius*) flour-incorporated bread, muffins, pasta and breakfast bars. All lupin flour-added products were rated within the acceptable half range where, in comparison to the control (wheat flour) lupin pasta, muffin and bread were rated lower in terms of texture, flavour and general acceptability. Interestingly, lupin flour addition did not reduce the acceptability of chocolate chip cookies (28% wheat flour replacement) and breakfast bars (100% wheat flour replacement) in comparison to control (wheat flour) indicating the potential use of lupin flour in these products. In contrast to this, enrichment of lupin kernel fibre significantly reduced the overall acceptability of breakfast bars including muffin and orange juice and did not change the overall acceptability of bread and pasta (Clark and Johnson 2002).

In biscuit manufacturing, lupin flour can be successfully incorporated into biscuits by replacing up to 20% of wheat flour to increase protein and dietary fibre contents (Jayasena and Nasar-Abbas 2011). Further increase in lupin content impaired flavour, taste, crispness and overall acceptability and also increased the hardness and fracturability of biscuits. However, the addition of lupin improved the colour of biscuits making it more attractive to the consumers. Same was observed in muffins where an improvement in colour was observed with up to 30% substitution of lupin flour with no significant change in taste, flavour, texture and overall acceptability (Nasar-Abbas and Jayasena 2012). Levent and Bilgiçli (2011) recommended 30% addition of lupin flour into gluten-free cake, and Ahmed (2014) suggested up to 10% substitution of lupin flour to wheat flour in cake to have minimal effect on overall acceptability.

Jayasena et al. (2010b) prepared lupin flour-added instant noodles to improve the protein and dietary fibre content. Addition of 20% lupin flour increased the protein content by 42% and dietary fibre by approximately 200% without affecting the sensory properties of the instant noodles. Textural characteristics including firmness, springiness, cohesiveness, gumminess and chewiness were not affected by lupin flour incorporation. There was no significant increase in the cooking losses up to 30% lupin incorporation. Lupin addition resulted in a distinct yellow colour in both cooked and uncooked noodles which increased the consumer likeness for colour. This may be attributed to the presence of lutein, zeaxanthin and β-carotene in lupin flour (Mohamed and Rayas-Duarte 1995). Similar results were observed by Rayas-Duarte et al. (1996) where the colour scores of spaghetti containing 15–30% lupin flour were much higher than those prepared by adding light buck wheat and amaranth. Mahmoud et al. (2012) and Doxastakis et al. (2007) reported incorporation of lupin protein into noodles and spaghetti (inclusion rate maximum 5% and 20%, respectively) to improve cooking loss, cooking time and cooking yield and also to have sensory values comparable to the control sample.

The main aim of adding lupin into cereal products has been to increase its nutritional properties, for example, to increase the protein and dietary fibre content without compromising consumer acceptability. Villarino et al. (2016) reviewed substitution of lupin flour in wheat bread (30–40%), biscuit (30%), muffin (60%) and chocolate chip cookies (28%) to significantly increase protein and dietary fibre content by 46–108% and 106–346%; 352 and 211%; 46 and 296%; and 51 and 316%, respectively. Furthermore, cereal proteins are deficient in lysine and have adequate amounts of sulphur-containing amino acids, whereas lupins are high in lysine and low in methionine. Thus, owing to their composition, lupin and wheat can be considered complementary with regard to their essential amino acids, and supplementing wheat with lupin can aid in satisfying the essential amino acid requirement (Dervas et al. 1999; Paraskevopoulou et al. 2012). For example, Doxastakis et al. (2007) reported LPI-containing spaghetti to have more available lysine than traditional semolina spaghetti. Therefore, addition of lupin into cereal products just not increases the protein content but also improves the quality of protein in the final product. In addition, several clinical studies have reported consumption of lupin-containing products to favourably modify the biomarkers of cardiovascular diseases and type 2 diabetes mellitus (Hall et al. 2005a, b; Lee et al. 2009, 2018; Bertoglio et al. 2011; Fornasini and Baldeon 2012; Harisa and Alanazi 2015; Bouchoucha et al. 2016a, b), a rising health concern worldwide.

Meat Analogs

To date, soybean ingredients (soybean flours, protein concentrates and isolates) are the main plant-based extenders used in comminuted meat products (Papavergou et al. 1999; El-sayed 2013; Danowska-Oziewicz and Kurp 2017). Formation of a functional protein matrix is essential in the production of comminuted meat products to ensure the final quality of the product (Alamanou et al. 1996; Drakos et al. 2007). Good emulsifying property and stability, water-binding capacity, gelation and cohesion of particles are some of the important characteristics required for the formation and stabilization of protein matrices (Alamanou et al. 1996). Some studies have reported the usefulness of lupin ingredients (flour, protein isolate and fibre) in making beef burger patties (El-sayed 2013), pork patties (Danowska-Oziewicz and Kurp 2017), frankfurters (Alamanou et al. 1996) and sausages (Papavergou et al. 1999; Archer et al. 2004). Lupins proteins possess good emulsifying and gel-forming properties (El-Adawy et al. 2001; Piornos et al. 2015) aiding in the stabilization of fat particles in comminuted meat products and thereby strengthening the structure of a processed/cooked product (Drakos et al. 2007). Furthermore, heating is reported to improve both foaming and emulsifying properties of lupin proteins (Pozani et al. 2002) and to improve the gel network structure of comminuted meat systems (Drakos et al. 2007). However, it should be noted that compared to soy protein isolates, lupin protein isolates form weaker gels with poor swelling index (Berghout et al. 2015).

Alamanou et al. (1996) studied the effect of lupin protein isolate addition on physicochemical properties, lipid oxidation and sensory quality of frankfurters. According to the study, addition of LPI increased the pH and the viscosity of the batter, improved water-binding capacity, reduced fat separation, increased processing yield and decreased purge accumulation indicating enhanced emulsion stability of batters. El-sayed (2013) replaced meat with lupine flour to be used as a binder in beef burger processing. According to the results, 5–7.5% of lupin flour incorporation into beef burger patties enhanced the physiochemical properties; total volatile basic-nitrogen, thiobarbituric acid value, pH value and water-holding capacity of the burgers. In comminuted meat production, meat batters should have good emulsion stability to avoid moisture loss and fat separation upon further processing. In addition, the ability to bind and retain meat juices during processing enhances mouthfeel and flavour in the final product. Addition of LPI (1–3%) into frankfurters reduced cooking loss by approximately 3.6% (Alamanou et al. 1996). Similar outcomes were reported by Danowska-Oziewicz and Kurp (2017) where addition of 3% lupin protein concentrate to pork patties reduced cooking loss by 3.33% compared to the control, and as to El-sayed (2013), addition of lupin flour into beef patties resulted in approximately 10% reduction in cooking loss compared to the control. However, in terms of firmness, studies have reported the contribution of lupin protein in the formation of protein matrix in comminuted meat products to be weaker than that of meat proteins (Alamanou et al. 1996; Danowska-Oziewicz and Kurp 2017). For example, increasing the lupin level in pork patties resulted in a significant decrease in the shear and compression force compared to the control containing only meat (Danowska-Oziewicz and Kurp 2017). Similarly, Alamanou et al. (1996) found increasing level of LPI to significantly affect the textural properties of frankfurters where a significant reduction was observed in hardness of the final product and inclusion of 2% LPI into fermented sausages resulted in lower firmness than the control (Papavergou et al. 1999).

El-sayed (2013) reported 5–7.5% of lupin flour incorporation into beef burger patties to have a similar effect on the sensory properties compared to the control sample containing only beef. In contrast, Danowska-Oziewicz and Kurp (2017) reported unpleasant flavour in lupin protein concentrate-added pork patties. The same have been reported by others where 3% inclusion of LPI in frankfurters (Alamanou et al. 1996) and 2% lupin seed flour in fermented sausages (Papavergou et al. 1999) resulted in unpleasant flavour and reduced consumer acceptability. Fruity, cheese-like, metallic, fatty, meat-like and hay-like flavour attributes have been already reported for lupin flour which might be the reason for the unacceptable flavour detected in lupin-containing food products (Bader et al. 2009; Danowska-Oziewicz and Kurp 2017). Though lupin flour resulted in unpleasant flavour, fermented sausages with LPI had similar sensory scores as the control (Papavergou et al. 1999). Several studies have reported fermentation to favourably modify the flavour profile of lupins (Schindler et al. 2011; Kaczmarska et al. 2018a, b). However, according to Alamanou et al. (1996) inclusion of 1–2% of LPI did not significantly affect the overall acceptability of frankfurters compared to control.

Furthermore, Archer et al. (2004) studied the application of lupin kernel fibre as a fat replacer in sausages. Compared to other insoluble fibres, lupin-kernel fibre, a product resulting after extraction of the protein, lipid and soluble carbohydrate fractions from lupin seeds, is reported to demonstrate unusually high water-binding capacity and viscosity properties (Turnbull et al. 2005), which can be used as a potential low-energy bulking agent and fat replacer in foods. In the study, one half of the fat in the sausage was replaced by fibre, and the lupin kernel breakfasts had 37% lower fat and 17% lower energy density compared to the full-fat diet. The lupin kernel diet had a more satiating effect than the full-fat diet and was rated neither acceptable nor unacceptable. However, similar to other meat products, the lupin kernel fibre-added sausage scored lower for flavour than the full-fat sausage.

Fermented Food Products

Studies have investigated the use of lupin in fermented foods, such as yoghurt, cheese, tofu and tempeh, as a replacement for cow's milk and soybean-based products (Petterson and Fairbrother 1996; Fudiyansyah et al. 2007; Jayasena et al. 2010a; Elsamani et al. 2014). To date, only preliminary laboratory-based investigations are performed regarding the development of fermented lupin-based products, and there are no reports on commercial preparations. Lupin milk is the sole ingredient in preparation of fermented foods. Unlike soybean milk, the production of lupin milk is quite challenging owing to the solubility and heat stability characteristics of lupin proteins. The main challenge in the production of lupin milk is that even exposure to mild heat treatments (e.g. pasteurization) to neutral aqueous lupin extracts tends to coagulate and sediment the protein. Australian researchers have encountered limited success in manufacturing lupin milk (Petterson and Fairbrother 1996). Other than that, off-flavour production during extraction and the retention of yellow carotenoid pigments in the milk poses great challenge in the production of lupin milk. Several authors have described the preparation of lupin milk using the alkaline and aqueous treatment methods (Jiménez-Martínez et al. 2003; John et al. 2007; Jayasena et al. 2010a; Elsamani et al. 2014). Generally, first lupins are debittered and dehulled, and the resulting cotyledons are ground with warm water, and the pH of the milk (filtrate) is adjusted to 6.5–7 using 0.1 M NaOH. Finally, the milk is heated in order to inactivate the lipoxygenase enzyme and then cooled to 4 °C for storage (Jiménez-Martínez et al. 2003; Jayasena et al. 2010a; Elsamani et al. 2014).

Elsamani et al. (2014) studied the effect of lupin milk addition on the physicochemical, microbiological and sensory properties of fresh and matured (75 days) soft cheese. Lupin incorporation increased the cheese yield compared to that of cow milk cheese, which was attributed to higher levels of fat and protein in lupin milk and low losses of total soluble solids in whey. The total bacterial count in cheese was not affected by lupin addition. In terms of sensory properties, except for 75% lupin-added cheese, no significant difference was recorded between pure milk and lupin-added fresh cheese with respect to taste and texture. Interestingly, incorporation of lupin milk in cheese significantly improved the flavour in both fresh and mature cheese, which may be due to production of lactic acid and flavour components such as acetaldehyde and volatile fatty acids. However, maturation significantly affected taste, texture and overall acceptability of lupin-containing cheese. Overall, incorporation of lupin milk at low concentration (25%) was found to significantly enhance the sensorial properties of both fresh and mature cheese compared to pure milk cheese (Elsamani et al. 2014). Same was observed in another study where with increasing lupin ratio in the cheese formula reduced overall acceptability, especially affecting the body and texture scores of the processed cheese (Awad et al. 2014).

In yoghurt production, lupin milk (Jiménez-Martínez et al. 2003; John et al. 2007), protein concentrates (Kuznetsova et al. 2014) and protein isolates (Hickisch et al. 2016) are used. As per Jiménez-Martínez et al. (2003), unflavoured, plain lupin yoghurt was scored as tasteless and unacceptable by the panelists. Furthermore, the yoghurt had to be supplemented with lactose to promote fermentation by lactic acid bacteria. Though the overall acceptance of strawberry-flavoured lupin-based yoghurt was similar to cow's milk yoghurt, it was scored poorly in terms of individual sensory attributes, colour, aroma, flavour and texture. However, beany flavour was absent in the lupin yoghurt. Hickisch et al. (2016) reported LPI to be an appropriate raw material for the production of a plant-based yoghurt alternative. Method of heating and microorganism type was found to play a critical role on the final quality of the product. Yoghurt prepared by UHT heating showed better rheological and textural properties and a lower tendency to syneresis compared to pasteurized yoghurt. Furthermore, John et al. (2007) reported yoghurt prepared from lupin milk to be superior to the commercial soybean-based yoghurt product in terms of sensorial properties. Despite the better gelling properties reported for soy proteins than lupins (Berghout et al. 2015), yoghurt produced from soy milk are reported to be firm and brittle (Donkor et al. 2007; Yang et al. 2012). It should be noted that LPIs exhibit weak gelling properties resulting in weak gels with a low number of bonds, which is considered more preferable in semisolid food applications (Berghout et al. 2015).

Tofu, tempeh, miso, natto and soy sauce are some of the common soy-based fermented products (O'Toole 2016; Mani and Ming 2017; Wang et al. 2019). Several authors have reported the application of lupin in the production of these products (Petterson and Fairbrother 1996; Jayasena et al. 2010a; Wolkers-Rooijackers et al. 2018). Since the quality of tofu is mainly determined by protein, lupin can be considered as a potential alternative for soybean owing to its almost similar protein content as soybean. Jayasena, Khu and Nasar-Abbas (Jayasena et al. 2010a) investigated the effect of lupin substitution on the quality characteristics of tofu. Tofu samples were prepared using a standard method by substituting lupin at 0, 20, 30, 40, 50 and 60% for soybean. Yield, chemical composition, texture, colour and sensory characteristics of the product were assessed. Lupin substitution significantly affected the yield, protein and fat content and the colour of the tofu samples, where yellowness ($L*$ and $a*$ values, $b*$ value) was significantly increased with increasing lupin level. However, lupin substitution did not affect the instrumental texture profile (hardness, cohesiveness, springiness and chewiness), and the authors recommended up to 40% lupin substitution in tofu without compromising the quality and sensory characteristics. Lupin tempeh preparation is almost simple as soybean tempeh preparation (Steinkraus 2018). Tempeh produced from lupin was rated above soy in terms of mouthfeel, flavour and colour (Fudiyansyah et al. 2007; Priatni et al. 2013; Steinkraus 2018). Tempeh is considered one of the main sources of vitamin

B12 for vegans (Wolkers-Rooijackers et al. 2018). Wolkers-Rooijackers et al. (2018) reported a significant increase in vitamin B12 level in lupin tempeh fermented using a mixed culture of *R. oryzae* and *P. freudenreichii*. Furthermore, Wickramasinghe (2017) reported for the first time the use of lupin in natto production. According to the results, the physicochemical and organoleptic properties of lupin natto and tempeh were similar and in some instances superior to natto and tempeh derived from soybean. It was also found the quality characteristics of lupin natto and tempeh to be mainly affected by fermentation time, source and microbial strain.

Future Trends in Processing and Product Development

Since the global demand for meat, dairy and fish products is on the rise and is no longer sustainable through animal products alone, there is a huge demand for plant-based protein alternatives. In addition, there has been a rising trend towards plant-based diet worldwide as a result of growing interest in the direction of vegetarian and vegan lifestyles along with rising concerns over animal welfare. This has also surged the demand for plant-based food products, and as a result, presently, plant-based food sector is growing at a double-digit rate. Soybean is the most widely used plant-based alternative for meat protein in foods. Lupin has been regarded as a promising alternative to soybean owing to its almost similar nutrient content and comparatively low level of antinutritional compounds (Arnoldi et al. 2007; Pastor-Cavada et al. 2009). One of the major advantages of lupins over soybean is its non-genetically modified (GM) status. Due to the concerns regarding GM products on health and the environment, lupins can be considered as a good alternative to fulfill the rising demand on natural food products. In addition to being used in soy-based food products, lupin ingredients, flour, fibre, protein concentrates and isolates are also used in different food systems such as bakery products, pasta, noodles, milk and meat products primarily to increase its nutritional and organoleptic properties. Given its high protein and fibre content, lupin flour is considered as an excellent raw material to supplement different food products (Doxastakis 2000; Johnson et al. 2017).

Techno-Functional and Sensorial Properties

The main challenges pertaining to the use of lupin in food products have been generally related to its techno-functionality and sensory properties. In most research work focused on supplementing lupin in cereal, milk and meat-based products, the most affected sensory attributes are flavour and texture. The undesirable aftertaste experienced in lupin-containing foods (e.g. grassy, beany, metallic, fatty, hay-like, meat-like and cheese-like) has been one of the main reasons limiting the use of lupin in food products (Bader et al. 2009). Since degradation of fat in lupin seeds mainly contributes to some of the undesirable flavours, the use of de-fatted lupin seeds and lupin protein isolates in foods and removal of free fatty acids, phospholipids and secondary plant metabolites can help to overcome this problem and enhance the sensory properties (Bader et al. 2011). In addition, bioprocessing, for example, fermentation and germination, has shown promising effect on masking the undesirable flavours originating from lupins and is also reported to enhance the flavour profile of the end product (Kaczmarska et al. 2018a, b).

Texture is the other important factor affected by the addition of lupin, which is attributed to the absence of gluten in lupins. Formation of the gluten matrix is one of the most important factors determining the textural properties of wheat-based foods (Villarino et al. 2016). There are several studies published with regard to addition of gluten-free flours into cereal-based foods and improvement of quality (Villarino et al. 2016), which may be applicable to lupin flour. For example, high-pressure treatment of non-gluten flours is reported to improve the quality of bread. High-pressure treatment facilitated denaturation of proteins leading to increased reactivity of sulphhydryl bonds and higher disulphide cross-linking thereby facilitating the development of the protein network (Villarino et al. 2016). This led to more acceptable breads. Apart from these issues, another major concern regarding the use of lupin ingredients in foods is the thermal characteristics of lupin protein. Lupin proteins are comparatively heat labile than soy proteins, and this poses a major obstacle in manufacturing processes thereby limiting the use of lupin protein isolates and concentrates in food applications. Modification of lupin ingredients to improve its functionality is a promising alternative. Technological parameters must be optimized in order to improve production and functionality of the proteins, in addition to making it cost-effective, sustainable and environmentally friendly. Further development in the field of food science and technology and material science will aid in the fractionation of proteins and other components with novel improved functionalities, modification of bioprocess-induced ingredients and formulation of food matrices to accomplish the successful incorporation of lupin in food products. In addition, protein extraction methods should also focus on recovery of lupin oil and fibre, which are also regarded as valuable food ingredients.

Product Development and Marketing

Although several studies have reported the potential food application of lupin ingredients, disappointingly, to date, none of these have reached into any significant commercialization. In Europe, lupin ingredients are used in food applications to some extent. However, compared to the food use of soy and pea ingredients, the use of lupin is much less common. This may be as a result of lack of knowledge on the food use of lupin or due to the nature of the food applications in practice. For example, though European consumers show a positive opinion on plant protein consumption, compared to soybean, many are unaware of the presence of lupin as a protein source (Lucas et al. 2015). Despite developing completely novel food products with lupins, reformulating traditional foods would be a more effective strategy to capture the attention of consumers initially. One such example is the use of lupin in gluten-free products (bakery products, pasta, breakfast cereals) as a means to enhance its protein content and quality, as the protein content in these products is often lower than standard goods. In future, a huge potential market demand can be expected for lupin-based products among consumers representing a specific group such as vegetarians, vegans and people with intolerance or allergy to gluten, soybean, milk or egg.

One approach to increase the use of lupin in foods is by the direct use of lupins (whole seed or splits) in minimally processed foods. However, this is a low value added market, and high-volume sales will be required to achieve any significant financial return, which is unlikely owing to the low consumption pattern of legumes worldwide, including lupins. Substitutional approach is another way of enhancing the food use of lupin in the export market. To make it successful, the products under consideration should be able to withstand high level of lupin substitution without any loss on its sensorial properties. In general, lupin and its ingredients are being used in minor quantities, that is, less than <5% of the total ingredients in bakery and gluten-free products (Lucas et al. 2015). According to studies published so far, the food categories which appear to have the greatest potential are the fermented food and snack food areas. A substitution rate of up to 40% have been successfully achieved in fermented soy-based products including tofu, tempeh, natto, soy sauces and miso (Petterson and Fairbrother 1996; Jayasena et al. 2010a). The snack food sector is an enormously growing industry worldwide. Since snack foods are enjoyed by all without any age, cultural or taste barriers, incorporation of lupins in the formulation of extruded and other snack foods could have unlimited potential and could be an effective tool in introducing lupins to a wide range of population. Moreover, replacement of lupin into animal products such as sausages, frankfurters, ice cream, cheese, yoghurt and dessert creams should also be considered as the current consumer food market is mainly interested in health properties and plant-based alternatives to meat, milk and eggs. While some work has been carried out on this regard with limited success, intense systematic research work is still required. Furthermore, to increase the food use of lupins, the availability and new uses of lupins should be promoted in other countries where lupin is not a common crop. However, it should also be noted that inclusion of lupin into food products should impart some unique function (health benefits) or quality attribute for the product to successfully embark on its journey and survive in the long run. Failure to do so will only incur extra raw material and processing costs.

Nutraceuticals

Another promising approach with huge market return is production of specialized products for niche markets. In this category, specific consumer groups with special needs can be targeted. The global nutraceutical markets in the USA, the EU, Japan and in many other countries are projected to show tremendous growth in the near future, especially targeting the areas of cardiovascular diseases, diabetes mellitus and neurological disorders. Centuries ago, lupins were consumed as a food by poorer sections of some societies. However, in contrast, based on the unique health-promoting properties exerted by lupins, the future role of lupins seems to bank mostly on the affluent part of societies, plagued by obesity, diabetes and cardiovascular diseases. Despite the reported health benefits of lupin and its components, lupin ingredients are hardly used as nutraceutical agents. For example, only a few lupin-containing food supplements and products with health claims are marketed in Europe (Lucas et al. 2015). Demonstrating the positive role of lupin-based foods towards ailments such as metabolic syndrome, diabetes, cardiovascular diseases and obesity through clinical trials will open up new avenues to lupin-based food products in the consumer market. To date, none of the international food safety organizations allow labeling claiming the health benefits of lupin ingredients. If the beneficial properties of lupin ingredients, either flour, protein concentrates/isolates or its fibre fractions, are confirmed through proper, valid and detailed clinical studies, its use as a functional or nutritional ingredient will have promising financial returns. And also proving its positive effects will draw greater interest towards products containing lupins where the product quality may be overlooked to some extent. For example, since the beneficial role of gamma-conglutin towards diabetes seems promising, a proper detailed study confirming its effect might promote producing gamma-conglutin-rich lupins through breeding programmes, and also products rich in gamma-conglutin can be marketed targeting a specific group of consumers.

References

Ahmed, A. R. (2014). Influence of chemical properties of wheat-lupine flour blends on cake quality. *American Journal of Food Science and Technology, 2*(2), 67–75. https://doi.org/10.12691/ajfst-2-2-4.

Alamanou, S., et al. (1996). Influence of protein isolate from lupin seeds (Lupinus albus ssp. Graecus) on processing and quality characteristics of frankfurters. *Meat Science, 42*(1), 79–93. https://doi.org/10.1016/0309-1740(95)00013-5.

Archer, B. J., et al. (2004). Effect of fat replacement by inulin or lupin-kernel fibre on sausage patty acceptability, post-meal perceptions of satiety and food intake in men. *British Journal of Nutrition, 91*(4), 591–599. https://doi.org/10.1079/BJN20031088.

Arnoldi, A., et al. (2007). Parameters for the evaluation of the thermal damage and nutraceutical potential of lupin-based ingredients and food products. *Molecular Nutrition & Food Research, 51*(4), 431–436. https://doi.org/10.1002/mnfr.200600246.

Arnoldi, A., et al. (2015). The health benefits of sweet lupin seed flours and isolated proteins. *Journal of Functional Foods, 18*, 550–563. https://doi.org/10.1016/j.jff.2015.08.012.

Australian Government. (2013). *The biology of Lupinus L. (lupin or lupine)*. Retrieved from http://www.ogtr.gov.au

Awad, R. A., Salama, W. M., & Farahat, A. M. (2014). Effect of lupine as cheese base substitution on technological and nutritional properties of processed cheese analogue. *Acta Scientiarum Polonorum Technologia Alimentaria, 13*(1), 55–64. https://doi.org/10.17306/J.AFS.2014.1.5.

Azeze, H., et al. (2016). Challenges on production and utilization of white lupin (Lupinus albus L.) in Ethiopia: A strategic orphan crop. *American Journal of Experimental Agriculture, 13*(4), 1–14. https://doi.org/10.9734/AJEA/2016/27930.

Bader, S., et al. (2009). Characterisation of odour-active compounds in lupin flour. *Journal of the Science of Food and Agriculture, 89*(14), 2421–2427. https://doi.org/10.1002/jsfa.3739.

Bader, S., et al. (2011). Influence of different organic solvents on the functional and sensory properties of lupin (Lupinus angustifolius L.) proteins. *LWT - Food Science and Technology, 44*(6), 1396–1404. https://doi.org/10.1016/j.lwt.2011.01.007.

Bähr, M., et al. (2014). Chemical composition of dehulled seeds of selected lupin cultivars in comparison to pea and soya bean. *LWT - Food Science and Technology, 59*(1), 587–590. https://doi.org/10.1016/j.lwt.2014.05.026.

Baldeón, M. E., et al. (2012). Hypoglycemic effect of cooked lupinus mutabilis and its purified alkaloids in subjects with type-2 diabetes. *Nutrición Hospitalaria, 27*(4), 1261–1266. https://doi.org/10.3305/nh.2012.27.4.5761.

Belski, R. (2012). Fiber, protein, and lupin-enriched foods: Role for improving cardiovascular health. *Advances in Food and Nutrition Research, 66*, 147–215. https://doi.org/10.1016/B978-0-12-394597-6.00004-5.

Berghout, J. A. M., Boom, R. M., & van der Goot, A. J. (2015). Understanding the differences in gelling properties between lupin protein isolate and soy protein isolate. *Food Hydrocolloids, 43*, 465–472. https://doi.org/10.1016/j.foodhyd.2014.07.003.

Bertoglio, J. C., et al. (2011). Hypoglycemic effect of lupin seed γ-conglutin in experimental animals and healthy human subjects. *Fitoterapia, 82*(7), 933–938. https://doi.org/10.1016/j.fitote.2011.05.007.

Bingemann, T. A., et al. (2019). Lupin: An emerging food allergen in the United States. *Annals of Allergy, Asthma and Immunology, 122*(1), 8–10. https://doi.org/10.1016/j.anai.2018.09.467.

Bouchoucha, R., Fradj, M., Ben, K., et al. (2016a). Anti-hyperglycemic and anti-hyperlipidemic effects of Lupinus albus in type 2 diabetic patients: A randomized double-blind, placebo-controlled clinical trial. *International Journal of Pharmacology, 12*(8), 830–837. https://doi.org/10.3923/ijp.2016.830.837.

Bouchoucha, R., Kacem, M., et al. (2016b). Effect of Lupinus albus on glycaemic control, plasma insulin levels, lipid profile and liver enzymes in Type 2 diabetics. *Journal of Food and Nutrition Research, 4*(9), 615–620. https://doi.org/10.12691/jfnr-4-9-9.

Burgos-Díaz, C., et al. (2019). Food-grade pickering stabilizers obtained from a protein-rich lupin cultivar (AluProt-CGNA®): Chemical characterization and emulsifying properties. *Food Hydrocolloids, 87*, 847–857. https://doi.org/10.1016/j.foodhyd.2018.09.018.

Cabello-Hurtado, F., et al. (2016). Proteomics for exploiting diversity of lupin seed storage proteins and their use as nutraceuticals for health and welfare. *Journal of Proteomics, 143*, 57–68. https://doi.org/10.1016/j.jprot.2016.03.026.

Calabrò, S., et al. (2015). Characterization and effect of year of harvest on the nutritional properties of three varieties of white lupine (Lupinus albus L.). *Journal of the Science of Food and Agriculture, 95*(15), 3127–3136. https://doi.org/10.1002/jsfa.7049.

Carvajal-Larenas, F. E., et al. (2016). Lupinus mutabilis: Composition, uses, toxicology, and debittering. *Critical Reviews in Food Science and Nutrition, 56*(9), 1454–1487. https://doi.org/10.1080/10408398.2013.772089.

Chiofalo, B., et al. (2012). The productive traits, fatty acid profile and nutritional indices of three lupin (Lupinus spp.) species cultivated in a Mediterranean environment for the livestock. *Animal Feed Science and Technology, 171*(2–4), 230–239. https://doi.org/10.1016/j.anifeedsci.2011.11.005.

Clark, R., & Johnson, S. (2002). Sensory acceptability of foods with added lupin (Lupinus angustifolius) kernel fiber using pre-set criteria. *Journal of Food Science, 67*(1), 356–362. https://doi.org/10.1111/j.1365-2621.2002.tb11410.x.

Confortin, T. C., et al. (2018). Importance of Lupinus albescens in agricultural and food-related areas: A review. *3 Biotech, 8*(10), 448. https://doi.org/10.1007/s13205-018-1474-x.

Curti, C. A., et al. (2018). Changes in the fatty acid composition in bitter lupinus species depend on the debittering process. *Food Chemistry, 263*, 151–154. https://doi.org/10.1016/j.foodchem.2018.04.118.

Czubinski, J., & Feder, S. (2019). Lupin seeds storage proteins composition and their interactions with native flavonoids. *Journal of the Science of Food and Agriculture, 99*, 4011–4018. https://doi.org/10.1002/jsfa.9627.

Czubinski, J., et al. (2019). Bioaccessibility of defatted lupin seed phenolic compounds in a standardized static in vitro digestion system. *Food Research International, 116*, 1126–1134. https://doi.org/10.1016/j.foodres.2018.09.057.

Danowska-Oziewicz, M., & Kurp, L. (2017). Physicochemical properties, lipid oxidation and sensory attributes of pork patties with lupin protein concentrate stored in vacuum, modified atmosphere and frozen state. *Meat Science, 131*, 158–165. https://doi.org/10.1016/j.meatsci.2017.05.009.

Dervas, G., et al. (1999). Lupin flour addition to wheat flour doughs and effect on rheological properties. *Food Chemistry, 66*(1), 67–73. https://doi.org/10.1016/S0308-8146(98)00234-9.

Dijkstra, A. J. (2015). Soybean oil. *Encyclopedia of Food and Health, 18*, 58–63. https://doi.org/10.1016/B978-0-12-384947-2.00638-3.

Donkor, O. N., et al. (2007). Rheological properties and sensory characteristics of set-type soy yogurt. *Journal of Agricultural and Food Chemistry, 55*(24), 9868–9876. https://doi.org/10.1021/jf071050r.

Doxastakis, G. (2000). Lupin seed proteins. In G. Doxastakis & V. Kiosseoglou (Eds.), *Novel macromolecules in food systems* (pp. 7–38). Thessaloniki: Elsevier Science B. V.. https://doi.org/10.1016/S0167-4501(00)80004-7.

Doxastakis, G., et al. (2002). Lupin, soya and triticale addition to wheat flour doughs and their effect on rheological properties. *Food Chemistry, 77*(2), 219–227. https://doi.org/10.1016/S0308-8146(01)00362-4.

Doxastakis, G., et al. (2007). Technological properties and non-enzymatic browning of white lupin protein enriched spaghetti. *Food Chemistry, 101*(1), 57–64. https://doi.org/10.1016/j.foodchem.2005.12.054.

Drakos, A., Doxastakis, G., & Kiosseoglou, V. (2007). Functional effects of lupin proteins in comminuted meat and emulsion gels. *Food Chemistry, 100*(2), 650–655. https://doi.org/10.1016/j.foodchem.2005.09.088.

Dueñas, M., et al. (2009). Germination as a process to increase the polyphenol content and antioxidant activity of lupin seeds (Lupinus angustifolius L.). *Food Chemistry, 117*(4), 599–607. https://doi.org/10.1016/j.foodchem.2009.04.051.

Duranti, M., et al. (2008). The major proteins of lupin seed: Characterisation and molecular properties for use as functional and nutraceutical ingredients. *Trends in Food Science and Technology, 19*(12), 624–633. https://doi.org/10.1016/j.tifs.2008.07.002.

El-Adawy, T., et al. (2001). Nutritional potential and functional properties of sweet and bitter lupin seed protein isolates. *Food Chemistry, 74*(4), 455–462. https://doi.org/10.1016/S0308-8146(01)00163-7.

Elsamani, M. O., et al. (2014). Biochemical, microbial and sensory evaluation of white soft cheese made from cow and lupin milk. *LWT - Food Science and Technology, 59*(1), 553–559. https://doi.org/10.1016/j.lwt.2014.04.027.

El-sayed, S. M. (2013). Influence of using lupine flour as binder on quality characteristics of beef burger. *Applied Sciences Research, 9*, 3809–3819.

Erbaş, M., Certel, M., & Uslu, M. K. (2005). Some chemical properties of white lupin seeds (Lupinus albus L.). *Food Chemistry, 89*(3), 341–345. https://doi.org/10.1016/j.foodchem.2004.02.040.

FAO. (2019). *FAOSTAT*. Retrieved from http://www.fao.org/faostat/en/#rankings/countries_by_commodity

Fontanari, G. G., et al. (2012). Cholesterol-lowering effect of whole lupin (Lupinus albus) seed and its protein isolate. *Food Chemistry, 132*(3), 1521–1526. https://doi.org/10.1016/j.foodchem.2011.11.145.

Fornasini, M. V., & Baldeon, M. (2012). Hypoglycemic effect of Lupinus mutabilis in healthy volunteers and subjects with dysglycemia. *Nutrición Hospitalaria, 27*, 425–433. https://doi.org/10.3305/nh.2012.27.2.5412.

Fraser, M. D., Fychan, R., & Jones, R. (2005). Comparative yield and chemical composition of two varieties of narrow-leafed lupin (Lupinus angustifolius) when harvested as whole-crop, moist grain and dry grain. *Animal Feed Science and Technology, 120*(1–2), 43–50. https://doi.org/10.1016/j.anifeedsci.2004.12.014.

French, R. J. (2015). Lupin: Agronomy. In C. Wrigley, H. Corke, K. Seetharaman, & J. Faubion (Eds.), *Encyclopedia of food grains* (2nd ed.). Amsterdam: Elsevier. https://doi.org/10.1016/B978-0-12-394437-5.00194-7.

Frick, K. M., et al. (2017). Quinolizidine alkaloid biosynthesis in lupins and prospects for grain quality improvement. *Frontiers in Plant Science, 8*, 1–12. https://doi.org/10.3389/fpls.2017.00087.

Fudiyansyah, N., et al. (2007). A nutritional, chemical and sensory evaluation of lupin (L. angustifolius) tempe. *International Journal of Food Science & Technology, 30*(3), 297–305. https://doi.org/10.1111/j.1365-2621.1995.tb01378.x.

Guarneri, F., Guarneri, C., & Benvenga, S. (2005). Identification of potentially cross-reactive peanut-lupine proteins by computer-assisted search for amino acid sequence homology. *International Archives of Allergy and Immunology, 138*, 273–277. https://doi.org/10.1159/000088864.

Guillamón, E., et al. (2010). Characterization of lupin major allergens (Lupinus albus L.). *Molecular Nutrition & Food Research, 54*(11), 1668–1676. https://doi.org/10.1002/mnfr.200900452.

Hall, R. S., & Johnson, S. K. (2006). Sensory acceptability of foods containing Australian sweet lupin (Lupinus angustifolius) flour. *Journal of Food Science, 69*(2), snq92–snq97. https://doi.org/10.1111/j.1365-2621.2004.tb15520.x.

Hall, R., et al. (2005a). Lupin kernel fibre-enriched foods beneficially modify serum lipids in men. *European Journal of Clinical Nutrition, 59*(3), 325–333. https://doi.org/10.1038/sj.ejcn.1602077.

Hall, R. S., Thomas, S. J., & Johnson, S. K. (2005b). Australian sweet lupin flour addition reduces the glycaemic index of a white bread breakfast without affecting palatability in healthy human volunteers. *Asia Pacific Journal of Clinical Nutrition, 14*(1), 91–97. Retrieved from http://www.epistemonikos.org/documents/98ab8d60de761fec6cae45257b0075012d6e19dd.

Harisa, G. I., & Alanazi, F. K. (2015). The beneficial roles of Lupineus luteus and lifestyle changes in management of metabolic syndrome: A case study. *Saudi Pharmaceutical Journal, 23*(6), 712–715. https://doi.org/10.1016/j.jsps.2015.02.006.

Hickisch, A., et al. (2016). Influence of lupin-based milk alternative heat treatment and exopolysaccharide-producing lactic acid bacteria on the physical characteristics of lupin-based yogurt alternatives. *Food Research International, 84*, 180–188. https://doi.org/10.1016/j.foodres.2016.03.037.

Huerta-Yépez, S., Tirado-Rodriguez, A. B., & Hankinson, O. (2016). Role of diets rich in omega-3 and omega-6 in the development of cancer. *Boletín Médico del Hospital Infantil de México, 73*(6), 446–456. https://doi.org/10.1016/j.bmhimx.2016.11.001.

Huyghe, C. (1997). White lupin (Lupinus albus L.). *Field Crops Research, 53*(1–3), 147–160. https://doi.org/10.1016/S0378-4290(97)00028-2.

Islam, S., & Ma, W. (2016). Lupine. In P. M. Finglas & F. Toldrá (Eds.), *Encyclopedia of food and health* (pp. 579–585). Amsterdam: Elsevier. https://doi.org/10.1016/B978-0-12-384947-2.00432-3.

Jayasena, V., & Nasar-Abbas, S. M. (2011). Effect of lupin flour incorporation on the physical characteristics of dough and biscuits. *Quality Assurance & Safety of Crops and Food, 3*(3), 140–147. https://doi.org/10.1111/j.1757-837X.2011.00100.x.

Jayasena, V., & Nasar-Abbas, S. M. (2012). Development and quality evaluation of high-protein and high-dietary-fiber pasta using lupin flour. *Journal of Texture Studies, 43*(2), 153–163. https://doi.org/10.1111/j.1745-4603.2011.00326.x.

Jayasena, V., Khu, W. S., & Nasar-Abbas, S. M. (2010a). The development and sensory acceptability of lupin-based tofu. *Journal of Food Quality, 33*(1), 85–97. https://doi.org/10.1111/j.1745-4557.2009.00290.x.

Jayasena, V., Leung, P. P. Y., & Nasar-Abbas, S. M. (2010b). Effect of lupin flour substitution on the quality and sensory acceptability of instant noodles. *Journal of Food Quality, 33*(6), 709–727. https://doi.org/10.1111/j.1745-4557.2010.00353.x.

Jimenez-Lopez, J. C., et al. (2018). Characterization of narrow-leaf lupin (Lupinus angustifolius L.) recombinant major allergen IgE-binding proteins and the natural β-conglutin counterparts in sweet lupin seed species. *Food Chemistry, 244*, 60–70. https://doi.org/10.1016/j.foodchem.2017.10.015.

Jiménez-Martínez, C., Hernández-Sánchez, H., & Dávila-Ortiz, G. (2003). Production of a yogurt-like product from Lupinus campestris seeds. *Journal of the Science of Food and Agriculture, 83*(6), 515–522. https://doi.org/10.1002/jsfa.1385.

John, S., Sipsas, S., & St. John, C. (2007). *Method to produce lupin protein-based dairy substitutes*. United States Patent Application Publication.

Johnson, S. K., et al. (2003). Sensory acceptability of white bread with added Australian sweet lupin (Lupinus angustifolius) kernel fibre and its glycaemic and insulinaemic responses when eaten as a breakfast. *Journal of the Science of Food and Agriculture, 83*(13), 1366–1372. https://doi.org/10.1002/jsfa.1552.

Johnson, S. K., et al. (2017). Lupins: Their unique nutritional and health-promoting attributes. In J. R. N. Taylor & J. M. Awika (Eds.), *Gluten-free ancient grains* (pp. 179–221). Amsterdam: Elsevier. https://doi.org/10.1016/B978-0-08-100866-9.00008-X.

Jung, S. (2009). Aqueous extraction of oil and protein from soybean and lupin: A comparative study. *Journal of Food Processing and Preservation, 33*, 547–559. https://doi.org/10.1111/j.1745-4549.2009.00400.x.

Kaczmarska, K. T., et al. (2017). Effect of germination and fermentation on carbohydrate composition of Australian sweet lupin and soybean seeds and flours. *Journal of Agricultural and Food Chemistry, 65*(46), 10064–10073. https://doi.org/10.1021/acs.jafc.7b02986.

Kaczmarska, K. T., et al. (2018a). Enhancing wheat muffin aroma through addition of germinated and fermented Australian sweet lupin (Lupinus angustifolius L.) and soybean (Glycine max L.) flour. *LWT, 96*, 205–214. https://doi.org/10.1016/j.lwt.2018.05.034.

Kaczmarska, K. T., et al. (2018b). Aroma characteristics of lupin and soybean after germination and effect of fermentation on lupin aroma. *LWT - Food Science and Technology, 87*, 225–233. https://doi.org/10.1016/j.lwt.2017.08.080.

Kapravelou, G., et al. (2013). Health promoting effects of Lupin (Lupinus albus var. multolupa) protein hydrolyzate and insoluble fiber in a diet-induced animal experimental model of hypercholesterolemia. *Food Research International, 54*(2), 1471–1481. https://doi.org/10.1016/j.foodres.2013.10.019.

Karamać, M., et al. (2018). Phenolic contents and antioxidant capacities of wild and cultivated white lupin (Lupinus albus L.) seeds. *Food Chemistry, 258*, 1–7. https://doi.org/10.1016/j.foodchem.2018.03.041.

Karnpanit, W. (2016) *Effect of cultivar and processing on anti-nutritional factors and bioaccessibility of minerals of Australian sweet lupin (Lupinus angustifolius L.)*. Western Sydney University. Retrieved from http://hdl.handle.net/1959.7/uws:38285

Kinder, D. H., & Knecht, K. T. (2011). Lupine (Lupinus caudatus L., Lupinus albus L.) seeds: History of use, use as an antihyperglycemic medicinal, and use as a food. In V. R. Preedy, R. R. Watson, & V. B. Patel (Eds.), *Nuts and Seeds in Health and Disease Prevention* (pp. 711–716). Amsterdam: Elsevier. https://doi.org/10.1016/B978-0-12-375688-6.10084-2.

Klupsaite, D., et al. (2017). The influence of lactic acid fermentation on functional properties of narrow-leaved lupine protein as functional additive for higher value wheat bread. *LWT, 75*, 180–186. https://doi.org/10.1016/j.lwt.2016.08.058.

Kohajdová, Z., Karovičová, J., & Schmidt, Š. (2011). Lupin composition and possible use in bakery—A review. *Czech Journal of Food Science and Technology, 29*(3), 203–211.

Kuznetsova, L., Zabodalova, L., & Baranenko, D. (2014). On the potential of lupin protein concentrate made by enzymatic hydrolysis of carbohydrates in dairy-like applications. *Agronomy Research, 12*(3), 727–736.

Lammi, C., et al. (2016). Peptides derived from soy and lupin protein as dipeptidyl-peptidase IV Inhibitors: In vitro biochemical screening and in silico molecular modeling study. *Journal of Agricultural and Food Chemistry, 64*(51), 9601–9606. https://doi.org/10.1021/acs.jafc.6b04041.

Lampart-Szczapa, E., et al. (2003). Chemical composition and antibacterial activities of lupin seeds extracts. *Nahrung/Food, 47*(5), 286–290. https://doi.org/10.1002/food.200390068.

Lee, Y. P., et al. (2006). Lupin-enriched bread increases satiety and reduces energy intake. *The American Journal of Clinical Nutrition, 84*, 975–980.

Lee, Y. P., et al. (2009). Effects of lupin kernel flour–enriched bread on blood pressure: A controlled intervention study. *The American Journal of Clinical Nutrition, 89*(3), 766–772. https://doi.org/10.3945/ajcn.2008.26708.

Levent, H., & Bilgiçli, N. (2011). Enrichment of gluten-free cakes with lupin (Lupinus albus L.) or buckwheat (Fagopyrum esculentum M.) flours. *International Journal of Food Sciences and Nutrition, 62*(7), 725–728. https://doi.org/10.3109/09637486.2011.572546.

Liu, K. (2015). Soybean: Overview. In *Encyclopedia of food grains* (2nd ed.). Amsterdam: Elsevier. https://doi.org/10.1016/B978-0-12-394437-5.00028-0.

Lucas, M. M., et al. (2015). The future of lupin as a protein crop in Europe. *Frontiers in Plant Science, 6*, 1–6. https://doi.org/10.3389/fpls.2015.00705.

Magalhães, S. C. Q., Fernandes, F., et al. (2017a). Alkaloids in the valorization of European Lupinus spp. seeds crop. *Industrial Crops and Products, 95*, 286–295. https://doi.org/10.1016/j.indcrop.2016.10.033.

Magalhães, S. C. Q., Taveira, M., et al. (2017b). European marketable grain legume seeds: Further insight into phenolic compounds profiles. *Food Chemistry, 215*, 177–184. https://doi.org/10.1016/j.foodchem.2016.07.152.

Magni, C., et al. (2004). Conglutin γ, a lupin seed protein, binds insulin in vitro and reduces plasma glucose levels of hyperglycemic rats. *Journal of Nutritional Biochemistry, 15*(11), 646–650. https://doi.org/10.1016/j.jnutbio.2004.06.009.

Mahmoud, E. A. M., Nassef, S. L., & Basuny, A. M. M. (2012). Production of high protein quality noodles using wheat flour fortified with different protein products from lupine. *Annals of Agricultural Sciences, 57*(2), 105–112. https://doi.org/10.1016/j.aoas.2012.08.003.

Mani, V., & Ming, L. C. (2017). Tempeh and other fermented soybean products rich in isoflavones. In *Fermented foods in health and disease prevention* (pp. 453–474). Amsterdam: Elsevier. https://doi.org/10.1016/B978-0-12-802309-9.00019-4.

Martínez-Villaluenga, C., Frías, J., & Vidal-Valverde, C. (2006). Functional lupin seeds (Lupinus albus L. and Lupinus luteus L.) after extraction of α-galactosides. *Food Chemistry, 98*(2), 291–299. https://doi.org/10.1016/j.foodchem.2005.05.074.

Martínez-Villaluenga, C., et al. (2010). Semolina supplementation with processed lupin and pigeon pea flours improve protein quality of pasta. *LWT - Food Science and Technology, 43*(4), 617–622. https://doi.org/10.1016/j.lwt.2009.11.001.

Mohamed, A. A., & Rayas-Duarte, P. (1995). Composition of Lupinus albus. *American Association of Cereal Chemists, 72*(6), 643–647.

Mosaad, M. N., et al. (2014). Hypoglycemic effect of lupin alkaloids and/or rosiglitazone in pregnant diabetic rats in relation to atherogenic index. *International Journal of Pharmacy and Pharmaceutical Sciences, 6*, 1–8.

Muñoz, E. B., et al. (2018). Gamma-conglutin peptides from Andean lupin legume (Lupinus mutabilis Sweet) enhanced glucose uptake and reduced gluconeogenesis in vitro. *Journal of Functional Foods, 45*, 339–347. https://doi.org/10.1016/j.jff.2018.04.021.

Musco, N., et al. (2017). Comparison of nutritional and antinutritional traits among different species (Lupinus albus L., Lupinus luteus L., Lupinus angustifolius L.) and varieties of lupin seeds. *Journal of Animal Physiology and Animal Nutrition, 101*(6), 1227–1241. https://doi.org/10.1111/jpn.12643.

Múzquiz, M., et al. (1989). A chemical study of lupinus hispanicus seed—Toxic and antinutritional components. *Journal of the Science of Food and Agriculture, 47*(2), 205–214. https://doi.org/10.1002/jsfa.2740470208.

Nasar-Abbas, S. M., & Jayasena, V. (2012). Effect of lupin flour incorporation on the physical and sensory properties of muffins. *Quality Assurance & Safety of Crops and Food, 4*(1), 41–49. https://doi.org/10.1111/j.1757-837X.2011.00122.x.

O'Toole, D. K. (2016). Soybean: Soymilk, Tofu, and Okara. In *Encyclopedia of food grains* (2nd ed., pp. 134–143). Amsterdam: Elsevier. https://doi.org/10.1016/B978-0-12-394437-5.00130-3.

Papavergou, E. J., Bloukas, J. G., & Doxastakis, G. (1999). Effect of lupin seed proteins on quality characteristics of fermented sausages. *Meat Science, 52*(4), 421–427. https://doi.org/10.1016/S0309-1740(99)00025-X.

Paraskevopoulou, A., et al. (2010). Dough rheology and baking performance of wheat flour–lupin protein isolate blends. *Food Research International, 43*(4), 1009–1016. https://doi.org/10.1016/j.foodres.2010.01.010.

Paraskevopoulou, A., Chrysanthou, A., & Koutidou, M. (2012). Characterisation of volatile compounds of lupin protein isolate-enriched wheat flour bread. *Food Research International, 48*(2), 568–577. https://doi.org/10.1016/j.foodres.2012.05.028.

Pastor-Cavada, E., et al. (2009). Analytical nutritional characteristics of seed proteins in six wild Lupinus species from Southern Spain. *Food Chemistry, 117*(3), 466–469. https://doi.org/10.1016/j.foodchem.2009.04.039.

Pavanello, C., et al. (2017). Effects of a lupin protein concentrate on lipids, blood pressure and insulin resistance in moderately dyslipidaemic patients: A randomised controlled trial. *Journal of Functional Foods, 37*, 8–15. https://doi.org/10.1016/j.jff.2017.07.039.

Petterson, D. S. (2015) Lupin: Overview. Encyclopedia of food grains 2nd edn: Elsevier Amsterdam Ltd. doi: https://doi.org/10.1016/B978-0-12-394437-5.00036-X.

Petterson, D. S., & Fairbrother, A. H. (1996). Lupins as a raw material for human foods and animal feeds. *Indonesian Food and Nutrition Progress, 3*(2), 35–41.

Pihlanto, A., et al. (2017). Bioactivities of alternative protein sources and their potential health benefits. *Food & Function, 8*(10), 3443–3458. https://doi.org/10.1039/C7FO00302A.

Piornos, J. A., et al. (2015). Functional and physicochemical properties of a protein isolate from Alu Prot-CGNA: A novel protein-rich lupin variety (Lupinus luteus). *Food Research International, 76*, 719–724. https://doi.org/10.1016/j.foodres.2015.07.013.

Pollard, N. J., et al. (2002). Lupin flours as additives: Dough mixing, breadmaking, emulsifying, and foaming. *Cereal Chemistry Journal, 79*(5), 662–669. https://doi.org/10.1094/CCHEM.2002.79.5.662.

Pozani, S., Doxastakis, G., & Kiosseoglou, V. (2002). Functionality of lupin seed protein isolate in relation to its interfacial behaviour. *Food Hydrocolloids, 16*(3), 241–247. https://doi.org/10.1016/S0268-005X(01)00094-7.

Priatni, S., et al. (2013). Quality and sensory evaluations of tempe prepared from various particle sizes of lupin beans. *Jurnal Teknologi dan Industri Pangan, 24*(2), 209–214. https://doi.org/10.6066/jtip.2013.24.2.209.

Ranilla, L. G., Genovese, M. I., & Lajolo, F. M. (2009). Isoflavones and antioxidant capacity of Peruvian and Brazilian lupin cultivars. *Journal of Food Composition and Analysis, 22*(5), 397–404. https://doi.org/10.1016/j.jfca.2008.06.011.

Rayas-Duarte, P., Mock, C. M., & Satterlee, L. D. (1996). Quality of spaghetti containing buckwheat, amaranth, and lupin flours. *Cereal Chemistry, 73*(3), 381–387.

Resta, D., et al. (2008). Evaluation of total quinolizidine alkaloids content in lupin flours, lupin-based ingredients, and foods. *Molecular Nutrition & Food Research, 52*(4), 490–495. https://doi.org/10.1002/mnfr.200700206.

Rosa Lovati, M., et al. (2012). Lupin seed γ-conglutin lowers blood glucose in hyperglycaemic rats and increases glucose consumption of HepG2 cells. *British Journal of Nutrition, 107*(1), 67–73. https://doi.org/10.1017/S0007114511002601.

Ruales, J., Pólit, P., & Nair, B. M. (1988). Nutritional quality of blended foods of rice, soy and lupins, processed by extrusion. *Food Chemistry, 29*(4), 309–321. https://doi.org/10.1016/0308-8146(88)90046-5.

Ruiz-López, M. A., et al. (2000). Chemical composition and antinutrient content of three lupinus species from Jalisco, Mexico. *Journal of Food Composition and Analysis, 13*(3), 193–199. https://doi.org/10.1006/jfca.1999.0887.

Rumiyati, R., James, A. P., & Jayasena, V. (2015). Effects of lupin incorporation on the physical properties and stability of bioactive constituents in muffins. *International Journal of Food Science & Technology, 50*(1), 103–110. https://doi.org/10.1111/ijfs.12688.

Sanz, M. L., et al. (2010). Lupin allergy: A hidden killer in the home. *Clinical & Experimental Allergy, 40*(10), 1461–1466. https://doi.org/10.1111/j.1365-2222.2010.03590.x.

Sathe, S. K. (2016). Beans: Overview. In *Encyclopedia of food grains* (2nd ed., pp. 297–306). Amsterdam: Elsevier. https://doi.org/10.1016/B978-0-12-394437-5.00033-4.

Sbihi, H. M., et al. (2013). Bitter and sweet lupin (Lupinus albus L.) seeds and seed oils: A comparison study of their compositions and physicochemical properties. *Industrial Crops and Products, 49*, 573–579. https://doi.org/10.1016/j.indcrop.2013.05.020.

Schindler, S., et al. (2011). Lactic fermentation to improve the aroma of protein extracts of sweet lupin (Lupinus angustifolius). *Food Chemistry, 128*(2), 330–337. https://doi.org/10.1016/j.foodchem.2011.03.024.

Sedláková, K., et al. (2016). Lupin as a perspective protein plant for animal and human nutrition—A review. *Acta Veterinaria Brno, 85*(2), 165–175. https://doi.org/10.2754/avb201685020165.

Siger, A., et al. (2012). Antioxidant activity and phenolic content in three lupin species. *Journal of Food Composition and Analysis, 25*(2), 190–197. https://doi.org/10.1016/j.jfca.2011.10.002.

Simopoulos, A. (2002). The importance of the ratio of omega-6/omega-3 essential fatty acids. *Biomedicine & Pharmacotherapy, 56*(8), 365–379. https://doi.org/10.1016/S0753-3322(02)00253-6.

Simopoulos, A. (2016). An Increase in the omega-6/omega-3 fatty acid ratio increases the risk for obesity. *Nutrients, 8*(3), 128. https://doi.org/10.3390/nu8030128.

Sironi, E., Sessa, F., & Duranti, M. (2005). A simple procedure of lupin seed protein fractionation for selective food applications. *European Food Research and Technology, 221*, 145–150. https://doi.org/10.1007/s00217-005-1151-2.

Sirtori, C. R., et al. (2004). Proteins of white lupin seed, a naturally isoflavone-poor legume, reduce cholesterolemia in rats and increase LDL receptor activity in HepG2 cells. *The Journal of Nutrition, 134*(1), 18–23. https://doi.org/10.1093/jn/134.1.18.

Sirtori, E., et al. (2011). Cross-reactivity between peanut and lupin proteins. *Food Chemistry, 126*(3), 902–910. https://doi.org/10.1016/j.foodchem.2010.11.073.

Steinkraus, K. (2018) Handbook of indigenous fermented foods (Revised and Expanded ed.). Boca Raon: CRC Press. https://doi.org/10.1201/9780203752821.

Sujak, A., Kotlarz, A., & Strobel, W. (2006). Compositional and nutritional evaluation of several lupin seeds. *Food Chemistry, 98*(4), 711–719. https://doi.org/10.1016/j.foodchem.2005.06.036.

Sweetingham, M., & Kingwell, R. (2008) Lupins—Reflections and future possibilities. In *Proceedings of the 12th International Lupin Conference*, 14-18 September 2008, Fremantle, Western Australia.

Tapadia, M., et al. (2019). Lupin seed hydrolysate promotes G-protein-coupled receptor, intracellular Ca2+ and enhanced glycolytic metabolism-mediated insulin secretion from BRIN-BD11 pancreatic beta cells. *Molecular and Cellular Endocrinology, 480*, 83–96. https://doi.org/10.1016/j.mce.2018.10.015.

Terruzzi, I., et al. (2011). Insulin-mimetic action of conglutin-γ, a lupin seed protein, in mouse myoblasts. *Nutrition, Metabolism and Cardiovascular Diseases, 21*(3), 197–205. https://doi.org/10.1016/j.numecd.2009.09.004.

Teshome, P. G., et al. (2012). Proximate composition and anti-nutritional factors of traditionally processed white lupine (Lupinous albus L.) fabaceae, grown in Ethiopia. *Ethiopian Journal of Biological Sciences, 11*(2), 133–146.
Tizazu, H., & Emire, S. (2010). Chemical composition, physicochemical and functional properties of lupin (Lupinus). *African Journal of Food Agriculture Nutrition and Development, 10*(8), 3029–3047.
Trugo, L. C., von Baer, E., & von Baer, D. (2016). Lupin: Breeding. In *Encyclopedia of Food Grains* (2nd ed., pp. 325–332). Amsterdam: Elsevier. https://doi.org/10.1016/B978-0-12-394437-5.00211-4.
Turnbull, C. M., Baxter, A. L., & Johnson, S. K. (2005). Water-binding capacity and viscosity of Australian sweet lupin kernel fibre under in vitro conditions simulating the human upper gastrointestinal tract. *International Journal of Food Sciences and Nutrition, 56*(2), 87–94. https://doi.org/10.1080/09637480500081080.
van de Noort, M. (2017). Lupin. In *Sustainable protein sources* (pp. 165–183). Amsterdam: Elsevier. https://doi.org/10.1016/B978-0-12-802778-3.00010-X.
Villarino, C. B. (2014). *Maximising the nutritional and sensory quality of lupin bread made using Western Australian bakers flour*. Bentley: Curtin University.
Villarino, C. B. J., et al. (2015). The effects of Australian sweet lupin (ASL) variety on physical properties of flours and breads. *LWT - Food Science and Technology, 60*(1), 435–443. https://doi.org/10.1016/j.lwt.2014.08.028.
Villarino, C. B. J., et al. (2016). Nutritional, health, and technological functionality of lupin flour addition to bread and other baked products: Benefits and challenges. *Critical Reviews in Food Science and Nutrition, 56*(5), 835–857. https://doi.org/10.1080/10408398.2013.814044.
Wang, D. H., et al. (2019). High levels of branched chain fatty acids in nātto and other Asian fermented foods. *Food Chemistry, 286*, 428–433. https://doi.org/10.1016/j.foodchem.2019.02.018.
Wickramasinghe, N. P. (2017). *Nutritional composition of (Australian sweet lupin) natto and tempeh: Effect of source and fermentation time*. Bentley: Curtin University.
Wolkers-Rooijackers, J. C. M., Endika, M. F., & Smid, E. J. (2018). Enhancing vitamin B 12 in lupin tempeh by in situ fortification. *LWT, 96*, 513–518. https://doi.org/10.1016/j.lwt.2018.05.062.
Wong, A., et al. (2013). Isolation and foaming functionality of acid-soluble protein from lupin (Lupinus angustifolius) kernels. *Journal of the Science of Food and Agriculture, 93*, 3755–3762. https://doi.org/10.1002/jsfa.6249.
Yang, M., Fu, J., & Li, L. (2012). Rheological characteristics and microstructure of probiotic soy yogurt prepared from germinated soybeans. *Food Technology and Biotechnology, 50*, 73–80.
Yoshie-Stark, Y., & Wäsche, A. (2004). In vitro binding of bile acids by lupin protein isolates and their hydrolysates. *Food Chemistry, 88*, 179–184. https://doi.org/10.1016/j.foodchem.2004.01.033.
Zhong, L., et al. (2018). Seed coats of pulses as a food ingredient: Characterization, processing, and applications. *Trends in Food Science and Technology, 80*, 35–42. https://doi.org/10.1016/j.tifs.2018.07.021.
Zhong, L., Wu, G., et al. (2019a). Characterization of polyphenols in Australian sweet lupin (Lupinus angustifolius) seed coat by HPLC-DAD-ESI-MS/MS. *Food Research International, 116*, 1153–1162. https://doi.org/10.1016/j.foodres.2018.09.061.
Zhong, L., Fang, Z., et al. (2019b). Extrusion cooking increases soluble dietary fibre of lupin seed coat. *LWT, 99*, 547–554. https://doi.org/10.1016/j.lwt.2018.10.018.

Moth Bean

T. Senthilkumar and M. Ngadi

Introduction

Moth bean (*Vigna aconitifolia* (Jacq.) Maréchal), also called as dew bean, haricot mat, Indian moth bean, mat bean, mattenbohne, math, moth gram, matki, or Turkish gram, originated in the Indian subcontinent and is mainly cultivated in India and Pakistan. Moth bean is cultivated in 1.5 million hectares with an annual approximate production of 0.4 million tons (Singhal 2003) in India. It is a drought-resistant crop and can be cultivated in arid and semiarid regions; arid regions of Rajasthan, India, have 85% of total moth bean cultivated area and produced 0.22 million tons of moth bean (55% of annual production) (S. Singh et al. 2017; Vir and Singh 2015). Legumes are a major source of protein for the developing world, and it is becoming more popular in the developed world due to changes in food habits.

The moth beans are brown, whitish brown, or yellowish brown in color, cylindrical or rectangular in shape, and elongate and subtruncate at the ends (Fig. 1). Like other legumes, the moth beans are composed of seed coat, cotyledon, and embryo. The main constituents of moth bean are carbohydrates and protein, which accounted for 61.9% and 21.9%, respectively (Table 1); the total carbohydrates and proteins in dry pulses are in the range from 24% to 68% and 20–30%, respectively.

The lipids and ash in moth beans are 3.5% and 3.3%, respectively (Table 1). The high protein content and less lipids make moth bean a part of healthy diet. Moth beans are a good source of minerals, such as phosphorus, magnesium, calcium, and iron. The moth beans are also rich in water-soluble vitamins such as thiamine, riboflavin, and niacin. The moth beans also pose some antinutritional factors such as trypsin inhibitors, polyphenols, hemagglutinins, phytic acid, and oligosaccharides (Mehta and Simlot 1982).

T. Senthilkumar (✉) · M. Ngadi
Mcgill University, Montreal, QC, Canada

A. Manickavasagan, P. Thirunathan (eds.), *Pulses*,
https://doi.org/10.1007/978-3-030-41376-7_11

Fig. 1 Moth bean seeds

Table 1 The nutritional composition of moth beans

S. No	Constituents	Per 100 g of dry moth beans
1	Carbohydrates (g)	61.9
2	Protein (g)	21.9
3	Lipids (g)	3.5
4	Fiber (g)	4.5
5	Ash (g)	3.3
6	Calcium (mg)	133
7	Phosphorus (mg)	356
8	Magnesium (mg)	183
9	Iron (mg)	11
10	Thiamine (mg)	0.50
11	Riboflavin (mg)	0.1
12	Niacin (mg)	1.7

Compiled from Kadam et al. (1985)

Moth beans are also used as feed for livestock. Green and dried moth bean plants are rich in nutrition and utilized by livestock (Ranjhan 1980). The production and consumption of moth beans are mostly confined to the Asian population. Only a little quantity is exported for the Asian population living all over the world.

Postharvest Processing

Harvesting and Storing

Moth bean crop is uprooted once the entire crops dry up and the pods turn light yellow. All the bean plants are stored as a heap and dried in sun for 3–5 days. After that, threshing is done by bullocks, threshers, or by the use of hand sticks. Thereafter, the seeds are again sun-dried until their moisture content reduced approximately to 8–10%. The dried seeds are stored in airtight earthen pots, gunny bags, and cloth bags. The seeds destined for seeding purposes were treated with endosulfan, whereas seeds for human consumption were not treated with any chemicals. It is estimated that 8–20% of seeds were lost during harvesting and storing of moth beans, so the harvesting and storing should follow proper procedures. The common storage pests in moth beans are *Callosobruchus chinensis*, and they are very active during the summer months from April to September.

Processing

Moth beans are usually consumed as whole seeds, dehusked seeds, and sprouts.

Milling

Moth bean seeds are milled to get dehusked splits, which are used in human diet. Dehulling is a very traditional method used widely in the developing world. The milling process improves the appearance, texture, cooking properties, and digestibility of seeds. The milling process is usually done using both manually and electric motor-operated disc shellers. The first step in milling is loosening of the husk by wet or dry method, or combining both. The wet method involves adding small amount of water to the seeds continuously and allowing the seeds to absorb the moisture for several hours, followed by draining the water and sun drying. The soaking and drying process shrinks the endosperm, which results in loosening of husk. The dry method involved adding vegetable oil to the seeds and drying in the sun. The oil penetrates in the husk and helps in the loosening of husk. After the loosening operation, the seeds were dehusked using manually or electric motor-operated disc sheller. The split seeds were cleaned using aspirator and sieve to get dehulled split seeds. The splitting operation results in loss of embryo and cotyledon edge breakage. This loss can be avoided by tempering the seeds after loosening of husk.

Grinding

Whole seeds or splits are dried and made into flour or added with moisture and grinded to make a batter. The grinding process is done using a mortar or grinder. The dry flour and wet-ground preparations were preferred equally by many countries. Handheld or mechanized mortars or grinders are available for wet batter preparations.

Germination

The germination of whole seeds resulted in sprouts, which increased the formation of riboflavin, niacin, and thiamine. The sprouted seeds are rich in amino acids and increase the protein digestibility. The sprouting of moth seeds was achieved by placing the whole seeds inside the muslin cloth for 25 h and tied properly. The moisture on the towel was maintained by sprinkling water intermittently.

Roasting

Moth beans are either directly cooked or subjected to other types of processing such as roasting and puffing. The roasted and puffed legumes are consumed as snacks. These snacks are very cheap and good for poor population. A study by Vijaylaksmi and Venkatrao (1977) showed that roasting did not result in reduction of various amino acids. It also improved the growth rate of rats, PER, and digestibility when compared with cooked moth bean.

Antinutritional Factors and Their Removal

Like other pulses, certain antinutritional factors like trypsin inhibitors, saponins, phytic acids, etc. are also found in moth bean. Studies carried out have indicated that by sprouting/cooking the seeds, these factors could be removed considerably. For instance, the trypsin inhibitors' activity was reduced up to 98% by cooking of sprouted seeds or pressure cooking of seeds. Similarly, saponin activity could be reduced by about 77% on cooking of sprouted seeds. Protein digestibility is generally known to be increased by about 20–50% on cooking of seeds.

Product Development

The moth bean is usually treated like poor man's food in the developing world. The moth bean is mostly consumed as sprouts, whole seed dhal, and split dhal.

Whole Seed Meal

Thick soup of whole seed meal is prepared by adding whole seeds in the boiling water and continuously stirred for proper mixing. The product is cooked at boiling temperature for a few minutes to make thick soup; a small amount of salt and pepper is added to make the product tasty. This is commonly prepared in the Indian subcontinent, and this whole meal soup is helpful in reducing the effects of cough and bronchitis among rural population (Kadam et al. 1985).

Dhal

The dehulled moth beans or whole moth beans, with cotyledons, are the most common form of moth bean consumption. The dehulled moth beans or whole moth beans are soaked in water for 30–60 min and fried in oil after adding turmeric powder and onion. Small quantities of chili powder and salt are also added to make it tastier. This product is consumed as a vegetable meal. The soaked moth beans, either dehulled or whole seeds, are used for the curry preparation. It is also used in a salted snack preparation called *bhujia*, which looks similar to macaroni (Kadam et al. 1985).

Sprouts

Moth beans are also germinated before they are used as human food because of its nutritional value. The seeds are soaked in water, and the soaked seeds are placed inside muslin cloth for 25 h to germinate overnight. The sprouted seeds are either used for the preparation of curry or fried in oil with onion, turmeric, and chili powder, and people also directly consume germinated moth beans. This preparation is often used as a vegetable in many parts of the Indian subcontinent. Preservation of moth bean sprouts is done by blanching in NaCl solution and drying at 60 °C (Dutta and Manjrekar 1983).

Medical Use

Moth bean soup is used for the treatment of cough and bronchitis in certain parts of India. In southern India, moth bean is consumed by women having abnormalities in their menstrual cycle. It is believed that moth bean regulates the menstrual cycle. It is also recommended in the diet of persons suffering from kidney-related issues. It is also used as an astringent, diuretic, and tonic. The moth bean soups and dhals are consumed by patients with fever (Kadam et al. 1985).

Papad

Moth bean flour is mixed with the desired amount of water to form a dough, and the dough is pressed hard to make a very thin loven bread, and it is either sun-dried or dried using industrial dryer; this dried product is commonly called as papad. Papad is also prepared from the mung bean flour and urad bean flour. Due to less moisture content, papad can be stored for long periods of time and can instantly be consumed as a snack by baking or frying in oil. Snacking on papad is the most common practice in India (Kumar 2002).

Bhujia

This is a thick paste of fine grinded flour of decorticated moth (mogar) beans mixed with appropriate quantity of spices to make a macaroni-like thread which is finely fried. It is commonly consumed among rural population in the Indian subcontinent (Kumar 2002).

Mangori

The dehulled moth beans are soaked in water for 30–60 min and wet-grinded as a thick paste with desired amounts of ginger, coriander, onion etc. added. The paste is, thereafter, separated into 2–4 g pieces and are dried in the sun. The end product is known as nuggets; these are stored for years for future needs as a substitute to regular vegetable, when they are not available during some seasons of a particular year (Kumar 2002).

Vada

It is prepared by mixing dhal flour with desired amounts of water, spices, ginger, and other essential ingredients; to cook vada, the dough with the spices is properly fried with edible oil. The consumption of vada is very common all over India, as it helps improve gastric troubles and disorders and constipation and also protects people from sunstrokes, which are frequent in summer months.

Kheech

It is prepared by boiling moth dhal and pearl millet seeds together in water to achieve a thick paste consistency. This tasty dish is served with butter. Kheech serves as an antiacid and a cooling agent for rural people from arid areas (Kumar 2002).

Roti

It is prepared by mixing the flour of pearl millet and moth bean which is flattened and roasted on a hot plate.

Rabri

A thick paste is made by mixing the flour of pearl millet and moth bean in buttermilk and is kept overnight. This increases the digestive capabilities. It also soothes stomach acidity and helps in releases of gas from the system. Further, it also works as a good sedative agent in hot and arid climates of the Indian desert. During summers it also acts as a cooling agent (Kumar 2002).

Other Products

Besides edible uses, moth bean is known as soil binder. Its dense mat-like spreading behavior of the canopy completely covers the ground, therefore reducing soil movement; it also shields soil from heat, preventing cracking and crust formation. It, therefore, reduces the soil organic matter and moisture losses.

Future Trends in Processing and Product Development

The moth beans are drought-resistant crops of the tropics. They are mostly grown in water scarcity areas and under the dry-farming system. In order to improve their production and utilization, detailed investigations on production, processing, and utilization are warranted. Such attempts will help to improve the protein supply in these countries. The chemical composition of the moth bean is rather comparable with other pulses. However, they have a significantly higher proportion of nonprotein nitrogen than other legumes. The proteins of both the legumes have not been isolated, characterized, or studied for the biochemical and nutritional properties. The biochemical significance of high nonprotein nitrogen in the seeds and detailed information on storage proteins of both the legumes merit further investigations. The higher content of minerals, particularly iron and molybdenum, in moth bean is of interest in the nutrition of pregnant women in these countries because of the prevalence of anemia in this group of the population.

Moth bean is mostly consumed in the form of *dhal* or sprouts. Improved milling technology needs to be employed to obtain higher yields of *dhal*. Several biochemical changes occur when seeds are converted into sprouts. Investigations are required to find the optimum conditions for the preparation of sprouts. Studies on preservation and storage of sprouts would improve the utilization of moth bean.

References

Dutta, R. V. S., & Manjrekar, S. P. (1983). *Preservation of germinated legumes by dehydration and bottling*. A report of the project work submitted in partial fulfilment of B. Pharm. to Birla Institute of Pillani, Rajasthan, India.

Kadam, S. S., Salunkhe, D. K., & Maga, J. A. (1985). Nutritional composition, processing, and utilization of horse gram and moth bean. *Critical Reviews in Food Science & Nutrition, 22*(1), 1–26. https://doi.org/10.1080/10408398509527407.

Kumar, D. (2002). *Production technology for moth bean in India*. Jodhpur: Indian Council of Agricultural Research, Central Arid Zone Research.

Mehta, S. L., & Simlot, M. M. (1982). Purification and characterization of trypsin inhibitor from moth bean (*Phaseolus aconitifolius*). *Indian Journal of Biochemistry & Biophysics, 19*, 403–407.

Ranjhan, S. K. (1980). *Animal nutrition and feeding practices in India* (2nd ed.). New Delhi: Vikas Publishing House.

Singh, S., Gupta, V., Yadava, N. S., & Singh, S. P. (2017). Yield and economics of moth bean [Vigna aconitifolia (Jacq.) Marechal] as influenced by different varieties and phosphorus levels. *International Journal of Current Microbiology and Applied Sciences, 6*(8), 831–835.

Singhal, V. (2003). *Indian agriculture* (pp. 196–197). New Delhi: Indian Economic Data Center.

Vijaylaksmi, D., & Venkatrao, S. (1977). Effect of processing and supplementation with sesame on the quality of the proteins of bengal gram, green gram and horse gram. In *FAO Expert Consultation on Grain Legume Processing, held at Central Food Technological Research Institute, November 14–18, Mysore, India.*

Vir, O., & Singh, A. K. (2015). Moth bean [Vigna aconitifolia (Jacq.) Maréchal] germplasm: Evaluation for genetic variability and inter characters relationship in hot arid climate of western Rajasthan, India. *Legume Research, 38*(6), 748–752. https://doi.org/10.18805/lr.v38i6.6718.

Mung Bean

G. Mohan Naik, P. Abhirami, and N. Venkatachalapathy

Introduction

Mung bean (*Vigna radiata L. Wilczek*) popularly known as green gram, believed to be native crop of India, is a tiny circular shaped bean in green color widely cultivated throughout Asia, including India, Pakistan, Bangladesh, Sri Lanka, Thailand, Laos, Cambodia, Vietnam, Indonesia, Malaysia, South China, and Republic of Formosa. This short-term legume can grow in varying environmental conditions, and later it expands it reach to the USA, Australia, and Africa. In general, mung bean is a source of high-quality protein which can be consumed as whole grains, dhal, or sprouted form and is an excellent complement to rice in respect to balanced human nutrition. In addition to being the prime source of human food and animal feed, it plays an important role in maintaining the soil fertility by enhancing the soil physical properties and fixing atmospheric nitrogen.

Structure of Mung Bean

Mung bean is an annual crop that is highly branched and is about 60–76 cm tall (Oplinger et al. 1990) with a slight tendency of twinning in upper branches. The central stem of this crop is roughly erect, but the side branches are semi-erect. The leaves of the plant are trifoliate, and it is deep-rooted. Clusters of 12–15 flowers are situated at the top of the plant, and eventually these flowers will develop into small cylindrical pods. The pods of this fully fertile and self-pollinated crop are linear,

G. Mohan Naik · P. Abhirami · N. Venkatachalapathy (✉)
Department of Food Engineering, Indian Institute of Food Processing Technology,
Ministry of Food Processing Industries, Government of India, Thanjavur, Tamil Nadu, India
e-mail: venkat@iifpt.edu.in

A. Manickavasagan, P. Thirunathan (eds.), *Pulses*,
https://doi.org/10.1007/978-3-030-41376-7_12

sometimes curve, round, and slender. The seeds enclosed within the pods are small and nearly globular. The three major components of dicotyledonous green gram seed are seed coat, cotyledon, and embryo accounting 12.1%, 85.6%, and 2.3% of the whole seed, respectively. Seed coat or testa is an outer covering which protects the embryo. The embryonic shoot above the cotyledon is epicotyl, and the embryonic root below the cotyledon is hypocotyl. Micropyle is a small pore on the seed that allows water absorption, and the hilum is a mark left on the seed coat by the stalk which attached the ovule to the ovary wall before it became a seed (Sefa-Dedeh and Stanley 1979).

Physical and Engineering Properties of Mung Bean

The knowledge of physical and engineering properties of mung bean is necessary for the designing and development of various separating, handling, storing, and drying systems, processes and controls in analysis, and determining the efficiency of the equipment required for the postharvest processing of this food legume. The relevant physical and engineering properties of green gram which include shape, size, mass, volume, bulk density, true density, porosity, angle of repose, and projected area are enlisted in Table 1. These properties are important for the quality of its deprived products such as texture of sprouts and consistency of dhal (Fig. 1).

Table 1 Physical and engineering properties of mung bean

Physical and engineering properties	Value
Length (mm)	4.9
Width (mm)	3.7
Thickness (mm)	3.6
Geometric mean diameter (mm)	4.3
Sphericity	0.82
Volume (mm^3)	33.2
Thousand seed weight (g)	35.6
Bulk density (kg/m^3)	756.81
True density (kg/m^3)	1335.4
Porosity (%)	40.8
Terminal velocity (m/s)	7.5
Angle of repose (degree)	27.6
Projected area (mm^2)	18.4

Reproduced from Dahiya et al. (2015)

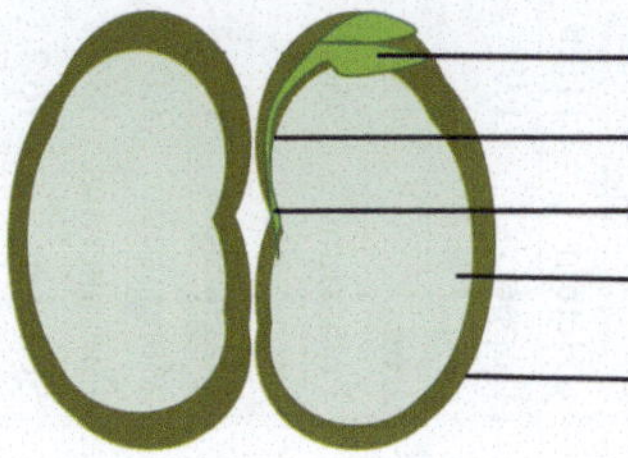

Fig. 1 Structure of mung bean

Nutritional Composition

Green gram can be a rich source of protein with higher digestibility and can serve to convalescing babies or malnutrition people. The nutrients are not distributed uniformly in major components such as seed coat, cotyledon, and embryo of the mung bean seed. The protein and lipids are found to be high in embryo, whereas the starch and crude fiber are concentrated in cotyledons and seed coats, respectively. The average moisture content present in the whole mung bean seed is 10.6 g/100 g of whole green gram with high protein (22.9 g), fat (1.2 g), total carbohydrate (61.8 g), crude fiber (4.4 g), and ash (3.5 g) per 100 g of sample (Adsule et al. 1986). The presence of antinutritional factors such as tannins (366.6 mg/100 mg), phytic acid (441.5 mg/100 g), hemagglutinin, trypsin inhibitors, proteinase inhibitors, and polyphenols (462.5 mg/100 g) were reported in mung bean, which affect the digestion and bioavailability of full nutrition (Mubarak 2005). The average chemical composition of the whole green gram is tabulated in Table 2.

Production Status

Pulses are one of the important groups of crops that plays a vital role in addressing the nutritional security all around the world. Pulses are cultivating throughout the world, and almost half of its production occurs in Asia especially in India. India is the largest producer of pulses and cultivated over 29 million hectares of area and recorded the highest ever production of 25.23 Mt during 2017–2018. In the case of mung bean, more than 3 Mt of green gram are produced in the world annually (FAOSTAT 2016). India contributes to the major share of mung bean in the world market with a production of 1.9 Mt in which Rajasthan with 42% area and 39% production outshined in the total mung bean production in the country (Ministry of Agriculture and Farmers Welfare, India 2018) which is followed by China (0.98 Mt), Myanmar (0.400 Mt), Indonesia (0.300 Mt), Thailand (0.210 Mt) and Pakistan (0.199 Mt). China is the largest exporter (Misiak et al. 2017), and India is the largest importer of mung beans.

Table 2 Chemical composition of the whole green gram seed

Carbohydrate (%)		Amino acid (g/16 g of nitrogen)		Lipid (% of total fat content)		Vitamin (mg/100 g dw)		Minerals (mg/100 g dw)	
Glucose	0.3	Alanine	4.1	Total saturated fatty acids	27.7	Thiamine	0.5	Calcium	113.4
Total soluble sugar	5.6	Arginine	5.8	Total unsaturated fatty acids	72.8	Riboflavin	0.3	Copper	1.0
Reducing sugar	1.8	Aspartic acid	13.0	Palmitic acids	14.1	Niacin	2.2	Iron	5.9
Nonreducing sugar	6.3	Cysteic acid	13.5	Stearic acid	4.3	Vitamin C	3.1	Potassium	956.6
Sucrose	1.3	Glutamic acid	18.3	Oleic acid	20.8	Pantothenic acid	1.9	Magnesium	162.4
Raffinose	1.1	Glycine	3.6	Linoleic acid	16.3	Nicotinic acid	1.6	Manganese	1.05
Stachyose	1.6	Histidine	3.2	Linolenic acid	35.7	–	–	Phosphorous	384.4
Verbascose	2.7	Isoleucine	4.3	Behenic acid	9.3	–	–	Sodium	16.7
Total dietary fiber	18.8	Leucine	7.6	–	–	–	–	Phytin phosphorus	171.3
Lignin	3.9	Lysine	6.5	–	–	–	–	Zinc	2.7
Cellulose	3.9	Methionine	1.2	–	–	–	–	–	–
Hemicellulose	4.7	Phenylalanine	5.4	–	–	–	–	–	–
Amylose	24	Proline	4.5	–	–	–	–	–	–
Starch	47	Serine	4.9	–	–	–	–	–	–
–	–	Threonine	3.2	–	–	–	–	–	–
–	–	Tryptophan	1.2	–	–	–	–	–	–
–	–	Tyrosine	2.7	–	–	–	–	–	–
–	–	Valine	5.1	–	–	–	–	–	–

Reproduced from Dahiya et al. (2015)

Postharvest Processing

Harvesting of the mung bean is usually done within 75–90 days when the pods turn dark or half to one third mature. The maturity of the pod is not uniform because of the extended period of flowering. Mature pods are usually handpicked. The moisture content present in the pods at the time of harvest is 13–15%, and to enhance the storage life, the moisture content should bring down to 12% by drying. In some newer varieties, pod matured uniformly, and the whole plant will be harvested and sun dried before threshing, and later seeds are separated from the pods by beating or trampling. Processing step brings the mung bean nearer to human consumption. Modernization of agriculture leads to farm mechanization which imparts combined harvesting and threshing equipment at farm level. Several farm machineries are available now for the harvesting of pulses which reduces the work load and time consumption.

Processing

Processing of mung bean is important in improving its nutritional value. It is concerned with its value addition. Milling of mung bean is one of the important processing methods employed to converting into dhal by decutilating and splitting. The process milling implies the removal of outer husk of mung bean and splitting the bean into two halves. Cleaning, grading, conditioning, dehusking, and polishing are the important unit operations in the mung bean milling process. The overall process flow is demonstrated in the given flowchart (Fig. 2) (Sahay and Singh 2004). Both wet milling and dry milling are preferred for commercial-scale production.

Unit Operations Involved in Green Gram Milling

Cleaning

Prior to the milling process, the mung beans received in milling plants should undergo a cleaning process in which the mung beans are cleaned from different organic and inorganic impurities such as chaffs, dust, stones, etc. with the help of screens, air drafts, or de-stoners. The cleaned grams are then graded according to their size in order to maintain the quality of the final product mostly by reel grader.

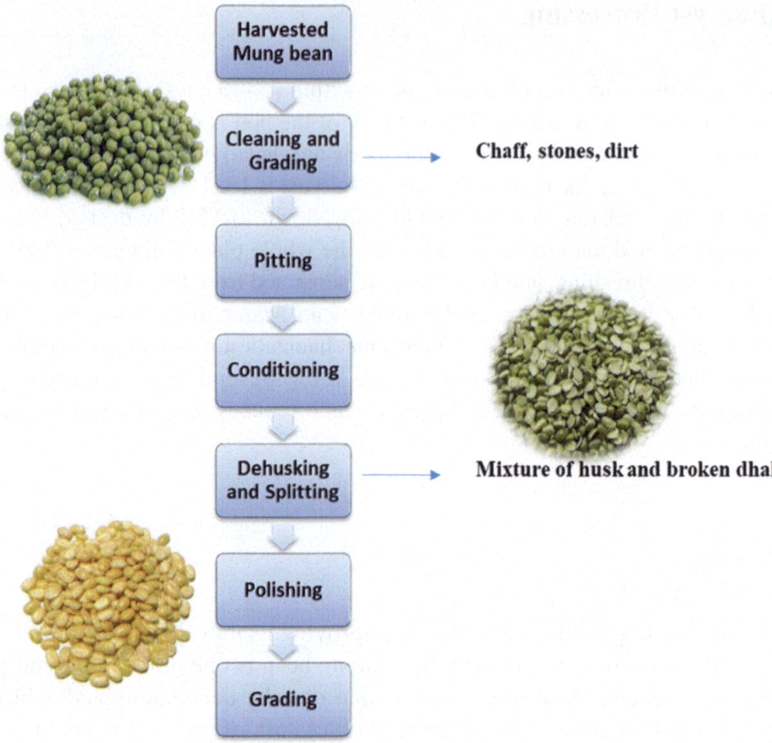

Fig. 2 Postharvest unit operation involved in mung bean processing

Conditioning

A gum/mucilaginous layer is present between the cotyledon and the outer kernel of the green gram. This gum adheres to the cotyledon to the seed coat. The thickness of the gum governs the degree of adherence, and it also influences the severity of conditioning. The main purpose of conditioning is to loosen the outer seed coat from cotyledon and make possible kernel separation. There are different ways of conditioning including water treatment, hydrothermal treatments, etc. with an alternate wetting and drying to reduce the milling losses. The drying time varies depending upon the type of milling method used for green gram processing.

Milling

Milling of mung bean involves the removal of outer husk and splitting the beans into two halves. Dehusking and splitting are important and major unit operations of mung bean milling. The conditioned mung beans are subjected to abrasive forces for the removal of outer husk. The dehusking is usually done with emery rolls, which furtherly splits in to two halves (dhal) by using vertical disc burr mill. Later the husk is aspirated off, and the split beans are separated by sieving. Since the kernel is tightly attached to the outer covering, alternate wetting and drying are followed for facilitating the removal of husk. Generally, there are two types of milling, namely, wet milling and dry milling. Both types of milling involves two basic steps: (1) preconditioning of mung bean for loosening the husk and (2) further milling of mung bean.

Wet Milling

Wet milling involves cleaning of mung bean followed by soaking in water for 4–12 h. After draining the water, soaked mung beans are mixed with red earth (2–3%) and heaped for overnight for effective moisture diffusion and equilibrium. Alternate sun drying (thin layers) and tempering of mung beans for 2–4 days have to be done prior to milling. Red earth should be separated from mung beans before milling. The milled beans can be graded according to the specifications. Unhusked beans can again pass for milling.

Dry Milling

In case of dry milling, pitting or scratching of outer surface will be done after the cleaning of mung beans followed by oil pretreatment and sun drying. The scratches are usually made by emery roller, and the oil (1.5–2.5 kg/t of green gram) mixed with mung beans is held for 12 h in a mixing vessel. Later oiled grains are spread for sun drying. After 3–4 days of drying, 3–5% water is sprayed on dried mung beans which are then subjected to milling. Whole dehusked beans recovered from the first milling shall undergo sun drying again after addition of water, followed by dehusking and splitting to obtain Grade (I) split dhal. In bigger dhal mills, dryers are used instead of sun drying.

Polishing

Polishing of dehusked and split beans is usually done to create a desirable shine for better consumer acceptability by treating it with small quantity of oil and water. Dhal is polished in different ways, such as nylon polish, oil/water polish, leather and makhmal polish. Generally polishing is done using soap stone, oil or water. The presence of oil and water improves the shine of the final product. The polished beans are graded according to the size to maintain the quality of final product.

Table 3 US grade and grade requirement for the class of mung bean

Maximum limits of							
		Foreign matters					
Grade	Moisture	Total (including stones)	Stones	Clean-cut weevil bored	Total damaged grains	Contrasting classes	Classes that blend
US No. 1	18.0	0.5	0.2	0.1	2.0	0.5	5.0
US No. 2	18.0	1.0	0.4	0.2	4.0	1.0	10.0
US No. 3	18.0	1.5	0.6	0.5	6.0	2.0	15.0

Reproduced from https://www.gipsa.usda.gov/fgis/standards/Bean-Standards.pdf (2017)

Table 4 AGMARK Grade specification for green gram (whole)

Maximum limits of tolerance (percent by weight)						
		Foreign matters				
Grade	Moisture	Organic	Inorganic	Other edible grains	Damaged grains	Weevilled grains percent by count
Special	10.0	0.10	Nil	0.1	0.5	2.0
Standard	12.0	0.50	0.10	0.5	2.0	4.0
General	14.0	0.75	0.25	3.0	5.0	6.0

Reproduced from https://dmi.gov.in (2005)

Grading

The cleaned grams are graded according to their size in order to maintain the quality of the final product mostly by reel grader. The polished beans are graded according to the size to maintain the quality of final product. Grading offers numerous advantages such as the following: it reduces the cost of marketing and transportation, assures the quality of produce, and farmers can get better price for their produce. US-based classification includes Grade I, Grade II, and Grade III. In this total defect percentage is different for each grade with same moisture content (Table 3). Directorate of Marketing Inspection (India) has drawn up specification of green gram for various quality factors under the Agricultural Produce (Grading and Marking), Act, 1937. The grade standards specified for green gram whole (Table 4), split-husked (Table 5) and split-unhusked (Table 6), are given below.

Product Development

Pulses are good sources of nutrients, vitamins, minerals, and other functional bioactive ingredients. Due to these factors, pulses are used as an important constituent of product in the development of different food formulations. The characteristic nature

Table 5 AGMARK Grade specification for green gram split (husked)

Maximum limits of tolerance (percent by weight)							
		Foreign matters					
Grade	Moisture	Organic	Inorganic	Other edible grains	Damaged grains	Weevilled grains percent by count	Brokens
Special	10.0	0.10	Nil	0.1	0.5	1.0	1.0
Standard	12.0	0.50	0.10	0.5	2.0	2.0	3.0
General	14.0	0.75	0.25	3.0	5.0	3.0	6.0

Reproduced from https://dmi.gov.in (2005)

Table 6 AGMARK Grade specification for green gram split (unhusked)

Maximum limits of tolerance (percent by weight)							
		Foreign matters					
Grade	Moisture	Organic	Inorganic	Other edible grains	Damaged grains	Weevilled grains percent by count	Brokens and fragments
Special	10.0	0.10	Nil	0.1	0.5	1.0	0.5
Standard	12.0	0.50	0.10	0.5	2.0	2.0	2.0
General	14.0	0.75	0.25	3.0	5.0	3.0	5.0

Reproduced from https://dmi.gov.in (2005)

of pulses is useful for the development of products like rich in protein, fiber, antioxidant, gluten-free, and low in fat and glycemic index (GI). The functional properties such as water holding capacity, oil holding capacity, and emulsification activity of pulse flours considerably contribute to the better quality of the product (Mazumdar et al. 2016).

A wide range of value-added products from green gram is available in the market as well as in household level. Most commonly green gram is widely consumed in the form of whole grain, split or whole dehusked dhal, and soaked, sprouted, and germinated grains. On the other hand, salted split dhal, flaked, chips and extruded green gram product is consumed as a snack; soaked, germinated and sprout is consumed as a green salad. The flour prepared by milling of whole grain, germination, or fermentation followed by drying is used in the development of functional foods for children, lactating mother, and infants, special formulated diet for patients, as well as ready-to-eat products. Generally, various bakery and confectionary products are produced by using refined or whole wheat flour as a base ingredient. In this context, replacement of wheat flour with alternative ingredient received considerable attention from the last three decades in food product development activities. Various literature is available for use of green gram or mung bean flour for the formulation of various value-added products like bread, cookies, biscuits, *dosa*, *idli*, functional beverages, and other ethnic foods (Fig. 3). Increase in nutritional awareness among the consumer and demand for health beneficial value-added products have been

Fig. 3 Processed products of mung bean

increased nowadays to fulfill the nutritional requirement. The incorporation of green gram in various forms could improve the nutritional status of the consumer by supplying adequate nutrients, thereby helping to alleviate the nutritional-related health issues.

Biscuit, Cookies, and Crackers

Biscuit and cookies belong to the category of baked products. Generally, wheat flour, sugar, and fat are used as the main ingredients. These are convenience foods consumed throughout the world in many forms such as cookies; short dough biscuit; hard sweet, snack crackers; soda crackers; and sweet biscuit. Cookies are prepared by the inclusion of a wide variety of dried nuts, chocolate chip, raisins, and dehydrated fruits. Biscuit and cookies have a longer shelf life because of lower water activity. They are nutritious and available in different functional forms (Davidson 2019). Shukla et al. (2016) developed the protein-enriched biscuit fortified with green gram flour. The experiment is carried out to replace the wheat flour with green gram flour (30%, 40%, 50% and 60%). It was observed that by incorporating green gram flour, the diameter and spread ratio were decreased with increase in addition of flour. The thickness and nutritional composition like protein, fat, and ash content were increased. The study revealed that incorporation of green gram flour with the replacement of wheat flour with a 30–60% level is acceptable. It not only improves the nutritional characteristics of the product but also improves the sensory quality.

Effect of incorporation of green gram flour on rheology, microstructure, and quality of cookies was investigated by Rajiv et al. (2012). The cookies were prepared by replacing wheat flour with green gram flour (10%, 20%, 30%, 40%, and 50%) in order to improve the proximate composition of the cookies. By increasing incorporation of green gram flour level extensibility, elasticity and peak viscosity of dough are decreased. The spread ratio of developed cookies ranged from 8.80 to 7.40. Cookies prepared by incorporating 40% of green gram flour and 60% of sodium stearoyl-2-lactylate as emulsifier were acceptable. Developed cookies were rich in protein, iron, calcium, and zinc, and total dietary fiber 1.25, 1.6, 2.0 and 2.3 times respectively than the control sample. Similarly, the addition of green gram flour of more than 50% yields hard texture and qualitatively unacceptable cookies. Tulse et al. (2014) studied the rheological and quality characteristics of cookies. The cookies were prepared by incorporation of co-milled wheat (70, 80, 90)/green gram (5, 10, 15)/barley gram (5, 10, 15). Water holding capacity and breaking strength increased; on the other hand hardness, spread ratio, extensibility, and stability of dough decreased by an increase in the proportion of green gram and barley flour. The cookies developed in the ratio 70:15:15 of co-milled wheat/green gram/barley were unacceptable and produce hard texture. The blend of co-milled wheat/green gram/barley at 80:10:10 cookies were good in textual quality and mouthfeel and rich in protein (12.30%) and dietary fiber (8%).

Savory crackers were developed by replacing the wheat flour partially or completely with green gram flour, and physicochemical and sensory attributes were investigated by Venkatachalam and Nagarajan (2017). Crackers were prepared at a ratio of wheat flour to green gram flour (100:0; 80:20; 60:40; 40:60; 20:80; 0:100) with additional ingredients like sugar, salt, sodium bicarbonate, vegetable oil, and water. All ingredients are mixed thoroughly and made as a dough. Afterward dough was sheeted and cut into small squares. Thereafter crackers were baked at 180 °C for 11 min in a conventional rotary oven. Products were removed and flavor, honey, and garlic were brushed on the surface of the crackers. The baking process was continued for another 4 min at the same temperature. It was pointed out that the addition of green gram flour improves the nutritional characteristics and functional properties of the crackers. The considerable changes in color value ($L*$, $a*$, $b*$, $\Delta E*$), ultrastructure, antioxidant activity, and sensory attributes were noticed in all formulations.

By the literature, it can be concluded that green gram flour is an excellent alternative against wheat flour up to an acceptable level. It not only improves the product sensory quality but also enriches the nutritional composition. Thereby it improves the nutritional status of the consumers.

Extruded Products

Extrusion is a high-temperature short-time process where a set of blended moist ingredients is enforced through a small opening in a die with a design specific to the food. Hot extrusion would be carried out by heating the product above 100 °C and in cold extrusion at ambient condition. It produces high-quality texturized products by modifying flour functionality and improves the digestibility and sensory characteristics (Patil and Kaur 2018). The extrusion study was carried out by blending green gram and rice flour by Chakraborty and Banerjee (2009). The expansion ratio of extruded product is decreased by an increase in moisture content; it shows that temperature and moisture content has a significant effect on product expansion ratio. The dough viscosity is decreased by an increase in the speed of screw which results in low-power consumption. Morphological characteristics showed gelatinization of starch and denaturation of protein by forming elongated and parallel air cells in extruded products. Bhattacharya (1996) investigated the effect of incorporation of green gram flour on extrudate characteristics. The operating conditions for double-screw extruder were as follows: screw speed (100–140 rpm); barrel temperature (100–175 °C); and barrel diameter (31 mm). The green gram to rice flour ratio was 1:1, and moisture content of the blended mixture was 18%. The expansion ratio and density of the extrudate varies between 1.31 and 2.53 and 149–1089 kg/m^3, respectively. At higher temperature level, torque was highest. Shear stress highly depended on process variables; on the other hand, temperature and speed of barrel directly influenced the quality characteristics of extrudates. Lower screw speed and higher barrel temperature yielded better quality of expanded products.

Protein-rich pasta was prepared by incorporation of green gram semolina to durum semolina in the ratio of 500:0, 400:100, 300:200, 200:300, and 100:400. The flour was mixed with water and formed a dough. The dough was extruded using a cold extruder and finally dried at 75 °C for 3 h. The viscosity of dough was decreased with an increase in the proportion of green gram semolina. Change in firmness, color, protein, and ash content of pasta was noticed. Pasta prepared by addition of 80% green gram semolina is unacceptable and sticky in nature, and higher cooking loss was observed. Incorporation of green gram semolina at the level of 60% yield better quality of pasta with respect to taste, texture, and quality. It can be concluded that green gram semolina offered a better quality of product as well as improved the nutritional composition (Jyotsna et al. 2013).

There is an increase in demand for ready-to-eat, ready-to-cook, and ready-to-serve convenience food from the last two decades. Pardeshi et al. (2013) developed ready-to-cook green gram nuggets by using cold extruder. The formulations were made by inclusion of wheat flour at the level of 0–30% in green gram flour. Steaming of the extrudate was done at 1.0 kg/cm^2 pressure followed by final drying (60 °C) up to desirable moisture level (7.73%). The optimized parameters of wheat flour (24%), steaming time (6 min), and initial moisture content (36.50%) yielded better quality of mung bean nuggets. The proximate composition of green gram nuggets was as follows: 65% carbohydrates; 22.50% protein; 1.5% fat, 3.5% ash, and 7.73% (wb) moisture. Ready-to-cook green gram nuggets could be stored up to 114 days at 30 °C in metalized polyester bags.

Bread

Bread and cake are staple foods prepared by using wheat flour, sugar, yeast, and shortening. Wheat flour containing 11.5–12.50% protein for bread and less than 10% for cake making is well suitable. Bread is made by mixing all ingredients in the form of dough, which has been baked at 225 °C in the oven. The protein content in the green gram ranges from 20% to 25% (Ganesan and Xu 2018). Thompson et al. (1976) investigated the preparation and application of mung bean flour in bread making. Initially dehulling of mung bean was carried out by steam conditioning at 7 psi for 5–7 s. After a 1-day resting period, dehulling was performed in pressure plate huller. Dehulled beans were milled into fine flour by using Buhler experimental mill. Proximate, functional, and rheological characteristics of mung bean flour were analyzed before the preparation of bread. It was noticed that 87% dehulling performance was achieved. The protein, fiber and ash content of mung bean range from 21.6–23.2%, 0.15–0.19% and 3.28–3.68% respectively. The bread was prepared by the straight dough method by adding mung bean flour (0–20%). After baking considerable changes in physical and sensory attributes were noticed. Inclusion of 20% mung bean flour in bread making produces lower loaf volume and a dark color compared to the control sample. The study points out that the replacement of 15% mung bean flour with wheat flour considerably produces the good quality of bread and the product approximately ten times rich in protein.

Noodles

Noodles are the convenience food prepared by either soft or hard wheat and consumed worldwide (Gulia et al. 2014). Protein-rich instant noodle was prepared, and its quality was analyzed by Jayarathne et al. (2006). High-protein noodles were prepared by using the optimal formulation: wheat flour (75 kg), soybean (10 kg), green gram (10 kg), and egg flour (5 kg). The flours were mixed with water, sodium bicarbonate, salt, sugar, and sodium tripolyphosphate. The dough was spread into a sheet and cut into noodles by sharp knife. Steaming and cooking are performed in a closed chamber at 100 °C for 90 s. The noodles were removed and allowed to cool at ambient condition. Frying was carried out for 90 s at 155–165 °C. The fried noodles were cooled and hermetically packed in oriented polypropylene pouches. The product was safe up to 6 months of the storage period; peroxide value (5.5) and moisture content (2.8%) of the products were within the acceptable standards. The nutritional composition of noodles was as follows: moisture 2.8%, ash 2.4%, crude protein 15%, fat 1.8%, and free fatty acids 0.6%.

Pickle

Pickling is a process of preservation of fruits and vegetables through controlled fermentation under acidic condition (pH < 4.5) by the addition of salt or brine, vinegar or acetic acid, sugar, oil, spices, and condiments (Erten et al. 2015). Protein-rich germinated green gram pickle was developed, and the preservation studies were carried out by Puranik et al. (2011). The germinated green gram pickle was prepared by the addition of salt, vinegar, garlic, oil, and other spices and condiments. The optimum level of ingredients for 50 g of pickle was oil 16 mL, vinegar 6 mL, garlic 7 g, and salt 4 g. The texture, flavor, and color of the product were observed within the acceptable limit with high protein and fiber content in its proximate composition. The product has a shelf life up to 2 months of storage period at the ambient condition with little or no change in its quality characteristics.

Future Trends in Processing and Product Development

Pulses are an integral part of our daily diet, rich in protein and fiber, and, hence, serve as an important source of the nutrient. Generally, green grams are harvested once the desirable maturity stage is attained. Adoption of improved harvesting, handling, drying, and storage methods is essential in order to obtain higher quality of processed food products. Apart from conventional sun drying, novel drying methods like microwave drying, radio-frequency drying, infrared drying, and other hybrid drying methods gain special attention in disinfection and drying of grain. Processing of green gram or pulses is the biggest challenge to obtaining the maximum dhal (split) yield. Prior to milling, pretreatment of green gram needs special attention to maximize the milling yield. Pretreatments like wet, hydrothermal, chemical, and enzymatic treatments are in practice. Currently, by using traditional, improved, or novel milling techniques, up to 75–80% of dhal yield was achieved. Still, research and innovation are essential to improve the milling yield, development of improved milling equipment, and drying technologies with low product loss and minimize the milling co-products.

Change in living lifestyle and demand for nutritious, convenience, and functional foods are increased. In the market convenience products like ready-to-eat, ready-to-cook, and ready-to-serve categories gain special interest toward the consumers. The value-added mung bean products like cookies, bread, biscuit, noodles, pasta, vermicelli, chips, salted split dhal, green gram flake, texturized products, and enriched flour mix are available in the market. In addition, incorporation of green gram flours produced by germination, sprouting, soaking, and fermentation serves as a perfect alternative toward wheat flour in the preparation of various bakery and confectionary products. Because of its high protein and fiber and low glycemic index, green gram flour is well suitable for therapeutic and novel food formulations. Development of improved processing methods, digestion and absorption studies, fortification, enrichment, level of antinutritional factor reduction, microbial safety, improved packaging, and storage methods with no or little effect on nutritional composition and product quality need further investigations.

Conclusion

Green gram is a leguminous crop that stores atmospheric nitrogen in root nodules, thereby considerably reducing the usage of chemical fertilizers in farm applications. It serves as an important source of protein, fiber and mineral in vegetarian diet. The wide range of milling technologies are available to separate the inner endosperm from outer husk. The efficiency and milling yield vary with the type of method used and preprocessing treatments. Postharvest processing and value addition are the major segments in effective utilization of mung bean. Processed products enriched and fortified with the addition of mung bean flour is very popular in the market. There is a growing demand for ready-to-eat, ready-to-cook, and ready-to-serve functional food due to awareness in consumers. Incorporation of green gram flour with wheat flours offers better quality of bakery products. Inclusion of sprouted, germinated, roasted flours in product development improves the product quality, nutritional composition, and sensory characteristics. The consumption of green gram and its value-added products improve the nutritional status of individuals; hence it serves an important source of nutrients and helps to eradicate the malnutrition in the world.

References

Adsule, R. N., Kadam, S. S., Salunkhe, D. K., et al. (1986). Chemistry and technology of green gram (*Vigna radiata* [L]. Wilczek). *Critical Reviews in Food Science and Nutrition, 25*(1), 73–105.

Bhattacharya, S. (1996). Twin-screw extrusion of rice-green gram blend: Extrusion and extrudate characteristics. *Journal of Food Engineering, 32*, 83–99.

Chakraborty, P., & Banerjee, S. (2009). Optimisation of extrusion process for production of expanded product from green gram and rice by response surface methodology. *Journal of Scientific and Industrial Research, 68*, 140–148.

Dahiya, P. K., Linnemann, A. R., Van Boekel, M. A. J. S., et al. (2015). Mung bean: Technological and nutritional potential. *Critical Reviews in Food Science and Nutrition, 55*(5), 670–688.

Davidson, I. (2019). *Biscuit, cookie and cracker production-process, production and packaging equipment*. Boca Raton, FL: CRC Press.

Department of Marketing Inspection. (2005). Retrieved May 28, 2019, from https://dmi.gov.in.

Erten, H., Boyaci-Gündüz, C. P., Ağırman, B., et al. (2015). Fermentation, pickling, and Turkish table olives. In Y. H. Hui, E. O. Evranuz, H. Erten, G. Bingöl, & M. E. J. Flores (Eds.), *Handbook of vegetable preservation and processing* (2nd ed., pp. 209–230). Boca Raton, FL: CRC press.

FAOSTAT (2016). Retrieved May 28, 2019 from https://www.fao.org/3/a-br622e.pdf.

Ganesan, K., & Xu, B. (2018). A critical review on phytochemical profile and health promoting effects of mung bean (*Vigna radiata*). *Food Science and Human Wellness, 7*, 11. https://doi.org/10.1016/j.fshw.2017.11.002.

Gulia, N., Dhaka, V., & Khatkar, B. S. (2014). Instant noodles: Processing, quality, and nutritional aspects. *Critical Reviews in Food Science and Nutrition, 54*(10), 1386–1399.

Jayarathne, U. S., Arampath, P. C., & Pemasiri, P. F. S. (2006). Development and quality evaluation of a protein-enriched instant noodle. *Tropical Agricultural Research, 18*, 1–11.

Jyotsna, R., Milind, Sakhare, S. D., et al. (2013). Effect of green gram semolina (Phaseolus aureus) on the rheology, nutrition, microstructure and quality characteristics of high-protein pasta. *Journal of Food Processing & Preservation, 38*, 1965–1972.

Mazumdar, D., Saikat, D., et al. (2016). Chapter 5. Utilization of pulses – Value addition and product development. In *pulses for sustainable food and nutrition security in SAARC region* (pp. 65–97). Bangladesh: SAARC Agriculture Centre.

Ministry of Agriculture and Farmers Welfare (2018). Pulses revolution from food to nutritional security-success report. Retrieved May 23, 2019 from https://www.farmer.gov.in/SucessReport2018-19.pdf.

Misiak, K., Gorna, B., Krol, E., et al. (2017). Yield and quality of mung bean (*Vigna radiata* (L) Wilczek) seeds produced in Poland. *Bulletin of University of Agricultural Sciences and Veterinary Medicine Cluj-Napoca. Horticulture, 74*(2), 149.

Mubarak, A. E. (2005). Nutritional composition and antinutritional factors of mung bean seeds (Phaseolus aureus) as affected by some home traditional processes. *Food Chemistry, 89*(4), 489–495.

Oplinger, E. S., Hardman, L. L., Kaminski, A. R., et al. (1990). Mungbean. In *Alternative feed crop manual*. Madison, WI: University of Wisconsin.

Pardeshi, I. L., Bhuskade, S. A., & Kalmegh, V. B. (2013). Development of cold extruded ready-to-cook Mung (Vigna radiata L.) nuggets. *Journal of Food Research and Technology, 1*(1), 21–28.

Patil, S. S., & Kaur, C. (2018). Review current trends in extrusion: Development of functional foods and novel ingredients. *Food Science and Technology Research, 24*(1), 23–34.

Puranik, V., Mishra, V., Singh, N., et al. (2011). Studies on development of protein-rich germinated green gram pickle and its preservation by using class one preservatives. *American Journal of Food Technology, 6*, 742–752.

Rajiv, J., Lobo, S., Jyothilaxmi, A., et al. (2012). Influence of green gram flour (Phaseolus aureus) on the rheology, microstructure and quality of cookies. *Journal of Texture Studies, 43*, 350. https://doi.org/10.1111/j.1745-4603.2012.00346.x.

Sahay, K. M., & Singh, K. K. (2004). *Unit operations of agricultural processing*. New Delhi: Vikas Publishing House Pvt. Ltd.

Sefa-Dedeh, S., & Stanley, D. W. (1979). The relationship of microstructure of cowpeas to water absorption and dehulling properties. *Cereal Chemistry, 56*(4), 379–386.

Shukla, R. N., Mishra, A. A., & Gautam, A. K. (2016). Development of protein enriched biscuit fortified with green gram flour. *Food Science Research Journal, 7*(1), 112–118.

Thompson, L. U., Hung, L., Wang, N., Rasper, V. F., & Gade, H. (1976). Preparation of mung bean flour and its application in bread making. *Canadian Institute of Food Science and Technology Journal, 9*(1), 1–5.

Tulse, S. B., Reshma, V., Rajiv, J., et al. (2014). Effect of co-milled wheat, green gram and barley on the rheological and quality characteristics of cookies. *Food Science and Technology International, 21*(7), 492–502.

United States Standards for Beans. (2017). Retrieved May 28, 2019, from https://www.gipsa.usda.gov/fgis/standards/Bean-Standards.pdf.

Venkatachalam, K., & Nagarajan, M. (2017). Physicochemical and sensory properties of savory crackers incorporating green gram flour to partially or wholly replace wheat flour. *Italian Journal of Food Science, 29*, 599–612.

Mungo Bean

Arumugam Sangeetha and Rangarajan Jagan Mohan

Introduction

In India, most of the people are vegetarian, and they depend on grains and beans. Every human requires a large amount of protein for a healthy life, and the mungo bean is a great source of protein. The mungo bean is also famous as a "Urad Dal" and is mainly used in foods for making dal. The scientific name of the mungo bean is *Vigna mungo* (Avinash and Patil 2018). Mungo [*Vigna mungo* (L.) Hepper; syn. *Phaseolus mungo* L.], which belongs to the family Fabaceae, is one of the most important pulse crops grown in Bangladesh. Mungo bean is an annual food legume that is also grown in Southern Asia such as in India, Pakistan, Bangladesh, Afghanistan, and Myanmar: it is a member of the Asiatic Vigna crop group. It is mainly a day-neutral warm season crop commonly grown in semiarid to subhumid lowland tropics and subtropics (Alam 2010). Mungo bean originated in India where it has been in cultivation from ancient times. India is the 15th largest producer and consumer of mungo bean in the world. The most suitable climate to cultivate mungo beans is 27–30 °C with heavy rainfall. This annual crop prefers loamy soil which has a high water preservation capability. The mungo bean grows normally in 90–120 days, and it also enriches the soil with nitrogen (DAC & FW 2015–2016). According to the International Market Research Companies (IMARC) group, the global mungo bean market reached a volume of 2.8 million tons in 2016, growing at a compound and annual growth

A. Sangeetha (✉) ·
Department of Nutrition and Dietetics, Jamal Mohamed College (Autonomous), Tiruchirappalli, Tamil Nadu, India

R. J. Mohan
Department of Nutrition and Dietetics, Jamal Mohamed College (Autonomous), Tiruchirappalli, Tamil Nadu, India

Dept of Food Product Development, Indian Institute of Food Processing Technology, Governmnet of India, Thanjavur, Tamil Nadu, India

A. Manickavasagan, P. Thirunathan (eds.), *Pulses*,
https://doi.org/10.1007/978-3-030-41376-7_13

rate (CAGR) of 6.7% during 2009–2016. IMARC's latest study, "Mungo Bean Market: Global Industry Trends, Share, Size, Growth, Opportunity and Forecast 2017–2022," provides a detailed analysis of the global black gram market. In this report of segments of the market on the basis of major geographic regions, India currently represents the largest producer of mungo beans, accounting for more than 70% of the global production. India is followed by Myanmar and Pakistan (IMARC 2017).

During 2015–2016, mungo bean accounted for an area of 3.19 million hectares in production, 1.95 million tonnes, and average productivity 596 kg per hectare. The production of pulses in general and mungo bean in particular has not been able to keep pace with the rapid increase in demand by the ever-increasing population. Lower production and productivity are mostly caused by several problems, with the mungo bean-growing farmers particularly having improper knowledge of practices. Therefore, it is necessary to assess the technological gap in production and also to know the problems and constraints in adopting improved mungo bean production technologies (Islam et al. 2011). Much of the Indian population (40%) is vegetarian, and pulses are important in providing protein (22%) and other essential nutrients to the large population of the country (Avinash and Patil 2018).

With high levels of nutrition, dietary fiber, proteins, and vitamins along with a strong flavor, mungo beans form an important part of a healthy diet. Driven by its health benefits, population growth, and changing dietary habits, the consumption of mungo bean has witnessed strong growth in the past few decades. Mungo bean differs from urad bean and other pulses in its peculiarity of attaining a mucilaginous pasty character when soaked in water.

In the south, it is consumed in variety of ways from north to south in preparation of different regular and popular dishes such as *vada*, *idli*, *dosa*, *halwa*, and *imarti* in combination with other food grains. It is also used as a nutritive fodder for milch cattle.

The mungo bean (*Vigna mungo*) is one of the most important food legumes grown and consumed in India, but till today only a few aspects of the nutritional composition have been studied (Grewal and Jood 2006; Khattak et al. 2007, 2008). Legumes, unfortunately, contain greater varieties of toxic constituents than any other plant family. The toxic compounds consist of some flavonoids, alkaloids, tannins, cyanogenic compounds, phytate, and trypsin inhibitors. For human consumption the legumes are processed by various methods that include soaking, boiling, sprouting, pressure-cooking, and fermentation, depending upon tradition and taste preferences. These processing treatments are also effective in eliminating the antinutritional factors. The effects of the various processing treatments on the nutritional and antinutritional composition of legumes, especially green and mungo bean, have also been studied (Chang and Xu 2008; Khattab and Arntfield 2009).

Among the various processing treatments, pressure-cooking was found to be the most effective in retention of the nutrients in cultivars of both legumes. For removal of antinutritional factors, both pressure-cooking and germination were found to be most effective among all the processing treatments (Kakati et al. 2010).

Nutritional Value of Mungo Bean

Mungo bean (*Vigna mungo* L.), raw mature seeds: nutritive value per 100 g

Principle	Nutrient value	Percentage of RDA (%)
Energy	341 kcal	17
Carbohydrates	58.99 g	45
Protein	25.21 g	45
Total fat	1.64 g	8
Cholesterol	0 mg	0
Dietary fiber	18.3 g	48
Vitamins		
Folates	216 µg	54
Niacin	1.447 mg	9
Pantothenic acid	0.906 mg	18
Pyridoxine	0.281 mg	22
Riboflavin	0.254 mg	20
Thiamin	0.273 mg	23
Vitamin A	23 IU	1
Vitamin C	0 mg	0
Electrolytes		
Sodium	38 mg	2.5
Potassium	983 mg	21
Minerals		
Calcium	138 mg	14
Copper	0.981 mg	109
Iron	7.57 mg	95
Magnesium	267 mg	67
Phosphorus	379 mg	54
Zinc	3.35 mg	30

Source: USDA National Nutrient database

Postharvest Processing

Postharvest technology emphasizes loss prevention and value addition to the raw food commodities through preservation and processing. Raw food materials are cleaned, graded, and then conditioned either for storage or for processing. Processing is done to make raw commodities edible through primary or secondary processing and ready to eat through secondary and tertiary processing.

At every stage of processing, value is added to the product. Estimated value additions to the raw food materials through primary and secondary/tertiary processing in India are 75% and 25%, respectively. Therefore, primary processing has a greater role in improving the economic benefits to the farmers. The process of industrialization in the country is changing the structure of the rural economy.

Concentrating the agricultural produce, processing, and value addition and its trade to urban areas will make farmers merely producers of raw materials. There is an urgent need, therefore, to change this scenario.

Seventy-five percent of pulses produced in India are processed; therefore, post-harvest technology is important in per capita availability. Pulse processing units vary in size from cottage industries to multi-story plants using pneumatic conveyors. The steps involved in making *dal* or *besan* at home or in mills are the following:

1. Cleaning (removing of foreign matter from pulse grain).
2. Dampening (soaking of the grain in water for desired time).
3. Tempering (keeping soaked grain for sun drying).
4. Splitting (grinding of grain to make *dal*).
5. Husking (removal of husk from *dal*).

Processing of *dal* is the second largest food processing industry after rice and flour milling. There are approximately 10,000 pulse mills with a processing capacity of 10–20 tons per day and an annual turnover of Rs. 45,000 crores. These mills are privately owned and operate on average for 200–250 days per year. The majority of *dal* mills use conventional technology with locally fabricated machinery that consumes large amounts of electricity and time and they are labor intensive. Output of the *dal* mills depends on the availability of raw material, capital, and energy, and also the capacity of the mill and the number of working days. The major portion of the pulses processed is milled by the *dal* mills with daily capacity ranging from 0.5 to 10 tonnes per day. Packing and storage of *dal* is related to loss of quantity as well as quality: the packing material is seldom of good quality.

Pulses are generally converted into *dal* by splitting the whole seed. More than 75% of the total legumes produced in the country is split into *dal*. Pulse processing is a small-scale industry including thousands of *dal* mills distributed throughout the country. Mills are mostly concentrated in producing areas such as Indore (Madhya Pradesh), Jalgaon and Akola (Maharashtra), and in other major cities such as Kolkata, Mumbai, Chennai, Hyderabad and Delhi. The Indian pulse milling process consists of several steps (Fig. 1). First, the pulses are cleaned and the stones and mud are removed. Then, the surfaces of the pulses are scratched so that when they are soaked in a mixture of water and vegetable oil, it is easier to remove the husks during the grinding process. Once the outer layer is removed, the pulses are split in half. To give a better finish, some processors polish the *dal*. Some pulses (mostly chickpea, *urd*, and *mung*) are milled to make flour *(besan)*. In the present economic structure, it is observed that a consumer pays Rs 100 for chickpea *dal* in India, under the most favorable conditions. The share of the farmer is only Rs 56, covering

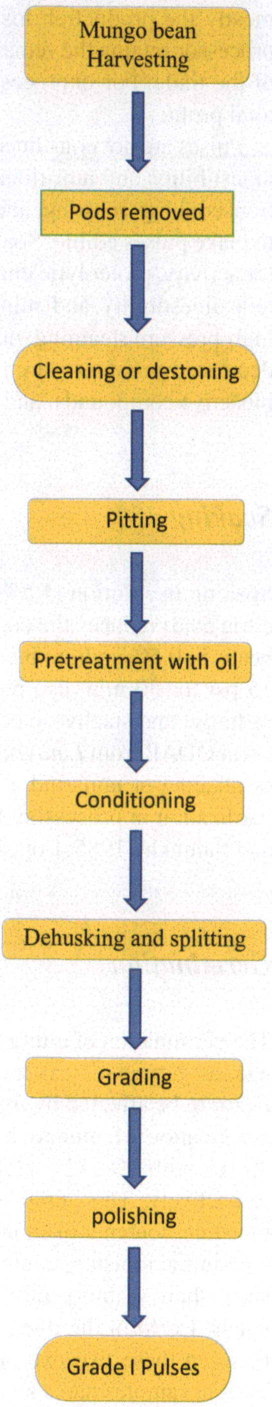

Fig. 1 Pulse milling process

mostly the production loss (Rs. 48), gaining only a marginal profit of 10%. The processor retains the remaining 44% for value adding and acting as the middleman of the trade. For unprocessed pulses, the grower gets about 0% of the share from total profit.

Pulses are not consumed raw; thus, these are processed to improve eating quality, digestibility, and nutritional and health significance. Dehusking, soaking, germination, cooking, roasting, and fermentation are processing techniques commonly used to make pulses edible. Soaking and germination help to eliminate the trypsin inhibitor activity, proteolytic enzyme inhibitors, phytates, and tannins, and increases protein digestibility and mineral bioavailability (iron, zinc). Conventional cooking, high-pressure steaming, microwave cooking, autoclaving, and splitting significantly decreased resistant starch and increased carbohydrates, fat, tocopherols, level of bioactive compounds, and antioxidant activity (Anjana 2016).

Soaking

Soaking in solution (1.5% sodium bicarbonate + 0.5% sodium carbonate + 0.75% citric acid) reduces the cooking time and improves protein quality, and soaking dry beans in 0.5% sodium bicarbonate solution for 18 h or pressure-cooking at 121 °C, 15 psi for 30 min also reduces flatulence-producing oligosaccharides, that is, the raffinose and stachyose content. Soaking is very important in removing the neurotoxin ODAP from *Lathyrus sativus* seeds: soaking them, particularly in boiled water or alkaline or tamarind solutions, is quite effective. Dose and duration need standardization in processing techniques (FAO of UN 2016; Gupta et al. 2011; Kadam and Salunkhe 1985; Lopez-Barrios et al. 2014).

Germination

The germination of mungo bean is used as an indicator for deterioration during storage, as germination is more sensitive to quality changes. Germination is the first factor to be affected by improper storage conditions. Figure 2 shows the changes in germination of mungo bean stored at three different temperatures: 20, 30, and 40 °C, with 9%, 12%, 15%, and 18% w.b. initial moisture content combinations, respectively. The initial germination was about 98.6% for all the various initial moisture contents of mungo bean stored at different temperatures. Grain samples with initial moisture content of 9%, 12%, 15%, and 18% w.b. stored at 20 °C did not show their viability until 25 weeks of storage; however, at 40 °C, germination reached zero in the third week of storage. Changes in germination at 30 °C were between these two extremes. At 40 °C, the 12%, 15%, and 18% initial moisture content samples had a significant decrease in germination after 1 week of storage.

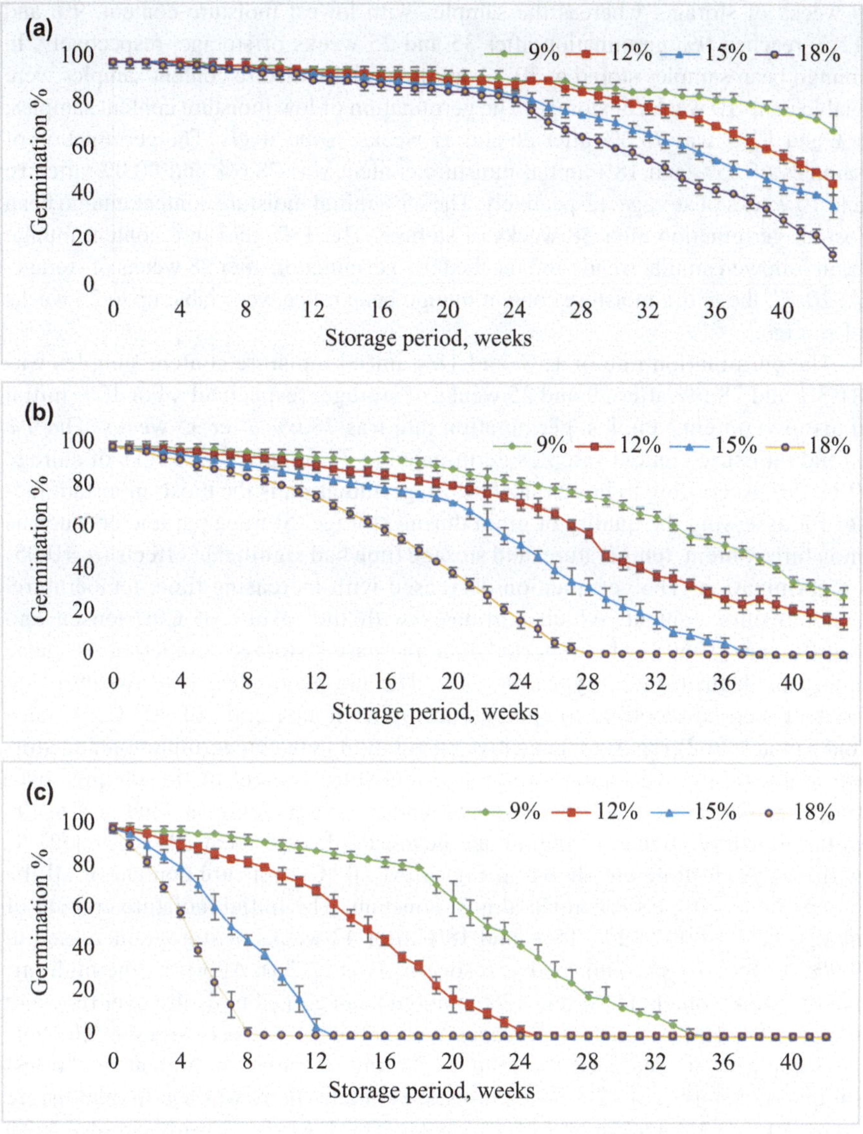

Fig. 2 Changes in germination of black gram with different initial moisture contents stored at (**a**) 20 °C, (**b**) 30 °C and (**c**) 40 °C with respect to storage period

Germination trends decreased to 0% germination after 35 weeks of storage for the samples with 9%, 12%, 15%, and 18% initial moisture content. The highest moisture content (18%) mungo bean reached 82.6% germination after 2 weeks of storage. Germination of 80% was reached in mungo beans of initial moisture content of 9%, 12%, and 15% after 18, 9, and 4 weeks of storage, respectively. The high moisture content samples, 15% and 18%, reached 0% germination after 13 and

9 weeks of storage, whereas the samples with lowest moisture content, 9% and 12%, reached 0% germination after 35 and 25 weeks of storage, respectively. In mungo bean samples stored at 30 °C, all the initial moisture content samples were viable up to 10 weeks of storage. The germination of low moisture content samples, 9% and 12%, was 78.6% after 26 and 21 weeks, respectively. The germination of samples of 15% and 18% initial moisture content was 78.6% and 80.0% after 16 and 10 weeks of storage, respectively. The 15% initial moisture content mungo bean lost all germination after 38 weeks of storage. The 18% moisture content mungo bean followed similar trends and reached 0% germination after 28 weeks of storage. At 20 °C, the entire moisture content mungo bean range was viable up to 25 weeks of storage.

The germination rate of 15% and 18% initial moisture content samples was 81.3% and 78.6% after 29 and 25 weeks of storage, respectively. For 12% initial moisture content samples, germination rate was 78.6% after 35 weeks. The 9% initial moisture content samples germinated at 78.6% after 42 weeks of storage (Fig. 2c). According to Pomeranz (1992), germination is the most important factor for assessing the quality of grain during storage. Storage parameters such as moisture content, temperature, and storage time had significant effects ($\alpha = 0.05$) on germination. The germination decreased with increasing time, temperature, and moisture content, which correlates with the results of Christensen and Kauffmann (1969), who reported that increased storage temperatures cause injury or death to most types of grain. The grain samples with low moisture content were susceptible to spoilage at higher temperature of 40 °C. Wallace and Sinha (1962) reported a negative correlation between germination and storage temperature. Minimum variation in moisture content of the samples over time was observed in all the initial moisture content samples. Changes in the initial moisture content of the mungo bean samples stored at 20, 30, and 40 °C with respect to time are shown in Fig. 3. At 20 °C, moisture content of all the mungo bean samples remained almost constant. The initial moisture content of mungo bean of 9%, 12%, 15%, and 18% after 43 weeks of storage increased to 9.9%, 12.8%, 16.3%, and 19.0%, respectively (Fig. 3a). At 30 °C, the high initial moisture content (15% and 18%) mungo bean gained moisture over time and increased to 16.2% and 18.1%, respectively, after 24 weeks of storage. The buffer samples were replaced to maintain the initial moisture content of the test sample. In samples of 12% initial moisture content, there was a gain in moisture content to 13.1% by the end of 43 weeks. The lowest moisture samples (9%) remained almost constant with time and increased to 10.3% moisture content at the end of 43 weeks of storage. Grain samples with initial moisture content of 9%, 12%, 15%, and 18% stored at 40 °C lost moisture over storage time from drying. The samples with lower initial moisture content of 9% and 12% decreased to 8.3% and 11.4% at the end of 34 and 23 weeks of storage, respectively. The moisture content of higher initial moisture samples (15% and 18%) also decreased with storage time and reached 14.8% and 17.7%, respectively, at the end of 12 and 8 weeks (Fig. 3c). According to Solomon (1951), controlling the relative humidity in biological experiments using chemical solutions such as

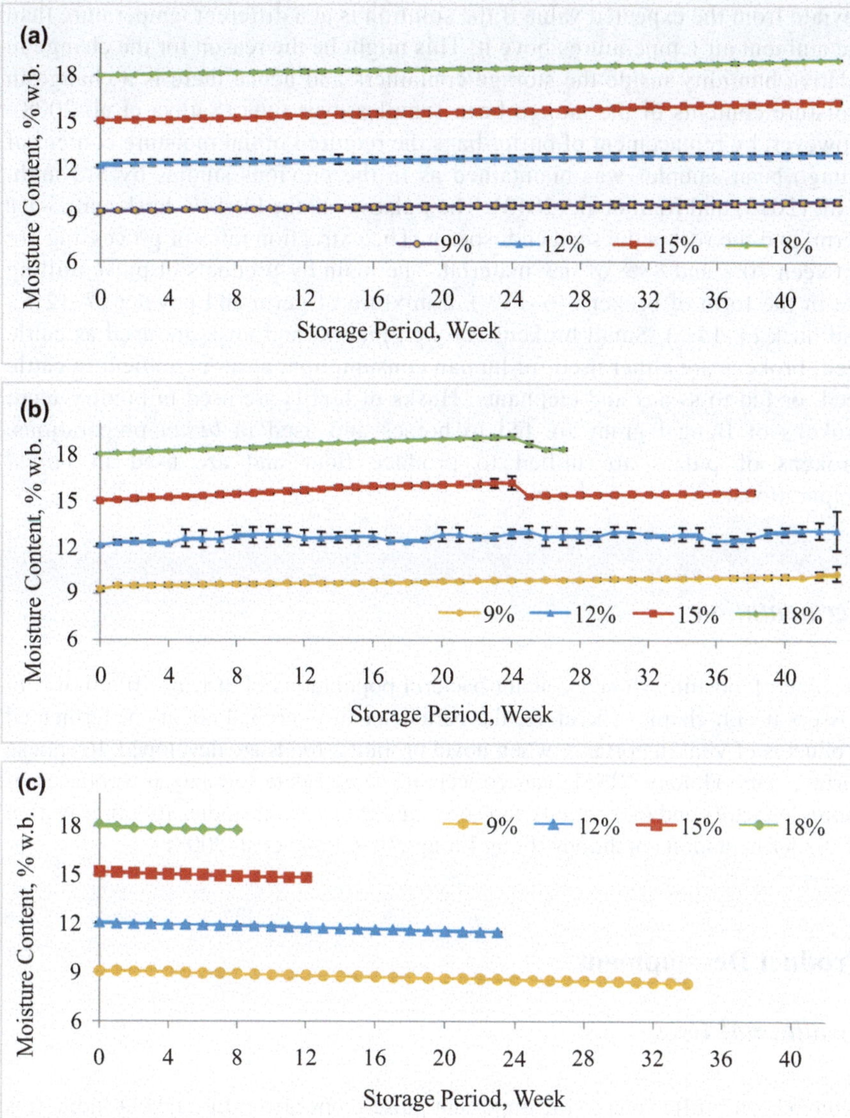

Fig. 3 Changes in moisture content of black gram with different initial moisture content stored at (**a**) 20 °C, (**b**) 30 °C, (**c**) 40 °C with respect to storage period

potassium hydroxide has been in practice for a long time. Errors in humidity control arise if the graded humidity solutions lose too much water through absorption of water vapor by materials enclosed with them or if they absorb water vapor from damp materials. Generally, solutions tend to give too low humidity at elevated temperatures. Furthermore, the equilibrium humidity will deviate from the expected value if the solution is at a different temperature than the ambient air temperature above it. This might be the reason for the change in relative humidity inside the storage containers and hence there is a change in moisture contents of the mungo bean samples over time (Sathya et al. 2008). However, by replacement of buffer bags the required initial moisture content of mungo bean samples was maintained as in the previous studies by Sravanthi et al. (2013) and Rani et al. (2013). The pulses contain 11–14% husk and 2–5% germ, and the rest is the seed endosperm. The extraction rates of processing are between 70% and 88% of raw material. The main by-products of pulse milling are in the form of brokens (6–13%), a mixture of germ and powder (7–12%), and husk (4–14%). Small brokens (broken pieces) and husk are used as cattle feed; brokens are either used for human consumption, as an ingredient in cattle feed, or fed to swans and elephants. Husks of lentils are used in poultry feed; brokens of Bengal gram are fed to horses and used in *besan* preparations. Brokens of pulses are milled to produce flour and are used in *papad* preparations.

Fermentation

Probiotic foodstuffs should contain bacteria populations of at least 10^6 cfu/mL to possess health claims. Therefore, the parameter of microbial counts of fermented product is of vital importance when novel probiotic foods are developed. Response surface methodology (RSM) can collect substantial data to analyze responses of multiple factors and interactions and so it is used in several studies for optimization of the fermentation conditions (Fung Liong 2010; Fung et al. 2008).

Product Development

Traditional Uses

Mungo bean (urd) is one of the important pulse crops grown throughout India. It is consumed in the form of 'dal' (whole or split, husked and unhusked) or parched. It is the chief constitute of 'papad' and also of 'bari' (spiced balls), which make a delicious curry. Urd differs from other pulses in its peculiarity of attaining, when ground up with water, a somewhat mucilaginous pasty character, giving additional body to the mass. In South India, the husked dal is ground into a fine paste and allowed to ferment and is then mixed with an equal quantity of rice flour to make 'dosa' and 'idli.' It is also fried to serve as a savoury dish. Urd dal is also used in the preparation of 'halwa' and 'imarti' (DPD 2012).

By-Product Utilization in the Milling Process

Various by-products that come from the milling of pulses include different fractions of pulses which need to be separated from the whole/split pulse. These by-products of *dal* milling such as husk, powder, and small brokens are usually sold as cattle feed. The pulse husk has been traditionally used as cattle feed because of the low bulk density, and it forms about 10% of the raw material that is sold as cattle feed at a lower price. The *dal* powder and small brokens, which are richer in nutrients, are sold at a higher price, also for cattle feed.

Value Addition in Supply Chain

The global consumer demand for high-quality pulses that are both fresh tasting and nutritious has created considerable interest. Investment in the development of new and improved postharvest storage, processing, and value addition methods in marketing pulses crops has been initiated. The competitive struggle for markets, which has resulted from more liberalized trade regimes, has required a much greater emphasis on efficient and effective postharvest handling, processing, storage, and distribution to access markets further and further away. Fresh packs and minimally processed products of pulses are developing gradually. The marketing concept for minimally processed pulses is based on the perceived consumer's desire for more natural, less processed, and high-quality, homemade-style preparations. Chilled ready-to-eat foods are very rapidly growing into a new segment of the market. Nutrition, coloration, dehulling, cooking properties, and moisture must be considered in developing quality standards in respect of such items. Efforts are underway to make value-added products such as breads and snack foods to increase prevailing pulses demand and to enhance their nutritional value and starch from pulses for use in the proper industries. For example, in Australia pulses have generally been used by the stock feed industry and any human consumption has been in a pre-cooked form such as baked beans; hence, pulses unfortunately have been seen by some, quite wrongly, as poor-quality food.

Most pulses are high in the oligosaccharides raffmose, stachyose, and verbascose, which cause flatulence. This problem combined with the long cooking times required to prepare them and the fact that they are usually only available in cans or as sauces has given pulses a poor marketing image as human food. Supermarket snack foods rarely have a pulse component. In contrast, Southeast Asian countries have top-selling snacks made from pulses.

Although this segment of the market is declining because of the popularity of Western-style chips, there is still a large market potential for such products. Pulses have a huge potential in extruded snack foods. They offer a good base for the extruded product as they produce a good flavor that is not overpowering when heated. They lend themselves to flavor addition and retain a good crunch if treated and stored properly. Pulses can also be used as "breakfast cereals" with added nuts or fruits. Breakfast bars are the home meal replacement option for breakfast. Both present an ideal opportunity for value adding with pulses. The addition of pulses that have been treated to remove any unusual flavors and softened for easy eating could differentiate a market cluttered with a range of products similar to one another. Pre-prepared meals are one of the fastest growing segments of the retail market. Manufacturers seek flavor, appearance, and an ability to retain texture and color when reheated from a frozen or cold state. Consumers want these properties as well as a healthy nutritious product. Many pulses can satisfy these requirements. They can add a valuable source of carbohydrates and proteins to a meal along with other positive nutritional benefits. Pulses that are ready to cook and eat quickly are another potential growth area. Pre-cooking and correct packaging is key to sales in this area. Pulses are a good source of carbohydrate, protein, fiber, and calcium, iron, thiamine, and riboflavin. With correct processing and packaging, pulse products can be sold as health supplements in supermarkets or health food stores. This is a specialty market, but with the appropriate scientific backing could be profitable. Pulses could also be used as face packs in beauty parlors to lend shine to skin.

While marketing pulses, regional, seasonal and varietal differences must be considered. Consumers in Tamil Nadu and Bihar states appreciate a deep yellow color in *dal*. For this, *dal* is mixed with yellow color additives (even nonpermitted dyes) in water solution, dried, and sold. This method of coloring is also used by millers to mask small patches of husk remaining on the *dal* from incomplete milling. In Western and Northwestern India, consumers prefer an oily-looking *dal,* for which an extra oil coating is given to the finished *dal* to impart an oily shine to the product.

Unfortunately, the pulse industry is competing in the market with many alternatives and therefore needs to differentiate itself. The key is to be imaginative and highlight the differences in pulses, rather than to make them look the same and have the same effect as an existing product.

The nutritional arena of pulses can make marked difference in their utilization and reduction in rampant malnutrition around the globe. The Food and Agriculture Organization (FAO) and United Nations have declared 2016 as "International Year of Pulses" (IYP) and seeds as "nutritious seeds for a sustainable future." The Indian Pulse and Grain Association has also portrayed pulses as the future ingredient to commensurate the nutritional and health benefits on the India Food Security Portal[1] by virtue of their richness in plant protein, energy, dietary fiber, and a wide variety of micronutrients and bioactive compounds.

Pulses and legumes terms are often used interchangeably. However, Codex Alimentarius Commission of FAO/WHO Food Standard programme[2,3] has given a clear distinction for pulses and legumes on the basis of their fat content. Pulses are dry seeds of leguminous plants and legumes that include oil seeds such as soybean and peanut. The crops that are harvested green are used as vegetables (green beans, peas, sprouts). There are about 11 primary pulses: (1) dry beans, (2) dry broad beans, (3) cow pea, (4) chickpea, (5) pigeon pea, (6) lentil, (7) bambara beans, (8) vetches, (9) lupins, and (10) minor pulses. These seeds are dicotyledonous; hence, they can be used as whole, decorticated, or dehusked (called "dal"). Since the time of Purans and Mahabharata, pulses have been an integral part of the Indian diet as dal-chawal, dal roti, and in popular snacks such as sattu, besan ke laddo, and besan sev. All these are regularly consumed in a wide variety of cuisines in different parts of the country. Sattu is considered an 'instant' food that has been made by preparing flour from roasted chickpea and barley or wheat since the time of the Rigveda (8000 BC) and is often used by common people living in adverse conditions. Papads are a traditional and very popular snack food primarily made from mungo bean flour with salt, oil, and spices. Papads are now under experimentation for value addition by adding different cereal flours and green leafy vegetables. Hummus is another popular product using chickpea that has been found to reduce the postprandial glucose responses to fourfold less than that of white bread.

Pulses are a source of constant supply of nutrition year round, and the nutritional value of pulses per 100 g is much higher than any other vegetarian food. The Indian Institute of Pulses Research, Kanpur, India, recently suggested that the pulses have tremendous scope to be popularized as a 'health food' or 'nutri-rich food.' The presence of antinutritional factors (ANF) limits the utilization of pulses, but the same factors also act as bioactive substances exhibiting significant favorable effects on health in reducing risks for coronary heart disease, diabetes, and obesity. Hence, food technologists are trying to explore pulses as functional foods and nutraceuticals. The inherent capacity of pulses to fix atmospheric nitrogen and thus increase soil fertility and decrease use of expensive chemical nitrogenous fertilizers make cultivation of pulses a good strategy for farmers also. Enhanced production of pulses can also create opportunities for local value-added processing, stimulate domestic demand, and provide off-farm employment and sources of income for the rural poor, especially women and youths.

Value addition in pulses can be more fruitful in the market. Motivated by consumer demand, there is a need for the development of foods with short cooking time, microbial safety, and high quality. Pulses, being a cheaper source of plant protein than nuts, milk, cheese, meat, and fish, can be used in bakery products such as pasta, breads, and snacks. Pulses provide ample opportunities to be used in such processed foods as bread and pasta as well as ingredients for 'designer' foods for snacks, baby foods, and sport foods. Pulses can fortify breakfast cereals, and microwaveable or partly prepared pulse-based meals can fulfil consumer demand for convenient meal solutions. The food service sector also prefers quick-cooking pulse products.

Germination of pulses is a simple and popular technique to enhance the palatability, digestibility, and nutritive value of seeds. Increased vitamins B and C, utilization of available proteins and carbohydrates, and decreased antinutritional factors and flatulence factors are useful outcomes of germination. Malting, roasting, and fermentation are other preferred methods for developing many local food products in different parts of the world.

Future Trends in Processing and Product Development

Functional properties such as solubility, water- and fat-binding capacity, and foaming are influenced by genetic makeup of the legumes and amino acid type, water and oil absorption, and protein solubility, which determine their utility in the development of bakery products, soups, and extruded products and ready-to-eat snacks. The ratio of amylase to amylopectin also determines the functional properties of pulse starch, such as in texture, rheological, and swelling properties that are important in food applications. High amylase starch retrogrades to greater extent than high amylopectin starches, resulting in higher degree of crystallinity, syneresis, and gel firmness (Radha et al. 1989; Kumar et al. 2015; Ramachandran 2014; Reddy 2006; Sasanam et al. 2011; Sat and Keles 2006; Vidyalankar and Charaka 1994).

Technological processing may evoke positive effects such as protein coagulation, starch swelling and gelatinization, texture softening, and formation of aroma components. Technological processing and biofortification at different levels on pea plants have improved nutrition composition, antioxidant capacity, chlorophyll content, and soluble sugars by severalfold, which can support developing attractive, convenient, ready-to-eat, and tasty legume-based food formulations as per consumer demand. Thermal treatment (pressure-cooking) increased the water absorption capacity in Bengal gram, mungo bean, and lentil while significantly decreasing fat absorption and foaming capacities. It also influenced the sensory scores for appearance, texture, flavor, and overall quality for Seviya and 'Chakli,' South Indian snacks.

It is quite clear that the studies reviewed so far in the field of value addition and quality attributes in the marketing of pulses, although these have covered several different aspects, are very limited in number. Most of these studies were either related to some aspect of marketing of pulses in general or concerned a particular state of the country. None of the studies was specific in discussing different ways of adding value in the marketing of these products, and little attempt has been made to study different quality characteristics that may increase consumer acceptability of the products. In view of the limitations of the studies and the developments that have recently taken place at national and international levels, a new set of studies are needed to explore emerging opportunities. There remains a gap that suggests renewed research on value addition and quality attributes in marketing of pulses to keep pace with changes.

In conclusion, to have value addition in mungo bean and other pulses and thereby make more profit available to farmers for their produce, different organizations have provided research efforts toward developing effective pre-milling treatments for loosening the seed coat of mungo bean grains to facilitate the milling process on one hand and to develop low-capacity dal mills to cater to the needs of crop-producing centers on the other. As a result of these efforts, some good designs of low-capacity dal mills have been developed by different R&D institutions and private sector units.

Acknowledgement The authors are thankful to the college Management Committee Members namely Dr. A.K. Khaja Nazeemudeen, Secretary and Correspondent, Hajee. M. J. Jamal Mohamed, Treasurer, Dr. K. Abdus Samad, Assistant Secretary and also to Dr. S. Ismail Mohideen, Principal for providing facilities and support.

References

Anjana, A. (2016). Value addition of pulse products in India. *Journal of Nutrition Health & Food Engineering, 5*(2), 590–592. https://doi.org/10.15406/jnhfe.2016.05.00166.

Avinash, C. S., & Patil, B. L. (2018). Trends in area, production and productivity of major pulses in Karnataka and India: An economic analysis. *Journal of Pharmacognosy and Phytochemistry, 7*(4), 2097–2102.

Christensen CM, Kaufmann HH (1969). Grain storage; the role of fungi in quality loss. University of Minnesota press, Minneapolis, MN.

Chang, K. C. S., & Xu, B. (2008). Effect of soaking, boiling, and steaming on total phenolic content and antioxidant activities of cool season food legumes. *Food Chemistry, 110*, 1–13.

FAO of UN. (2016). Pulses: Nutritious seeds for sustainable future. *A journey through all regions of the planet: Brazil, China, India, Mexico, Morocco, Pakistan, Spain, Tanzania, Turkey, USA, and recipes from some of the most prestigious chefs in the world.*

Fung, W. Y., & Liong, M. T. (2010). Evaluation of proteolytic and ACE-inhibitory activity of *Lactobacillus acidophilus* in soy whey growth medium via response surface methodology. *LWT - Food Science and Technology, 43*(3), 563e–567e.

Fung, W. Y., Woo, Y. P., & Liong, M. T. (2008). Optimization of growth of *Lactobacillus acidophilus* FTCC 0291 and evaluation of growth characteristics in soy whey medium: A response surface methodology approach. *Journal of Agricultural and Food Chemistry, 56*(17), 7910e–7918e.

Grewal, A., & Jood, S. (2006). Effect of processing treatments on nutritional and antinutritional contents on green gram. *Journal of Food Biochemistry, 30*, 535–546.

Gupta, M., Tiwari, B. K., & Norton, T. (2011). Value addition and international trade. In A. Pratap & J. Kumar (Eds.), *Biology and breeding of legumes*. Wallingford: CAB (eBooks).

Islam, M., Mohanty, A. K., & Kumar, S. (2011). Correlation growth yield and adoption of urdbean technologies. *Indian Research Journal of Extension Education, 11*(2), 20–24.

Kadam, S. S., & Salunkhe, D. K. (1985). Nutritional composition, processing, and utilization of horse gram and moth bean. *Critical Reviews in Food Science and Nutrition, 22*(1), 1–26.

Kakati, P., Deka, S. C., Kotoki, D., & Saikia, S. (2010). Effect of traditional methods of processing on the nutrient contents and some antinutritional factors in newly developed cultivars of green gram [*Vigna radiata* (L.) Wilezek] and Mungo bean [*Vigna mungo* (L.) Hepper] of Assam, India. *International Food Research Journal, 17*, 377–384.

Khattab, R. Y., & Arntfield, S. D. (2009). Nutritional quality of legume seeds as affected by some physical treatments. *LWT - Food Science and Technology, 42*, 1113–1118.

Khattak, A. B., Zeb, A., Bibi, N., Khalil, S. A., & Khattak, M. S. (2007). Influence of germination techniques on phytic acid and polyphenols content of chickpea (Cicer arietinum L.) sprouts. *Food Chemistry, 104*, 1074–1079.

Khattak, A. B., Zeb, A., & Bibi, N. (2008). Impact of germination time and type of illumination on carotenoid content, protein solubility and in vitro protein digestibility of chickpea (*Cicer arietinum* L.) sprouts. *Food Chemistry, 109*, 797–801.

Kumar, R., Bhayana, S., & Kapoor, S. (2015). The role of functional foods for healthy life: Current perspectives. *International Journal of Pharmacy and Biological Sciences, 6*(3), 429–443.

Lopez-Barrios, L., Gutierrez-Uribe, J. A., & Serna-Saldivar, S. O. (2014). Bioactive peptides and hydrolysates from pulses and their potential use as functional ingredients. *Journal of Food Science, 79*(3), R273–R283.

Mony, S. A., Haque, M. S., Alam, M. M., Hasanuzzaman, M., & Nahar, K. (2010). Regeneration of mungo bean (*Vigna mungo* L.) on changes of hormonal condition. *Notulae Botanicae Horti Agrobotanici Cluj-Napoca, 38*(3), 140–145. https://doi.org/10.15835/nbha3834744.

Pomeranz Y (1992). Biochemical, functional and nutritive changes during storage. In Storage of Cereal Grains and Their Products, ed. J.A. Anderson and A.W. Alcock, 55–141. St. Paul, MN: American Association of Cereal Chemists.

Radha, A., Narasinga, R., & Roy, D. N. (1989). Lectins, trypsin inhibitors, BOAA and tannins in legumes and cereals and the effects of processing. *Food Chemistry, 34*, 229–238.

Ramchandran, P. (2014). Food and nutrition security: Challenges in the new millennium. *The Indian Journal of Medical Research, 138*(3), 373–382.

Reddy, B. S. (2006). Indigenous technical knowledge on pulses: Storage and processing practices in Andhra Pradesh. *Indian Journal of Traditional Knowledge, 5*(1), 87–94.

Report given by International Market Research Companies, New York. (2017). *Mungo bean market: Global industry trends, share, size, growth, opportunity and forecast 2017–2022*. Retrieved from http://www.reportlinker.com/p04715134-summary/view-report.html

Report of Department of Agriculture, Cooperation & Farmers Welfare (DAC&FW). (2015–2016). *Directorate of Economics and Statistics. 2015–2016*.

Report of Directorate of Pulse Development, Department of Agriculture, Cooperation, Department of agricultural and Cooperation. (2012). *Pulses in India: Retrospect and prospects*. GOI: Ministry of Agriculture.

Sasanam, S., Paseephol, T., & Moongngarm, A. (2011). Comparison of proximate compositions, resistant starch content, and pasting properties of different colored cowpeas (*Vigna unguiculata*) and red kidney bean (*Phaseolus vulgaris*). *International Journal of Biological, Biomolecular, Agricultural, Food and Biotechnological Engineering, 5*(9), 553–557.

Sravanthi, B., Algusundaram, K., Deyas, D. S., & Chelladurai, V. (2013). Effect of storage conditions on red lentils. *Journal of Stored Products Research, 53*, 48–53. https://doi.org/10.1016/j.jspr.2013.01.004.

Sathya G., Jayasl D. S., White N. D. G. (2008). Safe storage guidelines for rye. Canadian Biosystems Engineering, P 50.

Solomon ME (1951). Control of humidity with potassium hydroxide, sulphuric acid, or other solutions. *Bull. Entomol. Res. 42*, 543–554.

Vidyalankar, J., & Charaka, S. (1994). *Part I. Motilal Banarsidass* (9th ed.). New Delhi: Red Books.

Wallace H. A. H., Sinha R. N. (1962). Fungi associated with hot spots in farm stored grain. *Can. J. Plant Sci. 42*, 130–141.

Pea

Martin Mondor

Introduction

Pea (*Pisum sativum* L.) is a major pulse grown worldwide, and Canada is the world's largest producer of this crop. In 2017–2018, Canada produced 4.112 × 10^9 kg and exported 3.085 × 10^9 kg of peas; for 2018–2019 and 2019–2020, respectively, the production forecast was 3.581 × 10^9 and 4.300 × 10^9 kg, and the export forecast was 3.200 × 10^9 and 3.100 × 10^9 kg (Agriculture and Agri-Food Canada 2019). Peas have high nutritional value with high protein, starch, and fiber contents but low levels of fat and sodium. The chemical composition varies depending on the growing conditions, year, and variety. Protein content (derived with a nitrogen conversion factor of 6.25) varies between 22.2% and 32.5% dry basis (d.b.), starch content varies between 33.4% and 52.3% d.b., fiber content varies between 19.8% and 31.4% d.b., and fat content varies between 0.5% and 4.0% d.b. (Nikolopoulou et al. 2007; Piecyk et al. 2012; Li and Ganjyal 2017). Furthermore, peas are gluten-free and not genetically modified and are low in allergens and glycemic index scores in comparison with many other cereals and pulses (Araya et al. 2003; Barac et al. 2010).

Given current concerns about food security and protein malnutrition, pea proteins are of tremendous interest for human food. Pea proteins have significant amounts of all the essential amino acids (Table 1). Holt and Sosulski (1979) suggested that peas satisfy adult human requirements for the essential amino acids except for the sulfur-containing ones (methionine + cysteine). In terms of molecular characteristics, pea proteins can be classified on the basis of their solubility in salt solution or water as globulins (salt-soluble) or albumins (water-soluble). Globulins represent approximately 50–70% of total pea proteins, whereas albumins represent 15–40% (Boye et al. 2010a). Salt-soluble globulins are composed of four fractions,

M. Mondor (✉)
St-Hyacinthe Research and Development Centre, Agriculture and Agri-Food Canada, St-Hyacinthe, QC, Canada
e-mail: Martin.mondor@canada.ca

A. Manickavasagan, P. Thirunathan (eds.), *Pulses*,
https://doi.org/10.1007/978-3-030-41376-7_14

Table 1 Amino acid content of pea (from Khattab et al. 2009)

Amino acids (g/100 g protein)	
Conditionally essential	
Arginine	7.93
Histidine	2.33
Essential	
Isoleucine	3.89
Leucine	7.84
Lysine	6.25
Methionine + cysteine	1.60
Phenylalanine	5.17
Threonine	4.46
Tryptophan	0.61
Valine	5.11
Nonessential	
Alanine	4.83
Aspartic acid + asparagine	11.16
Cystine	0.35
Glutamic acid + glutamine	18.46
Glycine	4.82
Proline	4.64
Serine	5.71
Tyrosine	3.34

namely, 2S, 7S, 11S, and 15S proteins, with the 7S (vicilin/convicilin) and 11S (legumin) fractions being the two most predominant (Chakraborty et al. 1979). Legumin has a hexagonal shape with a molecular weight varying between 330 and 410 kDa, whereas vicilin and convicilin are trimeric and have molecular weights of 150 kDa and 180–210 kDa, respectively (Mession et al. 2015). Several studies have shown that each fraction has different functional properties (Adebiyi and Aluko 2011; Rubio et al. 2014; Mession et al. 2015; Djoullah et al. 2018; Lam et al. 2018; Xiong et al. 2018a). As a result, the structure and properties of products containing pea proteins are expected to reflect their 11S/7S ratios as well as the processes used to produce the products. The functional properties of pea protein ingredients will dictate how they can be used in the formulation of food products. Key functional properties include solubility, water- and fat-adsorption capacities, emulsifying properties, foam-forming capacity and stability, and gelling properties (Shand et al. 2007; Aluko et al. 2009; Felix et al. 2010; Liang and Tang 2013; Stone et al. 2015b; Chen et al. 2019). All the aforementioned functional properties, as well as protein digestibility, will be affected by the processes used to produce the pea protein ingredients (Ma et al. 2011; Taherian et al. 2011; Nosworthy et al. 2017; Oliete et al. 2018; Qamar et al. 2019; Xiong et al. 2018b; Bogahawaththa et al. 2019; Burger and Zhang 2019).

Another important pea component is starch, which consists of two types of molecules: amylose and amylopectin (Fig. 1). Amylose consists mainly of

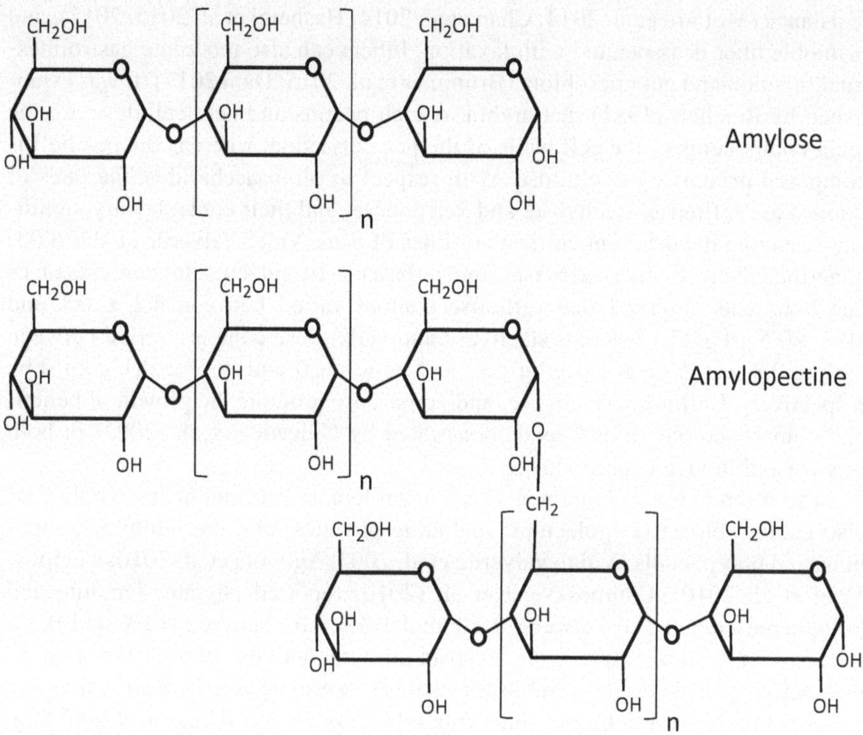

Fig. 1 Amylose and amylopectin structure

α-(1→4)-linked D-glucopyranosyl residues, but a slight degree of branching [9–20 α-(1→6) branch points per molecule] has been reported, and it has a molecular weight varying between 10^2 and 10^3 kDa. Amylopectin is composed of linear chains of (1→4)-α-D-glucose residues connected through (1→6)-α-linkages and has a molecular weight varying between 10^4 and 10^6 kDa (Ratnayake et al. 2002). The amylose/amylopectin ratio of pea starches differs significantly depending on the seed phenotype and genotype. An amylose content varying between 8% and 88% was reported by Ratnayake et al. (2002). Important functional properties of starch include solubility, swelling, gelatinization, and retrogradation. These properties as well as starch digestibility can be affected by processing (Yao et al. 2010; Leite et al. 2017). Recently, pea starch has been considered for the production of coatings (Mehyar et al. 2007; Chen et al. 2008, 2009; Corrales et al. 2009; Cano et al. 2015; Saberi et al. 2016a, b, 2017, 2018a, b).

In addition to their high protein and starch contents, peas are also an important source of dietary fiber, which consists of indigestible polysaccharides (e.g., cellulose, hemicelluloses, oligosaccharides, pectins, and gums), waxes, and lignin. Peas contain both soluble fiber (2–9 g/100 g) and insoluble fiber (10–15 g/100 g) (Tosh and Yada 2010). The health benefits of soluble fiber include the reduction of cholesterol level and favorable effects on glycemic control and insulin resistance (Mollard et al. 2014; Chan et al. 2014; Hashemi et al. 2015, 2017), and insoluble fiber is associated with laxation. Fibers can also modulate gastrointestinal function and gut microbiota (Brummer et al. 2015; Dahl 2017). It was established by Reichert (1981) that arabinose-rich pectins and hemicellulose are the main components of the cell walls of the pea cotyledon, whereas the pea hull is composed primarily of cellulose. With respect to oligosaccharides, the ones of interest are raffinose, stachyose, and verbascose, and their contents vary significantly among the different cultivars or lines of peas. Vidal-Valverde et al. (2003) quantified the raffinose, stachyose, and verbascose (α-galactoside) contents of 18 pea lines and observed that raffinose content varied between 4.1 ± 0.1 and 10.3 ± 0.5 g/kg d.b., whereas stachyose and verbascose contents varied between 10.7 ± 0.2 and 26.7 ± 1.0 g/kg d.b. and between 0 and 26.7 ± 0.1 g/kg d.b., respectively. Raffinose, stachyose, and verbascose promote the growth of beneficial colon bacterial strains, as demonstrated by Gulewicz et al. (2002) in both in vitro and in vivo experiments.

In addition to basic components such as protein, starch, and fibers, whole peas also contain bioactive molecules, including phytates, enzyme inhibitors, saponins, and polyphenols (Vidal-Valverde et al. 2003; Agboola et al. 2010; Campos-Vega et al. 2010). Campos-Vega et al. (2010) reported phytate, saponin, and polyphenol contents of between 0.2% and 1.3% d.b., between 0.1% and 0.3% d.b., and 0.25% d.b., respectively. Trypsin inhibitor activity, chymotrypsin inhibitor activity, and amylase inhibitor activity were detected, with values of 5.4–7.8 TIU (trypsin inhibitor units)/mg d.b., 740–10,240 IU/g, and 14–80 U/g, respectively, and hemagglutinin (lectin) activity of 5.1–15.06 U/kg was reported. All the aforementioned bioactive molecules may have detrimental or beneficial impacts on human health. Phytates are known to reduce mineral bioavailability owing to their high chelating potential. However, they have also been shown to have anticancer properties (Shamsuddin 2002), delay postprandial glucose absorption, and reduce the bioavailability of toxic heavy metals (Minihane and Rimbach 2002). Saponins are known for their hemolytic activity and their toxicity, although only a few saponins are toxic. There is also now some scientific evidence that saponins may possess anticancer activity and be beneficial for hyperlipidemia (Shi et al. 2004). Polyphenols can reduce mineral bioavailability and protein digestibility but also have antioxidant properties (Campos-Vega et al. 2010). It is also possible to reduce the level of polyphenols in yellow peas by cooking them (Han and Baik 2008). Lectins are known to reduce nutrient absorption, and enzyme inhibitors can negatively impact protein digestibility. However, lectins and enzyme inhibitors are removed by cooking (Campos-Vega et al. 2010). Recent work also demonstrated that pea lectin has anticancer properties (Islam et al. 2018). All the aforementioned bioactive molecules found in peas have long been considered antinutritional factors, but it is now clear that they may also confer health benefits.

Postharvest Processing

Peas are generally consumed after being cooked, in the form of whole seeds or decorticated splits. However, peas are also processed into ingredients, such as flour, protein-enriched ingredients, starch-enriched ingredients, and fibers, which can be used in the industry. Depending on the target application, different postharvest processes can be applied to peas. The most common processes are malting, dehulling, milling, fractionation, extraction, extrusion, and cooking. The basic principles of the aforementioned processes and their impact on the resulting pea ingredients are presented in the following subsections.

Malting

The malting process can be divided into three operations: steeping (the soaking of seeds in water), germination, and drying (Mondor et al. 2014). During the germination step, some of the reserve materials present in the seeds are degraded and are used partly for respiration and partly for synthesis of new constituents of the developing embryo. This process causes important changes in the composition of the seed, therefore modifying its biochemical, nutritional, and sensory characteristics (Vidal-Valverde et al. 2002). Several studies on the impact of pea seed germination on composition have been carried out (Kadlec et al. 2001; Vidal-Valverde et al. 2002; Martín-Cabrejas et al. 2003; Mondor et al. 2014; Villeneuve and Mondor 2014; Boye and Ma 2015; Ma et al. 2017; Ribéreau et al. 2018). One significant change that occurs during pea germination is the decrease in the amount of α-galactosides present in the seeds. Vidal-Valverde et al. (2002) reported that the total amount of α-galactosides in the pea variety Esla decreased from 5.19 ± 0.03% d.b. in raw seeds to 0.10 ± 0.01% and 0.12 ± 0.00% d.b. in germinated seeds after 6 days of germination with and without light, respectively. Kadlec et al. (2001) also reported a decrease in the content of α-galactosides in pea seeds during germination. Those authors observed a decrease of 59% after 3 days of germination. In addition to the amount of α-galactosides, the amount of phytates (inositol phosphates) present in pea seeds is also affected by germination treatment. Vidal-Valverde et al. (2002) reported decreases of 70% and 64% in the total amount of inositol phosphates present in germinated seeds after 6 days of germination with and without light, respectively. Germination treatments can also result in an increase in the concentration of some molecules of interest. For example, Vidal-Valverde et al. (2002) observed that the concentration of vitamin B_2 in germinated pea seeds increased by more than 100% after 6 days of germination. More recently, Mondor et al. (2014) studied the impact of supplementing breads at a 10% level with germinated or raw yellow pea flour on dough properties and bread-making potential. The results showed that dough mixing properties were barely affected and that both flours resulted in breads with similar specific volumes. However, the bread supplemented with 10% germinated pea flour had a slightly higher protein content than the bread supplemented with the raw pea flour did. In summary, malting can be applied as a pretreatment before the production of the flour or the various ingredients, in order to modify their composition.

Dehulling

Depending on the application, the pea seed coat, or hull, is left on or removed before consumption or processing. For the production of pea splits, the hull is removed, and then the cotyledon is split into its two halves called "splits." Two approaches can be used to remove hulls, namely, wet and dry milling. Generally, the dry method, which involves abrasion of the hull, is used because it is more efficient (Goyal et al. 2008). In an initial series of experiments, Black et al. (1998) studied the effects of pea genotype on dehulling efficiency. A total of 23 genotypes were considered. In a second series of tests, the impact of various pretreatments (soaking in water, soaking in sodium bicarbonate solution (10 g/kg), soaking in sodium chloride solution (10 g/kg), and preheating) on the dehulled fractions was studied with a blue field pea (CP 128952). For the soaking treatments, the peas (50 g) were soaked in 100 cm^3 of water or one of the salt solutions at 293.15 K for 600 s and then dried, whereas the preheating treatment was carried out in an oven at 343.15 K for 1800 s. The control sample underwent none of these treatments, and when the 23 genotypes were tested, no pretreatment was used. Dehulling efficiency varied between 71.16% and 85.72%, indicating that large variations are found between the genotypes. The pretreatments barely affected the dehulled fractions, with the splits representing 81.1%, 81.9%, 82.4%, 82.3%, and 81.0% for the control, the peas soaked in water, the peas soaked in sodium bicarbonate solution, the peas soaked in sodium chloride solution, and the preheated peas, respectively. Wang et al. (2008) studied the impact of dehulling on pea composition. Their results indicated that hull removal resulted in increases in protein, starch, K, P, phytic acid, stachyose, and verbascose contents and decreases in soluble dietary fiber, insoluble dietary fiber, total dietary fiber, Ca, Cu, Fe, Mg, and Mn contents.

Flour Production (Milling)

Because of a growing number of food applications, peas are now increasingly being processed into flour. Pea flour production involves grinding either whole peas or dehulled and split peas into small particles. Various milling methods, such as hammer, pin, roller, and stone mills, can be used. Hammer mills and pin mills fracture peas into small particles by exposing them to hammers or pins, whereas roller mills and stone mills fracture peas by compressing them between two hardened surfaces. Maskus et al. (2016) studied the impact of the milling method on the composition and the physical and functional properties of whole and split yellow pea flours. For composition, those authors characterized the protein, starch, and fiber contents of the flours as well as their particle size distribution. For physical and functional properties, the authors analyzed pasting properties, starch damage, foaming properties, oil- and water-absorption capacities, emulsifying properties, and color. It was observed that the milling method significantly affected the particle size distribution of the flours. The stone-milled flour was the one with the largest mean particles (595.6 ± 58.92 μm), followed by the flours produced by the hammer mill, coarse pin mill, and roller mill, which had similar mean particle sizes (274.2 ± 4.55 μm, 276.8 ± 20.72 μm, and 236.8 ± 3.68 μm, respectively) and by the flour produced by the fine pin mill, which had the smallest mean particle size, at 97.1 ± 16.81 μm. Significant differences were observed among the pea flours in terms of protein and starch contents, with values varying between 22.1 ± 0.01% and 23.5 ± 0.16% d.b. for protein and between 43.7 ± 0.05% and 49.6 ± 1.27% d.b. for starch. Despite the fact that these contents were statistically significantly different, it was not known whether these differences were large enough to be of practical significance. The milling method had an impact on pasting properties, oil- and water-absorption capacities, emulsion stability, and color, whereas foam capacity and emulsifying activity were not affected. Kaiser et al. (2019) studied the impact of hammer milling parameters on the characteristics of pea flour. Specifically, yellow split peas at 9% and 11% moisture were hammer milled with nine mill screen apertures (0.84–9.53 mm) at two rotor speeds (34 and 102 m/s). The authors reported that both rotor speed and screen aperture had a significant impact on flour particle size distribution, with the combination of a rotor speed of 102 m/s and a screen aperture of 0.84 mm resulting in the smallest median particle size (98 μm). Starch damage varied between 0.1% and 1.4% and was lower at 102 m/s than at 34 m/s for small screen apertures. The results also indicated that bulk density, redness, and yellowness were negatively correlated with flour particle size, whereas brightness was positively correlated.

Sometimes, to improve the nutritional value of the protein, flour can be treated enzymatically. Periago et al. (1998) used acid protease from *Aspergillus saitoi* to hydrolyze pea flour. This resulted in an increase in the free amino acid content in the hydrolyzed flour compared to the raw flour. The protein solubility for pH values below the isoelectric point of the protein (approximately pH 4.5) was improved by the enzymatic treatment, as were oil-absorption capacity, emulsification capacity, and gelation capacity, whereas foaming capacity was decreased. In vitro protein digestibility was also decreased by the enzymatic treatment. The authors concluded that the enzymatic hydrolysis treatment had a positive impact on most of the flour's functional properties, but the effect on the nutritional value of the flour was unclear.

Dry Fractionation (Air Classification)

Air classification is a dry fractionation process in which a finely milled flour is split into two size fractions (fraction enriched with large starch granules vs. fraction enriched with small proteins) using air flow to modify the particle size distribution (Reichert 1982; Rempel et al. 2019). When compared with wet extraction processes, air classification has the advantages of not using chemicals and water, not causing loss of insoluble proteins or protein denaturation, and requiring less energy (Pelgrom et al. 2013; Schutyser et al. 2015; Rempel et al. 2019). However, air classification results in a relatively low level of protein enrichment (Schutyser et al. 2015). Rempel et al. (2019) used an industrial-scale air classifier to separate three commercial pea flours at a cutoff point of 22 μm. Those authors studied the impact of the air classification process on the composition of the various fractions that were obtained. The three flours used in that study were a whole green pea flour, a split green pea flour, and an organic split yellow pea flour. For each flour, the same experimental procedure was carried out: feeding of the flour, milling, and air classification to obtain an initial coarse fraction and an initial fine fraction. The coarse fraction was then fed, milled, and air classified again into fine and coarse fractions. This procedure was repeated seven times to obtain a total of seven coarse fractions and seven fine fractions for each flour. The results indicated that nutrient shifting in the fine fraction depended on the processed flour. For the whole green pea flour, after seven passes, the shifting was 87.3%, 82.3%, 75.6%, 46.8%, and 12.4% wet basis (w.b.) for protein, fat, ash, dietary fiber, and starch, respectively, in comparison with values of 87.3%, 74.4%, 65.7%, 48.4%, and 8.7% w.b. for the split green pea flour and 84.7%, 94.5%, 70.7%, 57.1%, and 22.4% w.b. for the organic split yellow pea flour. The results showed that after repeated milling and air classification processes, protein, fat, and ash shifted to the fine fraction, whereas starch shifted to the coarse fraction. In terms of protein content, the most enriched fine fractions had a protein content varying between 42% and 50%. The coarse fractions that were the most enriched in starch content had a starch content varying between 60% and 67%. The total yields of the fine fraction were 39.6%, 34.9%, and 43.5% for the whole green pea flour, split green pea flour, and split yellow pea flour, respectively (Rempel et al. 2019).

Wet Extraction Processes

Wet extraction processes are common methods used to produce plant protein ingredients and include alkaline extraction-isoelectric precipitation (AE-IP), salt extraction-dialysis (SED), and micellar precipitation (MP) (Stone et al. 2015a; Lam et al. 2018). The AE-IP process consists of the extraction of the plant proteins at alkaline pH, followed by the removal of the insoluble matter and the precipitation of the proteins by adjusting the pH of the solution to the isoelectric point of the proteins (between pH 4 and 5). The proteins are then recovered and are resolubilized at pH 7 before being spray-dried. The resulting plant protein ingredient is composed mainly of globulins and usually has a protein content of over 90% d.b. This is the most common method of plant protein extraction reported in the literature. The SED process consists of the extraction of the plant proteins in a salt solution, followed by the removal of the insoluble matter and then dialysis to remove the salt before spray-drying. This process generally results in an ingredient containing both globulins and albumins. In the MP process, the plant proteins are first extracted in a salt solution, and then the insoluble residue is removed. Cold water is then added at a ratio of high-salt protein extract to water of 1:2 to 1:10 (v/v), resulting in the formation of micelles, which precipitate. The resulting ingredient has a micelle-like form and is stabilized by hydrogen bonds. It contains both globulins and albumins.

Stone et al. (2015a) studied the impact of each of these three wet extraction processes on the yield and functionality of the resulting pea ingredients. In terms of ingredient yield, those authors reported average yields of 15.7 ± 0.9% for the AE-IP process, 18.4 ± 1.4% for the SED process, and 7.1 ± 0.9% for the MP process, indicating that SED had the highest ingredient yield. Protein yield was found to be the highest for the SED process, at 71.9 ± 2.5%, followed by 69.1 ± 2.0% for the AE-IP process and 30.9 ± 1.8% for the MP process. However, the ingredient produced by MP had the highest protein content, at 85.2 ± 1.9% w.b., followed by ingredients produced by the AE-IP and SED processes, at 85.0 ± 1.5% and 75.6 ± 1.6% w.b., respectively. For functional properties, the ingredients produced by SED demonstrated the highest protein solubility (87.5 ± 1.3%), and those produced by MP demonstrated the lowest (45.9 ± 0.9%). The SED ingredients also had the highest oil-holding capacity (5.3 ± 0.3 g/g) and the lowest water-holding capacity (1.5 ± 0.4 g/g), whereas the MP ingredients had the highest water-holding capacity (3.4 ± 0.3 g/g). In general, the SED ingredients tended to have better foaming capacity, whereas the AE-IP ingredients produced foams that were more stable. Emulsifying capacity was greater for the SED ingredients than for those produced by AE-IP. All ingredients displayed high emulsion stability (approximately 98%). Sun and Arntfield (2010, 2011) applied an MP process followed by the resuspension of the proteins in distilled water to remove the unwanted salt by dialysis. Those authors studied the gelation properties of the resulting pea proteins at different pH levels (3–10) and NaCl concentrations (0–2 M) for a 14.5% (w/v) pea protein dispersion. The authors observed that the strongest gel stiffness was achieved at pH 4.0 and 0.3 M NaCl. A higher or lower pH value and a higher or lower NaCl concentration resulted in reduced gel stiffness. Gelation temperature was also influenced by the pH and the NaCl concentration. Higher pH values resulted in higher denaturation temperatures and higher enthalpies of denaturation, and higher NaCl concentrations inhibited pea protein denaturation, resulting in higher gelling points.

Extrusion

Extrusion is a process that consists of forcing a melted viscous material through a die where the vapor expands the material to a porous structure. Extrusion is one of the most versatile and efficient processes for producing starch-based foods and processing protein-fortified products (Ben-Hdech et al. 1991; Della Valle et al. 1994; Alonso et al. 2000; Hood-Niefer and Tyler 2010; Wang et al. 2012; Maskus and Arntfield 2015; Li et al. 2017; Kristiawan et al. 2018; Arribas et al. 2019). Kristiawan et al. (2018) studied the impact of extrusion parameters (flour moisture content, temperature, and specific mechanical energy) on the structural changes of extruded yellow pea flour. Specifically, flour moisture content varied between 18% and 35% (weight basis), temperature varied between 388.15 and 438.15 K, and specific mechanical energy varied between 50 and 1200 kJ/kg. Density, crystallinity, gelatinization enthalpy, solubility of the proteins in sodium dodecyl sulfate (SDS) and dithioerythritol (DTE), and solubility of the starch in water were measured to determine the structural changes in the extruded pea flour. The results indicated that increases in specific mechanical energy and in temperature resulted in a decrease in the density from 820 to 85 kg/m^3. Enthalpy and crystallinity results showed that for all extrusion conditions considered in that study, starch was amorphous. In addition, starch solubility in water increased up to 50%. Protein solubility in SDS decreased from 95% to 35% when the temperature was increased from 388.15 to 438.15 K, whereas protein solubility in DTE increased from 5% to 45% under the same conditions. The impact of specific mechanical energy on protein solubility followed similar trends. These results suggest the creation of protein network by S-S bonds, implicating larger SDS-insoluble protein aggregates, as a result of increasing temperature and specific mechanical energy, accompanied by the creation of covalent bonds other than S-S ones. The authors concluded that the interval from 388.15 to 419.15 K would be an appropriate temperature range for the design of extruded products.

Domestic Processing Methods

Peas are processed not only at the industrial scale but also at home by the consumer before consumption. Common domestic processing methods include blanching, soaking, and cooking. Zhao and Chang (2008) studied the impact of blanching peas and soaking them in water, followed by cooking with four different cooking methods and dehydration in a convection tray dehydrator. Peas (0.5 kg) were blanched in water (1:3, w/v) at 355.15 K for 300 s and then soaked in deionized water (ratio of peas to water of 1:5, w/v). Soaking temperatures of 295.15 and 355.15 K were chosen, with the soaking time varying between 1 and 24 h. After soaking, excess water was drained off, and the peas were cooked by indirect cooking, ordinary cooking, pressure cooking, or microwave cooking. This step was followed by dehydration of the cooked peas. Green and yellow peas were investigated in that study, and the impact of the treatments was assessed by measuring the firmness of the peas. For yellow peas, firmness varied between 183.8 ± 3.9 and 349.3 ± 13.0 kg force/100 g, whereas it varied between 160.8 ± 6.3 and 475.5 ± 20.6 kg force/100 g for green peas. For the cooking method, pressure cooking resulted in peas with the lowest firmness for both yellow and green peas. In another study, Eyaru et al. (2009) determined the effect of soaking, boiling, and pressure cooking on the starch fractions of yellow and green peas. The results indicated that soaking resulted in reduced starch fractions, possibly because of the leaching of soluble fractions. Ordinary cooking of soaked or unsoaked peas led to a significant decrease in resistant starch and an increase in rapidly digestible and slowly digestible starches and total starch. Pressure cooking led to an even greater reduction in resistant starch and a greater increase in rapidly digestible starch.

Product Development

Edible Applications

Pea ingredients have been used in the production of pasta and bakery products (Dalgetty and Baik 2006; Kamaljit et al. 2010; Petitot et al. 2010; Wang et al. 2012; Kaya et al. 2018), sausages (Kaack and Pedersen 2005; Pietrasik and Janz 2010; Vinauskienė et al. 2015), and low-fat chicken nuggets (Verma et al. 2015; Shoaib et al. 2018). However, pea ingredients are of interest not only for the production of food products for human consumption but also for use in animal diets (Cruz-Suarez et al. 2001; Owusu-Asiedu et al. 2002; Thiessen et al. 2003; Soto-Navarro et al. 2004; Bani et al. 2009; Davies and Gouveia 2010; Jiménez-Moreno et al. 2011; Sørensen et al. 2011; Röhe et al. 2017; Greenwell et al. 2018; Velayudhan et al. 2019).

Pasta, Breads, and Cookies

When pea ingredients are added to pasta or bread, the main objective is to increase the nutritional quality of the food product by increasing its protein and/or fiber content. However, the pasta- and bread-making processes as well as the characteristics of the final products depend on the substitution level and can be affected to varying degrees.

Pasta

Kaya et al. (2018) supplemented Turkish noodles with 2.5%, 5%, and 10% pea hulls and investigated the effects on the proximate composition, mineral composition, color, cooking properties, thiamine and riboflavin contents, texture, and sensory properties of the noodles. The crude ash, fiber, Ca, and Mg contents of the noodles significantly increased with pea hull substitution in comparison with the control. Lightness ($L*$) values were higher for the noodles in which pea hulls were substituted than for the control, but the substitution level did not have any effect. The water absorption and swelling volume of the noodles increased with pea hull substitution. Thiamine content was lower in the noodles with pea hulls than in the control, whereas riboflavin content was higher. The appearance, texture, and overall acceptability of the noodles were not significantly affected for substitution levels up to 10%. In another study, Wang et al. (2012) prepared pea starch noodles by twin-screw extrusion. Those authors studied the effects of processing variables (moisture content, screw speed, and barrel temperature) on the physicochemical, textural, and cooking properties of the noodles. The results indicated that cooking loss was reduced by an increase in the dough moisture content, whereas the other properties ($b*$ value, expansion ratio, cooking time, percentage of gelatinized starch, resistant starch content, firmness, and surface stickiness) were increased. Cooking loss and cooked weight were reduced by raising the barrel temperature, but the expansion ratio, percentage of gelatinized starch, resistant starch, firmness, and surface stickiness were increased. The $a*$ value and cooked weight were reduced as screw speed was increased, whereas the $b*$ value, expansion ratio, cooking time, percentage of gelatinized starch, resistant starch, firmness, and surface stickiness were increased. Petitot et al. (2010) looked at the impact of substituting 35% split pea flour for 35% durum wheat semolina on the minimum water content required for dough formation, particle size distribution of the dough after mixing, extrusion pressure, and pasta quality parameters (optimal cooking time, water uptake, cooking loss, and color) for pasta subjected to low-temperature drying. Pasta made from 100% durum wheat semolina was used as the control. The results indicated that the minimum amount of water required for dough formation was lower for the pasta with split pea flour than for the control pasta (39.4% vs. 49.0% d.b.). The particle size distribution of the dough was also affected by the substitution, with 41.1% fine particles (<1 mm) and 13.9% large particles (>4 mm) for the pasta with split pea flour in comparison with 68% and 1.6%, respectively, for the control. Also, extrusion pressure was lower for the wheat-pea dough (8.9×10^5 Pa) than for the durum wheat dough (10.7×10^5 Pa). The quality parameters of the pasta were also affected by the substitution: optimal cooking time was lower for the pasta with split pea flour than for the control pasta (8.5 ± 0.1 vs. 9.3 ± 0.3 min), water uptake was reduced by the substitution, from $192 \pm 7\%$ d.b. for the control to $166 \pm 6\%$ d.b. for the pasta with split pea flour, and cooking loss was increased by the substitution, from $5.6 \pm 0.4\%$ d.b. for the control to $6.8 \pm 0.8\%$ d.b. for the pasta with split pea flour. The substitution of pea flour for wheat flour also significantly decreased the brightness ($L*$ value) of the dried pasta and increased its redness ($a*$ value) without affecting its yellowness ($b*$ value). The authors also reported higher hardness and higher fracturability for the pasta with pea flour. The impact of the substitution on the various aforementioned parameters was attributed to the introduction of non-gluten proteins into the pasta network.

Breads and Cookies

The impact of incorporating pea flour into bread and cookies was also studied by Kamaljit et al. (2010), who incorporated pea flour into breads and cookies at 5% and 10% levels and evaluated the impact on the bread and cookie making processes as well as on the sensory characteristics of the resulting products. Two pea cultivars (Pb-87 and Pb-88) were considered in that study. For the breads, the substitution of pea flour for wheat flour resulted in an increase in the water absorption of the dough and a decrease in its stickiness. Specific volume was also lower for the breads with pea flour than for the control bread. For the cookies, a decrease in the cookie spread ratio was observed when the cookies were fortified with pea flour. Overall, the sensory score for the breads and cookies with 5% pea flour was similar to the sensory score for the controls. Pea cultivars barely affected the aforementioned production parameters and characteristics of the final products. Des Marchais et al. (2011) studied the impact of replacing commercial wheat flour at a 10% level (d.b.) with a pea protein isolate produced by the extraction of the pea proteins at pH 7.5 in water at room temperature, followed by purification by ultrafiltration/diafiltration using 50-kDa hollow fiber membranes (Taherian et al. 2011). Dough mixing properties and bread characteristics were determined. The results indicated that dough stability and development time were not affected by the substitution, although water absorption was increased by about 3%. The substitution of the protein isolate for wheat flour resulted in a bread with 20.9 ± 0.1% protein, a value that was significantly higher than the 14.5 ± 0.0% measured for the control bread. In contrast, the bread with the protein isolate had a specific volume of 4.3 ± 0.0 cm^3/g, in comparison with 5.4 ± 0.3 cm^3/g for the control. Despite the fact that the specific volume of the bread with the protein isolate was significantly lower than that of the control, the protein isolate bread was still considered acceptable from a consumer's point of view.

In addition to the preparation of breads containing pea flour, work on breads made with hulls and cotyledon fibers isolated from peas is also reported in the scientific literature. Dalgetty and Baik (2006) replaced wheat flour at levels of 3%, 5%, or 7% with pea hull fiber, insoluble fiber, or soluble fiber. The hull fiber fraction was separated from the flour obtained by milling the pea seeds, and the soluble and insoluble cotyledon fibers were isolated according to the procedure of Dalgetty and Baik (2003). The results indicated that water absorption was 64% for the dough for the control bread and that the substitution of hulls or fibers for wheat flour increased the water absorption of the dough in all treatments except the addition of 3% and 5% soluble fiber, which resulted in a decrease in water absorption. The mixing time, which was 233 s for the control, was increased by the addition of hulls and insoluble fibers but was not affected by the addition of soluble fibers. The breads made with hulls and fibers were also higher in moisture content than the control bread was.

Meat Products

There is also increasing interest within the food industry in using plant ingredients as substitutes in the production of meat products. The use of plant ingredients allows the production of meat products that are lower in fat, cholesterol, and calories. However, the direct substitution of nonmeat ingredients for fat may lead to texture problems, changes in sensory qualities after cooking, and cooking and purge losses (Kaack and Pedersen 2005; Pietrasik and Janz 2010; Verma et al. 2015; Vinauskienė et al. 2015). Therefore, before a plant ingredient can be considered for use as a substitute in the production of meat products, the ingredient's influence on the functionality, quality, and acceptability of the meat products must be investigated.

Low-Fat Bolognas

Pietrasik and Janz (2010) used pea flour, starch, and fiber for the production of low-fat bolognas. Low-fat bolognas (10% fat) were compared with low-fat formulations (10% fat) supplemented with wheat flour or one of the pea-based ingredients at a level of 4%. Conventional high-fat bologna (22% fat) was produced as the control. The results indicated that the low-fat bolognas had poor texture and binding properties in comparison with the high-fat bologna. However, the use of pea starch and fiber restored the texture profile values of the low-fat bolognas to match those of the high-fat bologna. In comparison with those for the low-fat bologna, cooking and purge losses were decreased by all the pea ingredients. Consumer acceptance of the low-fat bolognas supplemented with pea starch and fiber fractions was equivalent to consumer acceptance of the high-fat bologna. In general, low-fat formulations supplemented with pea starch and fiber resulted in little change in functionality, and those two ingredients performed as well as wheat flour did.

Low-Fat Chicken Nuggets

Verma et al. (2015) studied the impact of adding 8%, 10%, and 12% pea hull flour on the quality parameters of low-fat chicken nuggets. The results indicated that the emulsion stability of all the treatments was lower in comparison with the control, which did not contain pea hull flour, and was further decreased significantly as the level of hull flour increased. Color parameters were not affected by the addition of hull flour except at the 12% level, which decreased their values. Textural properties were also lower in the products containing pea hull flour than in the control. Sensory evaluation indicated that pea hull flour, when supplemented at a level of 8%, is a suitable ingredient for incorporation as a source of fiber into low-fat chicken nuggets without significant effects on various attributes. In another study on low-fat chicken nuggets, Shoaib et al. (2018) used pea protein isolates as meat extenders at different concentration levels (3%, 6%, 9%, and 12%). The protein content of the pea protein-enriched nuggets ranged between 32.84% and 39.31%, with the upper end of that range being higher than the value for the control (34.99%). The addition of pea proteins did not have any effect on the pH and ash content of the nuggets. However, the water-holding capacity of the nuggets was significantly increased by the addition of pea proteins, which resulted in a decrease in cooking loss. Cooking loss in nuggets enriched with pea proteins was found to be between 5.01% and 11.12%, in comparison with 12.43% for the control. There was no significant difference in the textural characteristics of the enriched nuggets. However, the pea protein isolates resulted in substantial flavor issues in the nuggets.

Animal Feed

Blue Shrimps

Different pea fractions have been used in animal diets. Cruz-Suarez et al. (2001) fed juvenile blue shrimp (*Litopenaeus stylirostris*) with a diet that included 30% whole, dehulled, extruded, dehulled-extruded, or micronized feed peas. The pea ingredients replaced a portion of the soybean meal and wheat that were in a control diet. The results indicated a significant weight gain after 28 days that varied between 269 ± 18% and 340 ± 29% for the shrimp fed with the diets containing the pea ingredients, in comparison with a 292 ± 14% weight gain with the control diet. However, only the micronized pea diet resulted in a weight gain that was significantly higher than the one observed for the control diet. This result was attributed to enhanced feed intake. The authors concluded that pea ingredients are very acceptable alternatives for blue shrimp diets.

Piglets

In another study, Owusu-Asiedu et al. (2002) fed 16-day-old weaned piglets with pea-based diets (containing raw, extruded, or micronized peas) to investigate the effect on apparent and standardized ileal amino acid digestibilities as well as on piglet growth rate. The pea-based diets were fed with or without enzyme (amylase and xylanase) supplementation. A soybean meal diet was used as the control diet. The results indicated that extrusion and micronization improved the amino acid digestibilities of pea-based diets supplemented with amylase and xylanase in comparison with the raw pea ingredient, in addition to improving the efficiency of feed utilization. However, feed intake and growth performance were not affected.

Rainbow Trouts and Common Carp

Thiessen et al. (2003) investigated the effect of feeding juvenile rainbow trout (*Oncorhynchus mykiss*) with pea-based diets (containing a raw/dehulled pea ingredient or an autoclaved air-classified pea protein). The pea ingredients replaced a portion of a soybean meal control diet (20% for the raw/dehulled pea ingredient and 25% for the autoclaved air-classified pea protein). The results showed no significant difference in feed intake, final weight, and specific growth rate measurements, indicating that pea ingredients are suitable ingredients for use in trout diet formulations at a level of 20–25%. Davies and Gouveia (2010) investigated the effect of feeding common carp (*Cyprinus carpio* L.) fry with pea-based diets (containing raw, autoclaved, or dry-cooked pea ingredients). A soybean meal diet was used as the control diet. The results indicated a significant weight gain after 49 days that varied between 241% and 289% for the carp fry fed the pea-based diets, in comparison with 302% for the control diet. The weight gains for the autoclaved and the dry-cooked pea-based ingredients were not significantly different than the weight gain for the control diet, whereas weight gain was found to be significantly lower for the raw pea diet than for the control diet. A similar trend was observed for the apparent nutrient and energy digestibility coefficients. The authors concluded that autoclaving and dry cooking can positively modify the nutritional value of pea seed meal, resulting in better growth performance, feed utilization efficiency, and apparent digestibility for common carp in comparison with raw pea seeds.

Broiler Chickens

More recently, Röhe et al. (2017) studied the effect of feeding different pea-based diets on intestinal morphology and functional glucose transport in the small intestine of broiler chickens. A soybean meal control diet was compared with three diets containing raw peas, fermented peas, or enzymatically predigested peas, each supplying 30% dietary crude protein. The results indicated that feeding pea-based ingredients to broilers resulted in lower glucose transport capacities than when broilers were fed the control diet. However, the authors concluded that additional studies are required to determine which components in peas are responsible for such effects and whether these effects have a beneficial or detrimental impact on gut function and animal health.

Nonedible Applications

Pea ingredients are of interest not only to the food industry but also for use in nonedible applications such as insect repellents and bioplastics.

Insect Repellents

In stored grains, insects are a problem because they cause extensive damage to the grains, reducing their viability and nutritional value (Semple et al. 1992). The two main methods to reduce losses of stored grains to insect pests are the application of synthetic insecticides and fumigation. However, the demand for natural alternatives to chemical insecticides is growing with the occurrence of insecticide-resistant insect strains, safety concerns about working with chemical insecticides, and the increasing demand for food without insecticide residues (Fields et al. 2001; Hou and Fields 2003). Many plant-derived chemicals are insecticidal to stored product pests. Studies on pea fractions demonstrated their potential as repellents against stored product insects (Bodnaryk et al. 1999; Fields et al. 2001; Hou and Fields 2003; Hou et al. 2004a, b; Pretheep Kumar et al. 2004). In an initial series of experiments, Fields et al. (2001) studied the repellent effect of pea fractions (protein, starch, and fiber) at different concentrations from 0.001% to 10% (w/w) on *Cryptolestes ferrugineus*, *Sitophilus oryzae*, and *Tribolium castaneum*. Those authors found that the starch fraction did not repel any of the insects, whereas the fiber fraction had repellent effects only for *C. ferrugineus* adults. The pea protein fraction was found to be the most effective, with repellent effects for both *C. ferrugineus* and *S. oryzae* (but not for *T. castaneum*). In addition, even after 4 weeks of exposure, *C. ferrugineus* and *S. oryzae* did not become habituated to the repellent action of the pea protein fraction. Hou and Fields (2003) studied the impact of grain species, temperature, and moisture content on the repellent effect of protein-rich pea flour. Those authors observed that the protein-rich pea flour was more effective in repelling *S. oryzae* for wheat and barley than for maize. One possible reason is that insect feed uptake is lower in maize than in wheat and barley, which may result in a lower uptake of protein-rich pea flour and therefore a lower repellent effect. However, the authors concluded that more studies were needed to determine the reasons for the difference observed between the three grain species. The authors also reported that increasing the temperature from 293.15 to 303.15 K or reducing the relative humidity from 85% to 65% increased the mortality of *S. oryzae* on wheat kernels. Another study by the same group (Hou et al. 2004a) investigated the impact of combining protein-rich pea flour at low concentrations (0–1 kg/m^3, w/w) with diatomaceous earth, neem, *Bacillus thuringiensis* (Berliner), malathion, and pyrethrum for controlling stored product beetles in wheat. The results indicated that protein-rich pea flour in combination with neem and in combination with malathion acted synergistically against *Tribolium castaneum* and against *S. oryzae*, respectively. The combination of diatomaceous earth or pyrethrum with protein-rich pea flour acted additively against *S. oryzae*, whereas all other combinations acted antagonistically. Lastly, Hou et al. (2004b) combined protein-rich pea flour with parasitoids to control stored product beetles. An additive effect of this combination was observed in large-scale testing, with a reduction in the population of *S. oryzae* on wheat kernels. All the aforementioned studies illustrated the potential of pea protein to control stored grain beetles.

Bioplastics

Another application of pea ingredients is pea-based bioplastics (Chen et al. 2008, 2009; Klüver and Meyer 2015; Perez et al. 2016; Perez-Puyana et al. 2016; Carvajal-Piñero et al. 2019). Chen et al. (2008) determined the impact of incorporating a pea starch nanocrystal (PSN) dispersion and native pea starch (NPS) granules into polyvinyl alcohol (PVA) films on the physical properties of the resulting films. The PSN or NPS was incorporated into the PVA solution at different levels that varied from 0% to 40% (weight basis). The light transmittance (Tr), tensile strength (σ_b), and elongation at break (ε_b) of the PVA, PVA/PSN, and PVA/NPS films were determined and compared in order to determine the impact of incorporating the pea ingredients on the properties of the PVA-based films. It was found that the Tr, σ_b, and ε_b of the PVA/PSN films containing 5% and 10% PSN were improved in comparison with the properties of the PVA films. However, the properties of the PVA/NPS films were lower than those of the PVA films. The smaller size of the PSNs in comparison with the NPS granules and the better dispersion of the PSNs in the PVA matrix resulted in stronger interactions with PVA and would explain the aforementioned observations. Other research by the same group (Chen et al. 2009) aimed to determine the impact of adding pea hull fibers and pea hull fiber-derived nanowhiskers (PHFNWs) to pea starch-based composite films. The pea hull fibers and PHFNWs were blended with the pea starch at different levels that varied from 0% to 30% (weight basis). Films were then prepared by means of a solution casting and evaporation process (Lu et al. 2006). The Tr, σ_b, and ε_b of the various films were characterized and compared in order to determine the effects of the use of pea hull fibers and PHFNWs as filler on the structure and properties of the pea starch composite films. The results revealed that the addition of PHFNWs to the pea starch improved the properties of the resulting films over both the pea starch film and the pea starch/pea hull fiber films. The improvement in the properties of the pea starch/PHFNW nanocomposite films was attributed to the homogenous dispersion of the PHFNWs within the pea starch and to the strong interactions between the matrix and the nanoscale filler. In addition to pea starch and fibers, pea proteins have also been considered for the development of pea protein-based bioplastics using injection molding (Perez et al. 2016; Perez-Puyana et al. 2016; Carvajal-Piñero et al. 2019) or extrusion (Klüver and Meyer 2015). Bioplastics containing 60% pea protein isolate and 40% glycerol were manufactured by means of a two-stage thermomechanical procedure described by Perez-Puyana et al. (2016) and involving an injection molding process. This ratio was found to be optimal to obtain good processability of the blends for producing bioplastics (Perez et al. 2016). The influence of molding time and injection pressure on the properties of the bioplastics was determined. The results indicated that both the Young's modulus values and Tr values decreased as molding time increased. In contrast, increasing the injection pressure led to a significant increase in the storage modulus and in the strain at break. The results also indicated that the different bioplastics had excellent water uptake capacities, with values of more than 100% for most of the bioplastics. This result indicates that these bioplastics could be used for various applications as absorbent material. Carvajal-Piñero et al. (2019) studied the impact of mixing speed and time on the microstructure and the rheological and mechanical behavior of bioplastics made from glycerol and pea protein isolate. The results indicated that intermediate mixing speeds (30 rpm) and short mixing times (60 and 600 s) led to bioplastics with good mechanical properties. Short and long mixing times and high mixing speeds (50 rpm) were found to be not suitable since they led to heterogeneity of the bioplastics. Klüver and Meyer (2015) processed pea protein isolates into bioplastics by extrusion. Glycerol was used as a plasticizer at a level of 67%. The thermoplastic behavior of the pea protein melt under extrusion conditions was confirmed by viscosity measurements. The results were well represented by the power law, with a power law index of 0.31. The authors concluded that pea protein isolates can be suitably extruded to produce pea protein-based bioplastics.

Future Trends in Processing and Product Development

Processing

In terms of pea processing, dry fractionation (air classification) and wet extraction processes (AE-IP, SED, and MP) are well established. However, processes such as membrane processing are in development and could be widely applied at the industrial scale in the near future. One such process is a variant of the AE-IP process and consists of replacing the isoelectric precipitation step with an ultrafiltration and/or diafiltration step to concentrate and purify the proteins (Fredrikson et al. 2001; Fuhrmeister and Meuser 2003; Makri et al. 2005; Boye et al. 2010b; Taherian et al. 2011; Mondor et al. 2012) (Fig. 2). Fuhrmeister and Meuser (2003) compared the performance of ultrafiltration plate modules with different molecular weight cutoffs (5, 10, 30, 50, or 100 kDa) in terms of pea protein retention and permeate flux. As expected, pea protein rejection decreased with an increase in the molecular weight cutoff, whereas permeate flux increased. The authors concluded that the 50-kDa membrane represented the best compromise between high protein rejection and high permeate flux. Taherian et al. (2011) produced four yellow pea protein isolates from yellow pea flours extracted at pH 7.5 and 298.15 K in water or in 0.06 M KCl solutions, followed by ultrafiltration and diafiltration at pH values of 7.5 and 7.5 or 6, respectively, using a 50-kDa hollow fiber membrane. Those authors compared phytic acid content and functional properties (solubility, foaming, emulsification ability, and gelling properties) between the pea protein isolates produced by

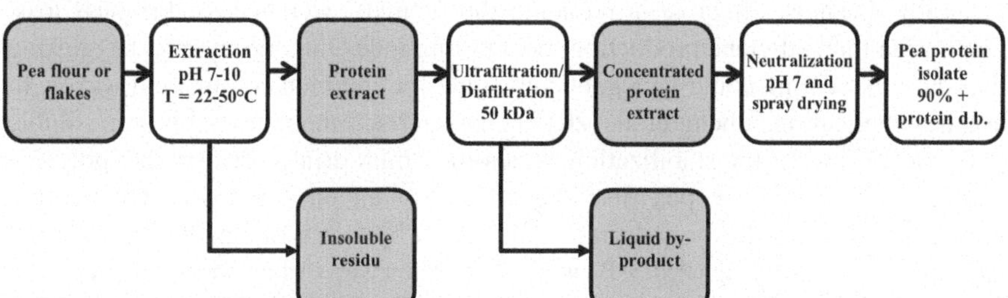

Fig. 2 Membrane purification process

membrane technologies and a commercial pea protein isolate. The results indicated that, depending on the ultrafiltration/diafiltration conditions, phytic acid content was reduced by 28–68% in comparison with the commercial isolate, and all the functional properties of the membrane-processed isolate were similar to or better than those of the commercial isolate. For example, the solubility of the pea proteins at pH 2.0 was approximately 20% for the commercial pea protein isolate and approximately 80% for the isolate produced by membrane technologies. In addition to resulting in isolates with improved functional properties, the use of membrane technologies in place of conventional isoelectric precipitation also results in higher isolate yield (Mondor et al. 2012) and in isolates with higher protein levels relative to the isoelectric precipitation process (Boye et al. 2010b). There is also a variant of the SED process, in which the salts present in the pea protein extract are removed by a combination of ultrafiltration and diafiltration (Tian et al. 1999).

Product Development

Edible Applications

In terms of product development, the use of pea-based ingredients in the production of pasta, breads, cookies, and meat products is now well established, and the demand in upcoming years should increase. However, in the future, pea-based ingredients have the potential to be used in other edible applications, such as edible coatings (Saberi et al. 2016a, b, 2017, 2018a, b) or stabilizers in drinks (Cheng et al. 2018). Saberi et al. (2016a, b, 2017) showed that biocomposite edible films with good physical, optical, and mechanical properties can be obtained by combining pea starch with guar gum. Saberi et al. (2018a) also blended pea starch and guar gum with a lipid mixture containing oleic acid and shellac (PSGG-Sh) to develop a coating that could be used on fruit. The new coating was sprayed uniformly on "Valencia" oranges that were then stored at either 278.15 or 293.15 K for up to 4 weeks. The quality parameters (weight loss, firmness, respiration rate, ethylene production, peel pitting index, and fruit decay rate index) of the PSGG-Sh-coated oranges were characterized each week, and the results were compared with the quality parameters of oranges coated with commercial wax and uncoated oranges. The novel coating was generally the most effective in reducing the oranges' weight loss, firmness loss, respiration rate, ethylene production, peel pitting index, and decay rate, suggesting that the new coating could be a good substitute for common commercial waxes. In another application, Cheng et al. (2018) characterized the potential of pea soluble polysaccharides for the stabilization of acidified milk drinks. Pea soluble polysaccharides extracted from pea fibers by means of an enzyme-assisted extraction method combined with spray-drying (PSPS-A) or by means of ethanol precipitation followed by oven-drying (PSPS-B) were used. The PSPS-A had an average molecular weight of 625 kDa, in comparison with 809 kDa for the PSPS-A. A milk drink (0.6% protein) was prepared by dissolving, in deionized water, 6% sucrose and skim milk powder. The skim milk dispersion was then added to the polysaccharide solution, and the pH value of the mixture was gradually acidified using 0.1 M citric acid and sodium citrate buffer. This step was followed by homogenization at room temperature at 4000 rpm for 120 s. The final concentration of milk protein was 0.3%. The stability of the acidified milk drinks was determined by measuring the precipitation rate by centrifugation at 4000 × *g* for 1200 s. Stable milk drinks (precipitation percentage less than 1%) were obtained with the addition of 0.15% PSPS-A or 0.1% PSPS-B over a pH range of 3.6–4.6. It was also observed that PSPS were effective in stabilizing acidified milk drinks over a wider pH range than soybean soluble polysaccharides containing the same functional groups.

Nonedible Applications

In terms of nonedible applications, the industrial production of bioethanol from pea starch or pea hulls could be envisaged in the near future (Fig. 3). Bioethanol can be considered a good alternative to fossil fuels as it can contribute to a cleaner environment (Li et al. 2009), and lignocellulosic residues, such as pea hulls, are promising raw materials for bioethanol production as they are residues and wastes and do not compete with primary food production. However, in order for pea hulls to be used for the production of fuel ethanol, it is important to get as much fermentable sugars from them as possible. Sarkar et al. (2015) studied the effect of alkaline pretreatments on the conversion of pea hulls into fermentable sugars. The pea hulls used in that study were composed of 82.3% fiber, 62.3% cellulose, 8.2% hemicelluloses, and 1.7% ash. The experiments consisted of preparing a slurry of pea hulls with distilled water and then adding different concentrations of hydrogen peroxide (4%, 7.5%, and 11%), adjusting the pH to 11.5 using NaOH (0.1 M) and incubating the slurries at different temperatures (298.15, 303.15, and 308.15 K) for different durations (18, 24, and 30 h). Sugar release from the pea hulls was determined by means of the anthrone method (Dreywood 1946). The results showed that sugar release increased with an increase in peroxide concentration and reaction temperature. However, an increase in reaction time resulted in the degradation of released sugar into residue. A predictive equation (quadratic model) for sugar yield was formulated on the basis of the experimental results and was used to find the conditions for

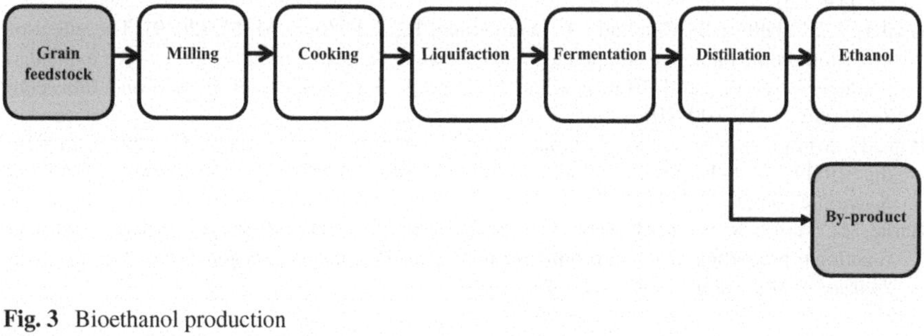

Fig. 3 Bioethanol production

which sugar release was maximized (9% peroxide hydrogen; 300.15 K; 18 h of incubation). The authors concluded that pea hulls are a promising renewable raw material for bioethanol production. In another study (Nichols et al. 2005), pea starch also showed great potential for the production of bioethanol. Starch-enriched fractions obtained by air classification from whole or dehulled peas were liquefied and saccharified using industrial α-amylase and glucoamylase to obtain a final ethanol concentration of 11.0% (w/v) in 48–52 h, with yields of 0.43–0.48 g of ethanol/g of glucose. That study demonstrated the feasibility of fermenting pea starch and suggested that whole or dehulled peas are a suitable feedstock for ethanol fermentation. In the same context, Prairie Green Renewable Energy is developing a world-scale ethanol and protein animal feed plant just south of Saskatoon, Saskatchewan, Canada. The plant will produce 196,000 m^3 of fuel-grade ethanol for the domestic market and about 2.28×10^8 kg of high-protein/high-amino acid animal feed per year (Prairie Green Renewable Energy 2016).

References

Adebiyi, A. P., & Aluko, R. E. (2011). Functional properties of protein fractions obtained from commercial yellow field pea (*Pisum sativum* L.) seed protein isolate. *Food Chemistry, 128*, 902–908.

Agboola, S. O., Mofolasayo, O. A., Watts, B. M., & Aluko, R. E. (2010). Functional properties of yellow field pea (*Pisum sativum* L.) seed flours and the *in vitro* bioactive properties of their polyphenols. *Food Research International, 43*, 582–588.

Agriculture and Agri-Food Canada. (2019, September 19). *Canada: Outlook for principal field crops*. Retrieved September 17, 2019, from http://www.agr.gc.ca/eng/industry-markets-and-trade/canadian-agri-food-sector-intelligence/crops/reports-and-statistics-data-for-canadian-principal-field-crops/canada-outlook-for-principal-field-crops-2019-07-19/?id=1563981168187#a4.

Alonso, R., Orúe, E., Zabalza, M. J., Grant, G., & Marzo, F. (2000). Effect of extrusion cooking on structure and functional properties of pea and kidney bean proteins. *Journal of the Science of Food and Agriculture, 80*, 397–403.

Aluko, R. E., Mofolasayo, O. A., & Watts, B. M. (2009). Emulsifying and foaming properties of commercial yellow pea (*Pisum sativum* L.) seed flours. *Journal of Agricultural and Food Chemistry, 57*, 9793–9800.

Araya, H., Pak, N., Vera, G., & Alviña, M. (2003). Digestion rate of legume carbohydrates and glycemic index of legume-based meals. *International Journal of Food Sciences and Nutrition, 54*, 119–126.

Arribas, C., Cabellos, B., Cuadrado, C., Guillamón, E., & Pedrosa, M. M. (2019). The effect of extrusion on the bioactive compounds and antioxidant capacity of novel gluten-free expanded products based on carob fruit, pea and rice blends. *Innovative Food Science and Emerging Technologies, 52*, 100–107.

Bani, P., Minuti, A., Ficuciello, V., Guerreschi, M., Astorri, G., & Galassi, G. (2009). *In vitro* digestibility of field pea as influenced by processing methods. *Italian Journal of Animal Science, 8*(Suppl. 2), 259–261.

Barac, M., Cabrilo, S., Pesic, M., Stanojevic, S., Zilic, S., Macej, O., & Ristic, N. (2010). Profile and functional properties of seed proteins from six pea (*Pisum sativum*) genotypes. *International Journal of Molecular Sciences, 11*, 4973–4990.

Ben-Hdech, H., Gallant, D. J., Bouchet, B., Gueguen, J., & Melcion, J.-P. (1991). Extrusion-cooking of pea flour: Structural and immunocytochemical aspects. *Food Structure, 10*, 203–212.

Black, R. G., Singh, U., & Meares, C. (1998). Effect of genotype and pretreatment of field peas (*Pisum sativum*) on their dehulling and cooking quality. *Journal of the Science of Food and Agriculture, 77*, 251–258.

Bodnaryk, R. P., Fields, P. G., Xie, Y., & Fulcher, K. A. (1999, September 21). *Insecticidal factor from field peas*. US Patent 5,955,082.

Bogahawaththa, D., Bao Chau, N. H., Trivedi, J., Dissanayake, M., & Vasiljevic, T. (2019). Impact of selected process parameters on solubility and heat stability of pea protein isolate. *LWT- Food Science and Technology, 102*, 246–253.

Boye, J. I., & Ma, Z. (2015). Impact of processing on bioactive compounds of field peas. In V. Preedy (Ed.), *Processing and impact on active components in food* (pp. 63–70). London: Elsevier.

Boye, J. I., Zare, F., & Pletch, A. (2010a). Pulse proteins: Processing, characterization, functional properties and applications in food and feed. *Food Research International, 43*, 414–431.

Boye, J. I., Aksay, S., Roufik, S., Ribéreau, S., Mondor, M., Farnworth, E., & Rajamohamed, S. (2010b). Comparison of the functional properties of pea, chickpea and lentil protein concentrates processed using ultrafiltration and isoelectric precipitation techniques. *Food Research International, 43*, 537–546.

Brummer, Y., Kaviani, M., & Tosh, S. M. (2015). Structural and functional characteristics of dietary fibre in beans, lentils, peas and chickpeas. *Food Research International, 67*, 117–125.

Burger, T. G., & Zhang, Y. (2019). Recent progress in the utilization of pea protein as an emulsifier for food applications. *Trends in Food Science and Technology, 86*, 25–33.

Campos-Vega, R., Loarca-Piña, G., & Oomah, B. D. (2010). Minor components of pulses and their potential impact on human health. *Food Research International, 43*, 461–482.

Cano, A. I., Cháfer, M., Chiralt, A., & González-Martínez, C. (2015). Physical and microstructural properties of biodegradable films based on pea starch and PVA. *Journal of Food Engineering, 167*(Part A), 59–64.

Carvajal-Piñero, J. M., Ramos, M., Jiménez-Rosado, M., Perez-Puyana, V., & Romero, A. (2019). Development of pea protein bioplastics by a thermomoulding process: Effect of the mixing stage. *Journal of Polymers and the Environment, 27*, 968–978. https://doi.org/10.1007/s10924-019-01404-3.

Chakraborty, P., Sosulski, F., & Bose, A. (1979). Ultracentrifugation of salt-soluble proteins in ten legume species. *Journal of the Science of Food and Agriculture, 30*, 766–771.

Chan, C. B., Gupta, J., Kozicky, L., Hashemi, Z., & Yang, K. (2014). Improved glucose tolerance in insulin-resistant rats after pea hull feeding is associated with changes in lipid metabolism-targeted transcriptome. *Applied Physiology, Nutrition, and Metabolism, 39*, 1112–1119.

Chen, Y., Cao, X., Chang, P. R., & Huneault, M. A. (2008). Comparative study on the films of poly(vinyl alcohol)/pea starch nanocrystals and poly(vinyl alcohol)/native pea starch. *Carbohydrate Polymers, 73*, 8–17.

Chen, Y., Liu, C., Chang, P. R., Anderson, D. P., & Huneault, M. A. (2009). Pea starch-based composite films with pea hull fibers and pea hull fiber-derived nanowhiskers. *Polymer Engineering and Science, 49*, 369–378.

Chen, M., Lu, J., Liu, F., Nsor-Atindana, J., Xu, F., Goff, H. D., Ma, J., & Zhong, F. (2019). Study on the emulsifying stability and interfacial adsorption of pea proteins. *Food Hydrocolloids, 88*, 247–255.

Cheng, M., Qi, J.-R., Feng, J.-L., Cao, J., Wang, J.-M., & Yang, X.-Q. (2018). Pea soluble polysaccharides obtained from two enzyme-assisted extraction methods and their application as acidified milk drinks stabilizers. *Food Research International, 109*, 544–551.

Corrales, M., Han, J. H., & Tauscher, B. (2009). Antimicrobial properties of grape seed extracts and their effectiveness after incorporation into pea starch films. *International Journal of Food Science and Technology, 44*, 425–433.

Cruz-Suarez, L. E., Ricque-Marie, D., Tapia-Salazar, M., McCallum, I. M., & Hickling, D. (2001). Assessment of differently processed feed pea (*Pisum sativum*) meals and canola meal (*Brassica* sp.) in diets for blue shrimp (*Litopenaeus stylirostris*). *Aquaculture, 196*, 87–104.

Dahl, W. J. (2017). Pea hull fiber: A dietary fiber to modulate gastrointestinal function and gut microbiota. *Cereal Foods World, 62*, 203–206.

Dalgetty, D. D., & Baik, B.-K. (2003). Isolation and characterization of cotyledon fibers from peas, lentils, and chickpeas. *Cereal Chemistry, 80*, 310–315.

Dalgetty, D. D., & Baik, B.-K. (2006). Fortification of bread with hulls and cotyledon fibers isolated from peas, lentils, and chickpeas. *Cereal Chemistry, 83*, 269–274.

Davies, S. J., & Gouveia, A. (2010). Response of common carp fry fed diets containing a pea seed meal (*Pisum sativum*) subjected to different thermal processing methods. *Aquaculture, 305*, 117–123.

Della Valle, G., Quillien, L., & Gueguen, J. (1994). Relationships between processing conditions and starch and protein modifications during extrusion-cooking of pea flour. *Journal of the Science of Food and Agriculture, 64*, 509–517.

Des Marchais, L.-P., Foisy, M., Mercier, S., Villeneuve, S., & Mondor, M. (2011). Bread-making potential of pea protein isolate produced by a novel ultrafiltration/diafiltration process, 11th International Congress on Engineering and Food (ICEF11). *Procedia Food Science, 1*, 1425–1430.

Djoullah, A., Husson, F., & Saurel, R. (2018). Gelation behaviors of denaturated pea albumin and globulin fractions during transglutaminase treatment. *Food Hydrocolloids, 77*, 636–645.

Dreywood, R. (1946). Qualitative test for carbohydrate material. *Industrial and Engineering Chemistry, Analytical Edition, 18*, 499.

Eyaru, R., Shrestha, A. K., & Arcot, J. (2009). Effect of various processing techniques on digestibility of starch in Red kidney bean (*Phaseolus vulgaris*) and two varieties of peas (*Pisum sativum*). *Food Research International, 42*, 956–962.

Felix, M., Perez-Puyana, V., Romero, A., & Guerrero, A. (2010). Development of thermally processed bioactive pea protein gels: Evaluation of mechanical and antioxidant properties. *Food and Bioproducts Processing, 101*, 74–83.

Fields, P. G., Xie, Y. S., & Hou, X. (2001). Repellent effect of pea (*Pisum sativum*) fractions against stored-product insects. *Journal of Stored Products Research, 37*, 359–370.

Fredrikson, M., Biot, P., Alminger, M. L., Carlsson, N.-G., & Sandberg, A.-S. (2001). Production process for high-quality pea-protein isolate with low content of oligosaccharides and phytate. *Journal of Agricultural and Food Chemistry, 49*, 1208–1212.

Fuhrmeister, H., & Meuser, F. (2003). Impact of processing on functional properties of protein products from wrinkled peas. *Journal of Food Engineering, 56*, 119–129.

Goyal, R. K., Vishwakarma, R. K., & Wanjari, O. D. (2008). Optimisation of the pigeon pea dehulling process. *Biosystems Engineering, 99*, 56–61.

Greenwell, H. L., Gramkow, J. L., Jolly-Breithaupt, M. L., MacDonald, J. C., & Jenkins, K. H. (2018). Effects of field pea supplementation on digestibility and rumen volatile fatty acid concentrations of beef-cattle diets containing high and low quality forages. *The Professional Animal Scientists, 34*, 631–641.

Gulewicz, P., Szymaniec, S., Bubak, B., Frias, J., Vidal-Valverde, C., Trojanowska, K., & Gulewicz, K. (2002). Biological activity of α-galactoside preparations from *Lupinus angustifolius* L. and *Pisum sativum* L. seeds. *Journal of Agricultural and Food Chemistry, 50*, 384–389.

Han, H., & Baik, B.-K. (2008). Antioxidant activity and phenolic content of lentils (*Lens culinaris*), chickpeas (*Cicer arietinum* L.), peas (*Pisum sativum* L.) and soybeans (*Glycine max*), and their quantitative changes during processing. *International Journal of Food Science and Technology, 43*, 1971–1978.

Hashemi, Z., Yang, K., Yang, H., Jin, A., Ozga, J., & Chan, C. B. (2015). Cooking enhances beneficial effects of pea seed coat consumption on glucose tolerance, incretin, and pancreatic hormones in high-fat-diet–fed rats. *Applied Physiology, Nutrition, and Metabolism, 40*, 323–333.

Hashemi, Z., Fouhse, J., Im, H. S., Chan, C. B., & Willing, B. P. (2017). Dietary pea fiber supplementation improves glycemia and induces changes in the composition of gut microbiota, serum short chain fatty acid profile and expression of mucins in glucose intolerant rats. *Nutrients, 9*, 1236.

Holt, N. W., & Sosulski, F. W. (1979). Amino acids composition and protein quality of field peas. *Canadian Journal of Plant Science, 59*, 653–660.

Hood-Niefer, S. D., & Tyler, R. T. (2010). Effect of protein, moisture content and barrel temperature on the physicochemical characteristics of pea flour extrudates. *Food Research International, 43*, 659–663.

Hou, X., & Fields, P. (2003). Effectiveness of protein-rich pea flour for the control of stored-product beetles. *Entomologia Experimentalis et Applicata, 108*, 125–131.

Hou, X., Fields, P., & Taylor, W. (2004a). Combination of protein-rich pea flour and pea extract with insecticides and enzyme inhibitors for control of stored-product beetles. *Canadian Entomologist, 136*, 581–590.

Hou, X., Fields, P., Flinn, P., Perez-Mendoza, J., & Baker, J. (2004b). Control of stored-product beetles with combinations of protein-rich pea flour and parasitoids. *Environmental Entomology, 33*, 671–680.

Islam, F., Gopalan, V., Lam, A. K.-Y., & Kabir, S. R. (2018). Pea lectin inhibits cell growth by inducing apoptosis in SW480 and SW48 cell lines. *International Journal of Biological Macromolecules, 117*, 1050–1057.

Jiménez-Moreno, E., Chamorro, S., Frikha, M., Safaa, H. M., Lázaro, R., & Mateos, G. G. (2011). Effects of increasing levels of pea hulls in the diet on productive performance, development of the gastrointestinal tract, and nutrient retention of broilers from one to eighteen days of age. *Animal Feed Science and Technology, 168*, 100–112.

Kaack, K., & Pedersen, L. (2005). Application of by-products from industrial processing of potato flour and yellow peas as ingredients in low-fat high-fibre sausages. *European Food Research and Technology, 221*, 313–319.

Kadlec, P., Rubecova, A., Hinkova, A., Kaasova, J., Bubnik, Z., & Pour, V. (2001). Processing of yellow pea by germination, microwave treatment and drying. *Innovative Food Science and Emerging Technologies, 2*, 133–137.

Kaiser, A. C., Barber, N., Manthey, F., & Hall, C., III. (2019). Physicochemical properties of hammer-milled yellow split pea (*Pisum Sativum* L.). *Cereal Chemistry, 96*, 313–323.

Kamaljit, K., Baljeet, S., & Amarjeet, K. (2010). Preparation of bakery products by incorporating pea flours as a functional ingredient. *American Journal of Food Technology, 5*, 130–135.

Kaya, E., Tuncel, N. Y., & Tuncel, N. B. (2018). Utilization of lentil, pea, and faba bean hulls in Turkish noodle production. *Journal of Food Science and Technology, 55*, 1734–1745.

Khattab, R. Y., Arntfield, S. D., & Nyachoti, C. M. (2009). Nutritional quality of legume seeds as affected by some physical treatments, Part 1: Protein quality evaluation. *LWT- Food Science and Technology, 42*, 1107–1112.

Klüver, E., & Meyer, M. (2015). Thermoplastic processing, rheology, and extrudate properties of wheat, soy, and pea proteins. *Polymer Engineering and Science, 55*, 1912–1919.

Kristiawan, M., Micard, V., Maladira, P., Alchamieh, C., Maigret, J.-E., Réguerre, A.-L., Emin, M. A., & Della Valle, G. (2018). Multi-scale structural changes of starch and proteins during pea flour extrusion. *Food Research International, 108*, 203–215.

Lam, A. C. Y., Can Karaca, A., Tyler, R. T., & Nickerson, M. T. (2018). Pea protein isolates: Structure, extraction, and functionality. *Food Review International, 34*, 126–147.

Leite, T. S., de Jesus, A. L. T., Schmiele, M., Tribst, A. A. L., & Cristianini, M. (2017). High pressure processing (HPP) of pea starch: Effect on the gelatinization properties. *LWT- Food Science and Technology, 76*, 361–369.

Li, C., & Ganjyal, G. M. (2017). Chemical composition, pasting, and thermal properties of 22 different varieties of peas and lentils. *Cereal Chemistry, 94*, 392–399.

Li, H., Kim, N.-J., Jiang, M., Kang, J. W., & Chang, H. N. (2009). Simultaneous saccharification and fermentation of lignocellulosic residues pretreated with phosphoric acid–acetone for bioethanol production. *Bioresource Technology, 100*, 3245–3251.

Li, C., Kowalski, R. J., Li, L., & Ganjyal, G. M. (2017). Extrusion expansion characteristics of samples of select varieties of whole yellow and green dry pea flours. *Cereal Chemistry, 94*, 385–391.

Liang, H.-N., & Tang, C.-H. (2013). pH-dependent emulsifying properties of pea [*Pisum sativum* (L.)] proteins. *Food Hydrocolloids, 33*, 309–319.

Lu, Y., Weng, L., & Cao, X. (2006). Morphological, thermal and mechanical properties of ramie crystallites—Reinforced plasticized starch biocomposites. *Carbohydrate Polymers, 63*, 198–204.

Ma, Z., Boye, J. I., Simpson, B. K., Prasher, S. O., Monpetit, D., & Malcolmson, L. (2011). Thermal processing effects on the functional properties and microstructure of lentil, chickpea, and pea flours. *Food Research International, 44*, 2534–2544.

Ma, Z., Boye, J. I., & Hu, X. (2017). *In vitro* digestibility, protein composition and techno-functional properties of Saskatchewan grown yellow field peas (*Pisum sativum* L.) as affected by processing. *Food Research International, 92*, 64–78.

Makri, E., Papalamprou, E., & Doxastakis, G. (2005). Study of functional properties of seed storage proteins from indigenous European legume crops (lupin, pea, broad bean) in admixture with polysaccharides. *Food Hydrocolloids, 19*, 583–594.

Martín-Cabrejas, M. A., Ariza, N., Esteban, R., Molla, E., Waldron, K., & López-Andréu, F. J. (2003). Effect of germination on the carbohydrate composition of the dietary fiber of peas (*Pisum sativum* L.). *Journal of Agricultural and Food Chemistry, 51*, 1254–1259.

Maskus, H., & Arntfield, S. (2015). Extrusion processing and evaluation of an expanded, puffed pea snack product. *Journal of Nutrition & Food Sciences, 5*, 378.

Maskus, H., Bourré, L., Fraser, S., Sarkar, A., & Malcolmson, L. (2016). Effects of grinding method on the compositional, physical, and functional properties of whole and split yellow pea flours. *Cereal Foods World, 61*, 59–64.

Mehyar, G. F., Han, J. H., Holley, R. A., Blank, G., & Hydamaka, A. (2007). Suitability of pea starch and calcium alginate as antimicrobial coatings on chicken skin. *Poultry Science, 86*, 386–393.

Mession, J.-L., Chihi, M. L., Sok, N., & Saurel, R. (2015). Effect of globular pea proteins fractionation on their heat-induced aggregation and acid cold-set gelation. *Food Hydrocolloids, 46*, 233–243.

Minihane, A. M., & Rimbach, G. (2002). Iron absorption and the iron binding and anti-oxidant properties of phytic acid. *International Journal of Food Science and Technology, 37*, 741–748.

Mollard, R. C., Luhovyy, B. L., Smith, C., & Anderson, G. H. (2014). Acute effects of pea protein and hull fibre alone and combined on blood glucose, appetite, and food intake in healthy young men—A randomized crossover trial. *Applied Physiology, Nutrition, and Metabolism, 39*, 1360–1365.

Mondor, M., Tuyishime, O., & Drolet, H. (2012). Production of pea protein concentrates by ultrafiltration: Influence of hollow-fibre module. *Innovative Food Science and Emerging Technologies, 14*, 135–138.

Mondor, M., Guévremont, É., & Villeneuve, S. (2014). Processing, characterization and bread-making potential of malted yellow peas. *Food Bioscience, 7*, 11–18.

Nichols, N. N., Dien, B. S., Wu, Y. V., & Cotta, M. A. (2005). Ethanol fermentation of starch from field peas. *Cereal Chemistry, 82*, 554–558.

Nikolopoulou, D., Grigorakis, K., Stasini, M., Alexis, M. N., & Iliadis, K. (2007). Differences in chemical composition of field pea (*Pisum sativum*) cultivars: Effects of cultivation area and year. *Food Chemistry, 103*, 847–852.

Nosworthy, M. G., Franczyk, A. J., Medina, G., Neufeld, J., Appah, P., Utioh, A., Frohlich, P., & House, J. D. (2017). Effect of processing on the *in vitro* and *in vivo* protein quality of yellow and green split peas (*Pisum sativum*). *Journal of Agricultural and Food Chemistry, 65*, 7790–7796.

Oliete, B., Potin, F., Cases, E., & Saurel, R. (2018). Modulation of the emulsifying properties of pea globulin soluble aggregates by dynamic high-pressure fluidization. *Innovative Food Science and Emerging Technologies, 47*, 292–300.

Owusu-Asiedu, A., Baidoo, S. K., & Nyachoti, C. M. (2002). Effect of heat processing on nutrient digestibility in pea and supplementing amylase and xylanase to raw, extruded or micronized pea-based diets on performance of early-weaned pigs. *Canadian Journal of Animal Science, 82*, 367–374.

Pelgrom, P. J. M., Vissers, A. M., Boom, R. M., & Schutyser, M. A. I. (2013). Dry fractionation for production of functional pea protein concentrates. *Food Research International, 53*, 232–239.

Perez, V., Felix, M., Romero, A., & Guerrero, A. (2016). Characterization of pea protein-based bioplastics processed by injection moulding. *Food and Bioproducts Processing, 97*, 100–108.

Perez-Puyana, V., Felix, M., Romero, A., & Guerrero, A. (2016). Effect of the injection moulding processing conditions on the development of pea protein-based bioplastics. *Journal of Applied Polymer Science, 133*, 43306. https://doi.org/10.1002/app.43306.

Periago, M. J., Vidal, M. L., Ros, G., Rincón, F., Martínez, C., López, G., Rodrigo, J., & Martínez, I. (1998). Influence of enzymatic treatment on the nutritional and functional properties of pea flour. *Food Chemistry, 63*, 71–78.

Petitot, M., Boyer, L., Minier, C., & Micard, V. (2010). Fortification of pasta with split pea and faba bean flours: Pasta processing and quality evaluation. *Food Research International, 43*, 634–641.

Piecyk, M., Wołosiak, R., Drużynska, B., & Worobiej, E. (2012). Chemical composition and starch digestibility in flours from Polish processed legume seeds. *Food Chemistry, 135*, 1057–1064.

Pietrasik, Z., & Janz, J. A. M. (2010). Utilization of pea flour, starch-rich and fiber-rich fractions in low fat bologna. *Food Research International, 43*, 602–608.

Prairie Green Renewable Energy. (2016). *Not your typical ethanol plant*. Retrieved April 25, 2019, from http://pgre.ca/difference/.

Pretheep Kumar, P., Mohan, S., & Balasubramanian, G. (2004). Effect of whole-pea flour and a protein-rich fraction as repellents against stored-product insects. *Journal of Stored Products Research, 40*, 547–552.

Qamar, S., Bhandari, B., & Prakash, S. (2019). Effect of different homogenisation methods and UHT processing on the stability of pea protein emulsion. *Food Research International, 116*, 1374–1385. https://doi.org/10.1016/j.foodres.2018.10.028.

Ratnayake, W. S., Hoover, R., & Warkentin, T. (2002). Pea starch: Composition, structure and properties – A review. *Starch/Stärke, 54*, 217–234.

Reichert, R. D. (1981). Quantitative isolation and estimation of cell wall material from dehulled pea (*Pisum sativum*) flours and concentrates. *Cereal Chemistry, 58*, 266–270.

Reichert, R. D. (1982). Air classification of peas (*Pisum sativum*) varying widely in protein content. *Journal of Food Science, 47*, 1263–1268.

Rempel, C., Geng, X., & Zhang, Y. (2019). Industrial scale preparation of pea flour fractions with enhanced nutritive composition by dry fractionation. *Food Chemistry, 276*, 119–128.

Ribéreau, S., Aryee, A. N. A., Tanvier, S., Han, J., & Boye, J. I. (2018). Composition, digestibility, and functional properties of yellow pea as affected by processing. *Journal of Food Processing & Preservation, 42*, e13375.

Röhe, I., Goodarzi Boroojeni, F., & Zentek, J. (2017). Effect of feeding soybean meal and differently processed peas on intestinal morphology and functional glucose transport in the small intestine of broilers. *Poultry Science, 96*, 4075–4084.

Rubio, L. A., Pérez, A., Ruiz, R., Guzmán, M. A., Aranda-Olmedo, I., & Clemente, A. (2014). Characterization of pea (*Pisum sativum*) seed protein fractions. *Journal of the Science of Food and Agriculture, 94*, 280–287.

Saberi, B., Vuong, Q. V., Chockchaisawasdee, S., Golding, J. B., Scarlett, C. J., & Stathopoulos, C. E. (2016a). Mechanical and physical properties of pea starch edible films in the presence of glycerol. *Journal of Food Processing & Preservation, 40*, 1339–1351.

Saberi, B., Thakur, R., Vuong, Q. V., Chockchaisawasdee, S., Golding, J. B., Scarlett, C. J., & Stathopoulos, C. E. (2016b). Optimization of physical and optical properties of biodegradable edible films based on pea starch and guar gum. *Industrial Crops and Products, 86*, 342–352.

Saberi, B., Thakur, R., Bhuyan, D. J., Vuong, Q. V., Chockchaisawasdee, S., Golding, J. B., Scarlett, C. J., & Stathopoulos, C. E. (2017). Development of edible blend films with good mechanical and barrier properties from pea starch and guar gum. *Starch/Stärke, 69*, 1600227.

Saberi, B., Golding, J. B., Marques, J. R., Pristijono, P., Chockchaisawasdee, S., Scarlett, C. J., & Stathopoulos, C. E. (2018a). Application of biocomposite edible coatings based on pea starch and guar gum on quality, storability and shelf life of 'Valencia' oranges. *Postharvest Biology and Technology, 137*, 9–20.

Saberi, B., Golding, J. B., Chockchaisawasdee, S., Scarlett, C. J., & Stathopoulos, C. E. (2018b). Effect of biocomposite edible coatings based on pea starch and guar gum on nutritional quality of "Valencia" orange during storage. *Starch/Stärke, 70*, 1700299.

Sarkar, A., Kate, A. E., Kumbhar, B. K., & Singh, A. (2015). Effect of alkaline pretreatment parameters on saccharification of waste pea hulls. *Journal of Biobased Materials and Bioenergy, 9*, 433–438.

Schutyser, M. A. I., Pelgrom, P. J. M., van der Goot, A. J., & Boom, R. M. (2015). Dry fractionation for sustainable production of functional legume protein concentrates. *Trends in Food Science and Technology, 45*, 327–335.

Semple, R. L., Hicks, P. A., Lozare, J. V., & Castermans, A. (1992). *Towards integrated commodity and pest management in grain storage: A training manual for application in humid tropical storage systems*. Rome: Food and Agricultural Organization of the United Nations.

Shamsuddin, A. M. (2002). Anti-cancer function of phytic acid. *International Journal of Food Science and Technology, 37*, 769–782.

Shand, P. J., Ya, H., Pietrasik, Z., & Wanasundara, P. K. J. P. D. (2007). Physicochemical and textural properties of heat-induced pea protein isolate gels. *Food Chemistry, 102*, 1119–1130.

Shi, J., Arunasalam, K., Yeung, D., Kakuda, Y., Mittal, G., & Jiang, Y. (2004). Saponins from edible legumes: Chemistry, processing, and health benefits. *Journal of Medicinal Food, 7*, 67–78.

Shoaib, A., Sahar, A., Sameen, A., Saleem, A., & Tahir, A. T. (2018). Use of pea and rice protein isolates as source of meat extenders in the development of chicken nuggets. *Journal of Food Processing & Preservation, 42*, e13763.

Sørensen, M., Morken, T., Kosanovic, M., & Øverland, M. (2011). Pea and wheat starch possess different processing characteristics and affect physical quality and viscosity of extruded feed for Atlantic salmon. *Aquaculture Nutrition, 17*, e326–e336.

Soto-Navarro, S. A., Williams, G. J., Bauer, M. L., Lardy, G. P., Landblom, D. G., & Caton, J. S. (2004). Effect of field pea replacement level on intake and digestion in beef steers fed by-product-based medium-concentrate diets. *Journal of Animal Science, 82*, 1855–1862.

Stone, A. K., Karalash, A., Tyler, R. T., Warkentin, T. D., & Nickerson, M. T. (2015a). Functional attributes of pea protein isolates prepared using different extraction methods and cultivars. *Food Research International, 76*, 31–38.

Stone, A. K., Avramenko, N. A., Warkentin, T. D., & Nickerson, M. T. (2015b). Functional properties of protein isolates from different pea cultivars. *Food Science and Biotechnology, 24*, 827–833.

Sun, X. D., & Arntfield, S. D. (2010). Gelation properties of salt-extracted pea protein induced by heat treatment. *Food Research International, 43*, 509–515.

Sun, X. D., & Arntfield, S. D. (2011). Dynamic oscillatory rheological measurement and thermal properties of pea protein extracted by salt method: Effect of pH and NaCl. *Journal of Food Engineering, 105*, 577–582.

Taherian, A. R., Mondor, M., Labranche, J., Drolet, H., Ippersiel, D., & Lamarche, F. (2011). Comparative study of functional properties of commercial and membrane processed yellow pea protein isolates. *Food Research International, 44*, 2505–2514.

Thiessen, D. L., Campbell, G. L., & Adelizi, P. D. (2003). Digestibility and growth performance of juvenile rainbow trout (*Oncorhynchus mykiss*) fed with pea and canola products. *Aquaculture Nutrition, 9*, 67–75.

Tian, S., Kyle, W. S. A., & Small, D. M. (1999). Pilot scale isolation of proteins from field peas (*Pisum sativum* L.) for use as food ingredients. *International Journal of Food Science and Technology, 34*, 33–39.

Tosh, S. M., & Yada, S. (2010). Dietary fibres in pulse seeds and fractions: Characterization, functional attributes, and applications. *Food Research International, 43*, 450–460.

Velayudhan, D. E., Mejicanos, G. A., & Nyachoti, C. M. (2019). Evaluation of pea protein isolates as a protein source for broilers. *Poultry Science, 98*, 803–810.

Verma, A. K., Banerjee, R., & Sharma, B. D. (2015). Quality characteristics of low fat chicken nuggets: Effect of salt substitute blend and pea hull flour. *Journal of Food Science and Technology, 52*, 2288–2295.

Vidal-Valverde, C., Frias, J., Sierra, I., Blazquez, I., Lambein, F., & Kuo, Y.-H. (2002). New functional legume foods by germination: Effect on the nutritive value of beans, lentils and peas. *European Food Research and Technology, 215*, 472–477.

Vidal-Valverde, C., Frias, J., Hernández, A., Martín-Alvarez, P. J., Sierra, I., Rodríguez, C., Blazquez, I., & Vicente, G. (2003). Assessment of nutritional compounds and antinutritional factors in pea (*Pisum sativum*) seeds. *Journal of the Science of Food and Agriculture, 83*, 298–306.

Villeneuve, S., & Mondor, M. (2014). Processing and bread-making potential of proteins isolated from malted and non-malted pea seeds by ultrafiltration/diafiltration. *Food Bioscience, 8*, 33–36.

Vinauskienė, R., Morkūnaite, R., & Leskauskaitė, D. (2015). The influence of pea products on the functional properties of frankfurters. *CyTA Journal of Food, 13*, 282–292.

Wang, N., Hatcher, D. W., & Gawalko, E. J. (2008). Effect of variety and processing on nutrients and certain anti-nutrients in field peas (*Pisum sativum*). *Food Chemistry, 111*, 132–138.

Wang, N., Maximiuk, L., & Toews, R. (2012). Pea starch noodles: Effect of processing variables on characteristics and optimisation of twin-screw extrusion process. *Food Chemistry, 133*, 742–753.

Xiong, T., Ye, X., Su, Y. T., Chen, X., Sun, H., Li, B., & Chen, Y. (2018a). Identification and quantification of proteins at adsorption layer of emulsion stabilized by pea protein isolates. *Colloids and Surfaces B: Biointerfaces, 171*, 1–9.

Xiong, T., Xiong, W., Ge, M., Xia, J., Li, B., & Chen, Y. (2018b). Effect of high intensity ultrasound on structure and foaming properties of pea protein isolate. *Food Research International, 109*, 260–267.

Yao, W., Liu, C., Xi, X., & Wang, H. (2010). Impact of process conditions on digestibility of pea starch. *International Journal of Food Properties, 13*, 1355–1363.

Zhao, B., & Chang, K. C. (2008). Evaluation of effects of soaking and precooking conditions on the quality of precooked dehydrated pea, lentil and chickpea products. *Journal of Food Processing & Preservation, 32*, 517–532.

Pigeon Pea

G. Jeevarathinam and V. Chelladurai

Introduction

Pigeon pea (*Cajanus cajan* L.), commonly called red gram and tur in South Asia, is one of the major legumes consumed in South Asian countries. Pigeon pea is rich in protein, and it can grow under poor rainfall and in rainfed conditions. The pigeon pea is used as a food crop (dried peas, flour, or green vegetable peas) as well as for forage and cover crops (Fig. 1). India is the major producer of pigeon pea, producing about 67.24% of the world pigeon pea crop every year. India also has 79.65% of the world pigeon pea production area, 4.92 million hectares. Pigeon pea is commonly grown in Asia, Africa, Latin America, and the Caribbean islands. Even though India ranks number 1 in pigeon pea production, the Caribbean island countries Saint Vincent and Grenadines, and Trinidad and Tobago, rank first and second, respectively, in pigeon pea productivity. In Asia, India, China, Myanmar, and Nepal produce pigeon pea, and in Africa, the major pigeon pea growing countries are Kenya, Uganda, Malawi, Mozambique, and Tanzania (FAOSTAT 2014). In most of India pigeon pea is grown as a rainfed crop, and the states of Maharashtra and Karnataka share the major part of pigeon pea production. The annual production of India from 2011 to 2012 is given in Fig. 2.

In India, pigeon pea ranks as the second largest crop after chickpea in the production area at Kharif season. Pigeon pea contains 20–22% protein, and is also an excellent source of such amino acids as lysine, tryptophan, and methionine. Pigeon pea also contains a good amount of minerals (iron, magnesium, phosphorus),

G. Jeevarathinam
Department of Food Technology, Bannari Amman Institute of Technology, Sathyamangalam, India

V. Chelladurai (✉)
Department of Agricultural Engineering, Bannari Amman Institute of Technology, Sathyamangalam, India
e-mail: chelladurai@bitsathy.ac.in

A. Manickavasagan, P. Thirunathan (eds.), *Pulses*,
https://doi.org/10.1007/978-3-030-41376-7_15

Fig. 1 Pigeon pea crop (**a**) and harvested and threshed pigeon pea (**b**)

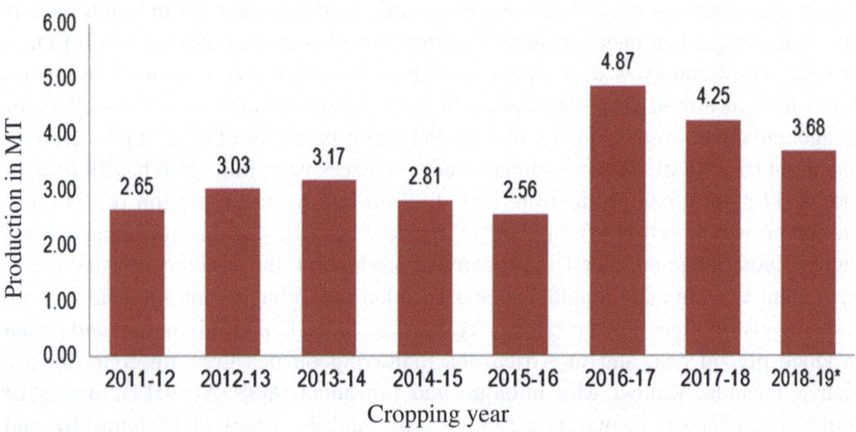

Fig. 2 Annual production of pigeon pea in India. *MT* million tonnes

riboflavin, folacin, and niacin. In India and other parts of South Asia, pigeon pea is usually consumed with cereal grains (rice and wheat), and this combination provides most of the nutrients required for a healthy diet (Akporhonor et al. 2006; Ramcharan and Walker 1985). In most parts of South Asia, pigeon pea is commonly used to make curry or dhal, which is consumed with rice- or wheat-based products such as roti or chappathi. The nutritional composition of pigeon pea is given in Table 1.

Pigeon pea provides a wide range of products, including dried seeds, pods and immature seeds used as green vegetables, leaves and stems used as fodder, and the dry stems that are used as fuel. It also increases soil fertility by nitrogen fixation as well as by leaf drop and nutrient recycling (Snapp et al. 2002). It is an essential pulse crop that performs well in poor soils and areas where there is insufficient or

Table 1 Nutritive value of pigeon pea

Composition	Nutritive value
Protein (%)	22.3
Fat (%)	1.7
Minerals (%)	3.5
Fiber (%)	1.5
Carbohydrate (%)	57.6
Calcium (mg/100 g)	73
Phosphorus (mg/100 g)	304
Iron (mg/100 g)	5.8

inadequate moisture availability (Kimani 2001). Depending on the distribution pattern, the plant can withstand low moisture and performs well in areas with an annual rainfall of less than 1000 mm. For crops such as maize, sorghum, or groundnuts, pigeon pea can be introduced as an intercrop without significantly reducing the main crop yield. The grain has a high nutritional value with a high protein content ranging from 21% to 25%, making it very important for many poor families who are unable to afford dairy- or meat-based diets and also to improve food security and health.

Postharvest Processing

Harvesting

Green pigeon pea pods are harvested for diverse purposes. Green pigeon is sold mostly to the urban market and other readily marketable locations. The pods of the green pigeon pea should be harvested before they begin to lose their green colour. The bright green and fully developed seed is preferred, whereas the appearance at this stage fluctuates depending upon the cultivars. The green pods that are being used for domestic purposes at a small scale can be picked 'by hand' using a sickle. For industrial use or for large-scale processing such as canning, mechanical harvesting methods are used for cutting plants and vines, followed by drying and threshing. It is even possible to continuously harvest the pods for consumption if ratoonable cultivars are developed as it is also a perennial crop. With the availability of ample rainfall and irrigation facilities, it can be grown as shrubs or trees for a period of 4–5 years. Under these preferable conditions, the harvest can be done at each growth after the plants reach the harvesting or maturity stage. This crop when harvested in its green colour serves as a superior vegetable. Currently, however, it is being cultivated in only in a few areas in India and is a very important market commodity. The likelihood for a widening consumption rate by the consumers of this vegetable in our country is notable as unique and healthy cultivars are being grown and consumer pressure also encourages increasing the yield. Before the breaking stage of the pods when they turn yellow, the dry seeds are harvested from pigeon pea. Pods that have reached the maturity stage continue to ripen even if the plants are cut, and very few pods shatter when dry. At this time, heavy crop losses occur. Green pigeon pea can be threshed by manual and mechanical methods. The harvested crop is left to dry in the threshing yard in the sun for 1 week, depending on weather conditions. In manual threshing, the vines and pods are beaten with sticks to separate the seed. In some places, animals are caused to walk on the pigeon pea crop to thresh it, but in most areas mechanical threshers are used (Pandey 1998).

Threshing (Shelling)

The pods that are developed are harvested in the field and will be shelled or threshed to separate the green pigeon peas from the walls of the pods, depending on the cultivar characteristics, with a great difference in the recovery of peas through shelling, and it is also notable that recovery will be higher in some cultivars compared to others. Shelling of green pigeon peas is done by mechanical means or manual means depending upon the volume of the processor's product where shelling recovery is an essential part of the process. The benefits of hand shelling are very low capital investment is required, an attractive product results, and high yield results in comparison to machine shelling. Frozen product packers and packers of peas that are sold in polythene bags prefer the process of hand shelling.

Storage

After the pigeon pea is threshed, to reduce the moisture content, the grains should be dried in the hot sun, which can reduce the moisture content of the grain as much as 10%. The grains should be stored in gunny sacks or in dry clean storage areas.

Methods of Storage

Poor or poorly developed storage conditions can cause losses despite abundant loss prevention methods followed by the farmers, which can have only a biased result. Thus, the method of storage is vital in storage loss reduction. Storage structures are usually made of such materials as mud, wood, plastic, sand, steel, concrete, and jute bags. At the farm level, these structures are used to store pigeon pea. Some farmers use communal mud bins. Jute bags are widely used in the market and in urban dhal mills. The storage structures used for seeds or grains intended for consumption differ from the usual type. Reduction of storage losses is widely accomplished by seed treatments, considered to be most important, in which contact insecticides such as dichloro-diphenyl-trichloroethane (DDT), β-hexachlorocyclohexane (BHC), and malathion are applied to the seeds. These toxic chemicals are being used to protect bulk stocks of seeds from insects and pathogens: 0.2% ethylene dibromide is used as a fumigant, and 0.2% malathion mixed with tricalcium phosphate 0.2% is found to be more potent. However, this type of chemical fumigation causes viability loss as the chemicals can react with the enzymes present in seeds. In the traditional practice, seeds were coated for protection from insects, and this age-old practice is being followed in several villages in India. In this process, the stored seeds are coated with an edible oil as a thin film. As this type of treated seeds is not usually favoured by pulse beetles, the treatment can be applied as a conservative storage method. Also, such oils as mustard, sunflower, safflower, castor, cotton, neem (*Azadirachta indica*), and karanj or honge (*Pongamia glabra*) are being investigated for controlling pulse beetle infestations. Results show that 1.0% honge and neem oils are effective as surface protectants against attack by *Callosobruchus chinensis*. This study of beetle infestation confirmed that honge oils can protect for 319 days and neem oil can protect for 161 days. It is noted that when pigeon pea is stored in the dhal form less pulse beetle infestation occurs. By preference, pigeon pea should be processed and stored as dhal. To decrease loss during storage and for pigeon pea stocks to be safe to consume, chemical formulations should be avoided (Chavan 2012; Feng et al. 2015).

Storage Pests

In India, the pulse beetles afflicting pigeon pea are the most significant storage pest, widely referred to as dhoras. Of the three bruchid species, *Callosobruchus chinensis* (L.), *Callosobruchus maculatus* (F.), and *Callosobruchus analis* (F.), *C. maculatus* is the most important. When the pods are in the ripening stage in the field, these insects intermittently begin infestation. The infected grains are subsequently carried into storage after harvesting, thereby causing losses. In the stored pulses, the development rate, mortality rate, and oviposition of some species of Bruchinae (Bruchidae; now Chrysomelidae) breeding were explained by Howe and Currie in 1964. It is been observed that *C. maculatus* and *C. chinensis* tend to complete their life cycle at 30 °C in several weeks and it is also noted that relative humidity in stored pigeon pea is 70%. Susceptibility toward pulse beetle attacks is observed to differ among cultivars. Among those distinct cultivars, seed infestation caused by *C. maculatus* was between 7.0% and 28.7%. Beetle incidence is highest in JA 8 and lowest in ICPL 7 (Howe and Currie 1964). Under laboratory conditions, when studying the relative susceptibility of the 33 cultivars of pigeon pea to *C. chinensis*, the least susceptible ones are ICPL 148 and ICPL 151 and that most highly susceptible is 79-74 and ICPL 289. No cultivars have 100% resistance to infestation, and the susceptibility to infestation varies significantly among cultivars. In different pulses the seed surface, seed coat thickness, and seed size are linked with the resistance mechanism to the different species of *Callosobruchus*. The study indicated that *C. maculatus* and *C. chinensis* showed preference of oviposition in the whole grain, compared to dhal, which indicates that the seed coat provides the ovipositional stimulus. The beetle does not show preference to some characters of seed size, colour, and texture; thickness of the seed coat alone influenced beetle incidence.

Storage Losses

At the storage stage, about 90% of pulse postharvest loss takes place, in which about 50% is caused by insects and pests; mass, quality, nutritive value, and hygienic deterioration are among the storage losses that occur. A recent survey stated that when comparing the legumes, large losses occur only in pigeon pea. Among legumes 32.6% of damage is caused in pigeon pea, 18.5% in cowpea, 14.9% in bean, 9.9% in mung bean, and 4.8% in chickpea; pigeon pea is noted as the most vulnerable to pest growth when compared to cowpea and chickpea. It was also noted that storage losses in pigeon pea were caused by the growth and development of *C. maculatus*, which causes insect infestation of about 14–64%. Temperature, humidity, moisture content of stored grains, air concentration in the storage structure, health and hygiene, and pesticide use are some of the farm level factors that influence storage losses. Temperature, moisture, and oxygen are the requisite physical factors to be regulated. Insects, moulds, bacteria, and yeast are noted to be biological factors in storage loss.

Processing Green Seeds

Shelling, freezing, and canning are major processing methods after harvesting. Canning and freezing consist of several operations such as cleaning, blanching, and filling cans or polyethylene bags.

Cleaning

Appropriate cleaning methods are based on hand and machine shelling. The cleaning process removes all the damaged seeds and foreign matter. Cleaning precedes freezing small quantities during hand shelling. Seeds are cleaned with blanching water, but before blanching the seeds should be cleaned with cold water so that the blanching water is kept as clean as possible. Sellers themselves will clean all fresh market produce, which should be washed. For cleaning and washing, seeds that have been mechanically shelled are transferred to conveyors to be dropped onto large mesh screens where pod pieces, damaged seeds, and small seeds are removed by air blasting. The mesh screen filters all the pods and foreign matter. The seeds are then washed with cold water in various combinations and with various types of flotation washers.

Canning

Canning of pigeon pea is followed using traditional methods in various states in India. As the demand for canned pigeon peas has recently increased in developing countries, canned pigeon peas for export markets are increased, becoming a common export business in countries such as the Dominican Republic where 80% of the green pigeon pea harvested annually is canned and exported. Those cultivars highly suitable for the process of canning have been developed through the maturity of pigeon peas; agroclimatic environmental factors highly influence the quality of the vegetable in that the mature green seeds of pigeon pea serve better than the starchy yellow seeds. For the process of canning, cultivars with seeds bright green in colour and uniform, and cultivars with a large quantity of pods, are preferred. Consumers prefer green seeds with higher soluble sugar content but the genotype for this type of trait is not yet developed. For producing the best quality of canned pigeon peas, harvesting the green seeds at the mature stage is the most important process to be followed, but the nonsynchronized flowering characteristic of this plant makes harvesting of pigeon peas at the maturity stage difficult. Some of the factors that similarly depend on the maturity of the pigeon pea are volume, viscosity, drained mass, the colour of brine, and uniformity of colour.

Blanching

In the process of canning and freezing, blanching is one of the fundamental applications of the heat treatment process, essentially done for colour fixation, flavour improvement, and volume reduction. It helps in the improvement of the texture, which could allow placing a larger quantity of peas into the cans. Blanching also helps in the removal of mucous substances and also free starch for the purpose of obtaining clear brine. It also removes intercellular gases, which can be strained during the process of heating. There are two basic methods of blanching wherein the common method is to obtain a clear brine. In this type, the seeds are heated for 5 min in hot water at 185 °F (85 °C). After the heating process, the seeds are immediately cooled to about 80 °F (26.7 °C) from 185 °F. Steam blanching, used to decrease seed shrinking and nutrient loss, is another blanching method. This method consumes more energy and requires a larger investment, and so this method is not been practiced and accepted in any of the developing countries as an alternative method. Finally, the seeds are checked for removal of foreign matter and for any off-colour that may have escaped notice in other blanching methods before canning and freezing take place. It is notable that the canning and freezing process also involves all the same methods that take place in blanching.

Filling, Closing, and Cooling Cans

The can is filled with seeds with 2% of brine solution in varying amounts after blanching and cooling at 195 °F (90.5–93.3 °C) with no added sugar or other additives. Without using any mechanical exhaust, the brine is retained at its boiling point for closure of small cans. The larger type cans are supplemented with a vacuum-creating medium before closure as the brine at its near-boiling point does not create an ample vacuum. Thus, additional medium for the creation of a vacuum is needed. The cans should be immediately processed thermally after closing to inhibit the growth of thermophilic bacteria that will cause spoiling of the stored product later at high temperatures.

Freezing

The automatic continuous system and the labour-intensive batch system are the two types of freezing methods. In the automatic continuous system, the blanched and cooled seeds are transported by conveyor to a fluidized bed freezer that operates below freezing temperature (−10 to −20 °F) (−23.3 to −28 °C) where individual seeds are quickly frozen. Seeds that have been specially treated with wax initially are hand-packed into cartons once frozen; the seeds are wax treated to prevent product dehydration when stored at 0 °F (−17.8 °C). In a labour-intensive batch system type, the blanched seeds are dropped into cooled water tanks after the hot water blanch. After cooling, seeds are hand-packed into polyethylene bags. Later, the seeds will be placed for freezing in trays in a batch freezer at −10 to −20 °F (−23.3 to −28.9 °C) for 4–10 h depending on freezer design, package size, and initial temperature. The frozen bags then will be placed into corrugated containers for storage at 0 °F (−17.7 °C).

Dehulling

In many countries, in the initial process the hull is removed for processing grain legumes, followed by splitting the seeds into dicotyledonous material. In India, the process of dehulling the green pigeon pea is the initial method to convert the whole seed into dhal. The process of dehulling is performed in one of two ways: loosening the husk from the cotyledons, and removing the husk from the cotyledons followed by splitting them by roller machine or stone chakki.

Dehulling Methods

Dehulling pigeon pea is a traditional practice in India. The hand pounding method common in earlier days was later replaced by the stone chakkis method. The many traditional methods used are broadly classified into two categories: wet method and dry method. Water soaking, dehulling, and sun drying are steps involved in the wet method and in other methods that involve practices such as oil/water application, sun drying, and dehulling. According to a survey in India, pigeon pea for commercial purposes is dehulled into dhal on a very large scale by mechanical means in mills, where the products are first graded, followed by mild abrasion by the roller machine; called the tempering operation, this step improves the water-absorbing capacity of the seeds by scratching the seeds to loosen the testa. After treating with oil and water, the materials are dried in the sun, spread in a drying yard; if necessary, the material should be stirred occasionally, followed by dehusking by the roller machine. After dehusking is complete, several products are obtained and separated: dehusked split (dhal), dehusked un-splits (pearled), and un-dehusked material of split and un-split seeds. If more dhal is needed, the entire process can be repeated. The next step is a small-scale processing method using the stone chakki, which contains two grinding stones; when followed by the villagers in their homes, the operations probably vary from region to region. For example, in states such as Maharashtra, Uttar Pradesh, and Madhya Pradesh, the peas are soaked in water for about 2–14 h. In other places, the material is treated with oil before dehulling. In domestic practice, people hand pound and remove the seed coat after splitting the peas using chakki and treating with oil or water. In Uttar Pradesh, pigeon pea is heated in an iron pan, with or without sand, before the process of grinding. There are more economical methods, and the machines have been recently improved. Machines such as attrition-type mills (plate mills), a type of method that has not been widely followed, assist dehulling when the hull is not attached to the cotyledons. When the hull is firmly attached to the cotyledons, abrasive-type mills are used, incorporating carborundum that gradually abrades the seed coat (Sahay and Singh 1994).

Milling

Domestic Milling

Domestic milling involves cleaning and grading of seed, pretreatment or conditioning of seed, and splitting the cotyledons.

Cleaning and Grading of Seed

Sieves are used to discard dry husks, inert materials, and small shrivelled seeds.

Conditioning of Seed

(a) **Dry conditioning**

An energy-coated roller is used for pitting after cleaning and grading of pigeon pea. Pitted seeds are manually combined with warm oil (about 1% linseed or mustard oil), followed by 2–5 days of sun drying. Before the last day of sun drying, 2–5% water is sprinkled on the seed, which is thoroughly mixed and heaped for tempering overnight.

(b) **Wet conditioning**

This method involves 2–4 days of alternating wetting and drying of seed. Kurien et al. (1986) proposed uniform adjustment of moisture within the seed by exposing it to hot air (300 °C) for several minutes. This step provides complete removal of the husk (99%) in a single pass with limited scouring of the peripheral seed surface when abrasive (pearling) action is performed in stone rollers while dehusking. This method gives 10–15% more dhal compared to the traditional commercial method and reduces the time and cost of processing.

Saxena et al. (1981) proposed treatment of pigeon pea seed with sodium bicarbonate (0.3–1.0%), which resulted in recovery of 68% dhal and 88% dehusking efficiency. Reddy (1981) found 75% dhal recovery and 96% dehusking by pitting seed followed by sodium bicarbonate treatment, 12-h tempering, and oil treatment [250 g oil (100 kg seed)$^{-1}$], followed by another 12-h tempering in a modern rice pearler before dehusking. Further, Srivastava et al. (1988) reported that soaking pigeon pea seed in sodium bicarbonate (6%) solution results in high dehusking efficiency. Pigeon pea roasting on dry sand for 5 min at 100 °C with an initial moisture content of 12% was also recommended.

Dehusking and Splitting

Kurien (1971) reported that either disk shellers or roller machines are used for dehusking and splitting. A disk sheller used for wet processing operates on the attrition principle and is useful to simultaneously remove the husk and split the cotyledons. However, this process causes excessive breakage. Dry processing is performed by using a roller machine that works on the abrasion principle and is efficient for dehusking when used alone. When used simultaneously for dehusking and splitting, rounding of the edges of the split cotyledons causes additional powdering and a decrease in dhal commercial value.

Polishing of Dhal

A cone polisher similar to the one used in a rice mill is used to polish dhal. Rubber rollers were used to reduce scouring (Narain et al. 1986). A pin-like machine similar to a pin mill provides good splitting with less breakage. Dehusking and splitting are normally done at the same time. Figure 3 shows the traditional milling of pigeon pea.

Dhal Milling Using Pantnagar Technology

Narain et al. (1986) reported a new technology for making dhal at the Govind Ballabh Pant University of Agriculture and Technology, Pantnagar, India (Fig. 4) that involves the following steps.

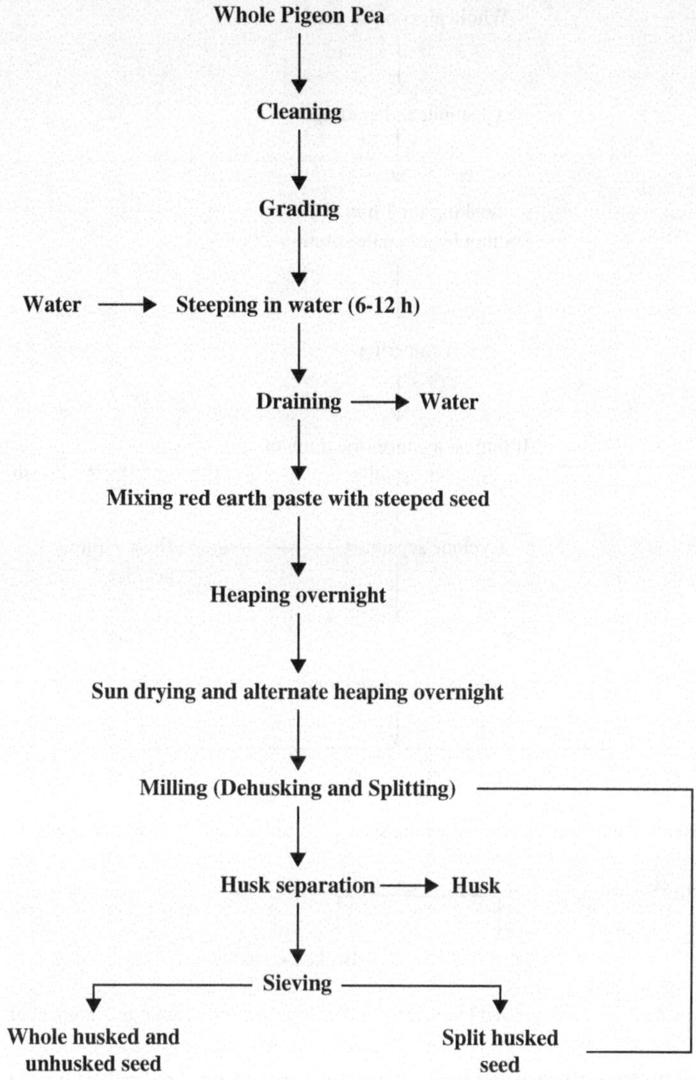

Fig. 3 Flowchart for traditional milling of pigeon pea by the wet method. (From Narain et al. 1986)

1. **Pretreatment**

   In pretreatment, pigeon pea seed is mixed with 6.0% solution of sodium bicarbonate, followed by 5–6 h tempering and drying to 10% moisture content (wet basis).

2. **Machine system**

   The system consists of a concentric double-cylinder auger feeder and discharge machine. To provide a rough surface, the inner cylinder is made of emery. Part of the length carried by the emery cylinder flows toward the outlet to provide a restricted passage with its plane normal to the axis of the auger. Residing time

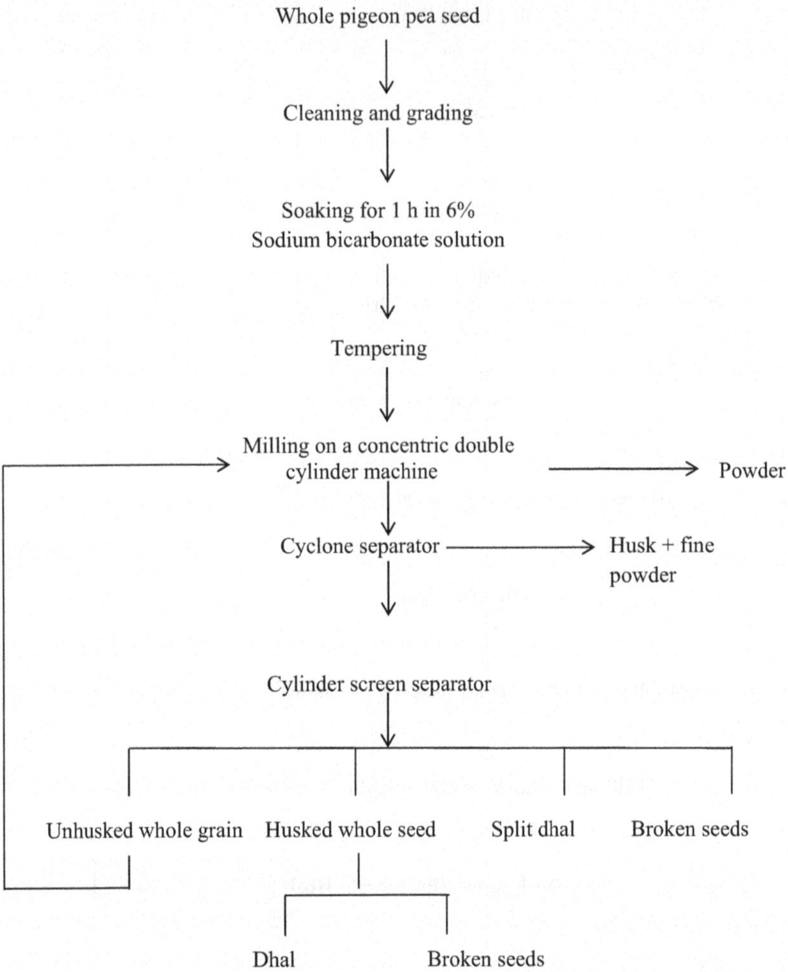

Fig. 4 Flowchart of the improved Pantnagar dhal milling process. (Source: Narain et al. 1986)

of seed in the milling zone is regulated by controlling the discharge opening size through an automatic loading device. The milling system is fitted with an elevator for feeding the dehusking roller, a cyclone separator for separating the husk, and a dhal grader for separating dehusked seed and broken seeds.

In dhal recovery and dehusking, the Pantnagar technology was reported to be on a par with the CFTRI technology. It is not weather dependent and has a high dehusking performance with a single pass (more than 97%). Less energy, 45 kWh t^{-1} is needed when compared to 85 kWh t^{-1} for the CFTRI process, and 60 kWh t^{-1} as compared to the traditional process (Narain et al. 1986). Singh and Jambunathan (1990) discussed different modifications of dhal milling in India. A survey showed that dhal recovery ranges from 60% to 85% depending on the milling procedure and presoaking treatments.

Dry Milling Method of Pigeon Pea

This method is normally followed in Madhya Pradesh and Uttar Pradesh. The pulses are subjected to pitting in a roller followed by an oil treatment (applying 0.5–2.0% linseed oil or any edible oil). After this the pulses are spread in the drying yard for sun drying for 2–4 days. During the intervening nights, the pulses are tempered by heaping and covering. Once again, pulses are moistened uniformly with about 5% water after sun drying and held overnight as such on heaps for moisture equilibrium. Then, for splitting and dehusking these pulses are allowed to pass from the roller. About 50% of the pulses are dehusked and broken in the first operation. After this, aspiration removes the husks, and split dhal is segregated from the mixture of husked and unhusked whole pulses. The mixture is moistened again and sun dried and then dehusked followed by splitting. This process of alternate wetting and drying is repeated until all the remaining pulses are converted to split dhal. The average yield of dhal varies from 68% to 75% (Fig. 5).

Wet Milling Method for Pigeon Pea

In this milling method, the grains are soaked in water for 3–12 h. The soaked pulses are thoroughly mixed with red earth at about 5%. The mixture is held overnight in heaps. Then, the whole mixture is dried for 2–4 days in the sun until the grain husk is shriveled and loosened. Between these days, the pulses are tempered overnight. The red earth is separated from the pulses by sieving. The dried grains are dehusked and split into the disk sheller. The dhal and other fractions are separated. Approximately 95% of the pulses are dehusked and split in a single milling process. Further, the remaining product is pretreated and milled to turn into dhal. The red earth makes it easier to raise the rate of drying and in loosening the husk. This method involves about 5–7 days for processing of a batch of pulses (Fig. 6).

CFTRI Method of Pigeon Pea Milling

There are some other approaches, such as the Central Food Technological Research Institute (CFTRI) process that removes oil mixing and for husk loosening. To condition clean and graded grains, dry heat treatment by two passes through an LSU drier with hot air is used. The grains are tempered for 6 h in a tempering bin after each pass through the dryer. The preconditioned pulses are conveyed to the pearler or dehusker where almost all pulses are dehusked in a single operation. The gota (whole grain dehusked) is separated from split pulses and the mixture of husk, brokens, etc. Water is added to gota at a controlled rate, then collected and for about 1 h allowed to remain in the same condition. Some of the moistened gota form lumps of varying sizes. To separate them, these lumps are fed to the lump breaker. These gota are passed to the LSU dryer where they are exposed to a few hours of hot air. For splitting purposes, the gota is dried to a proper level of moisture. The hot,

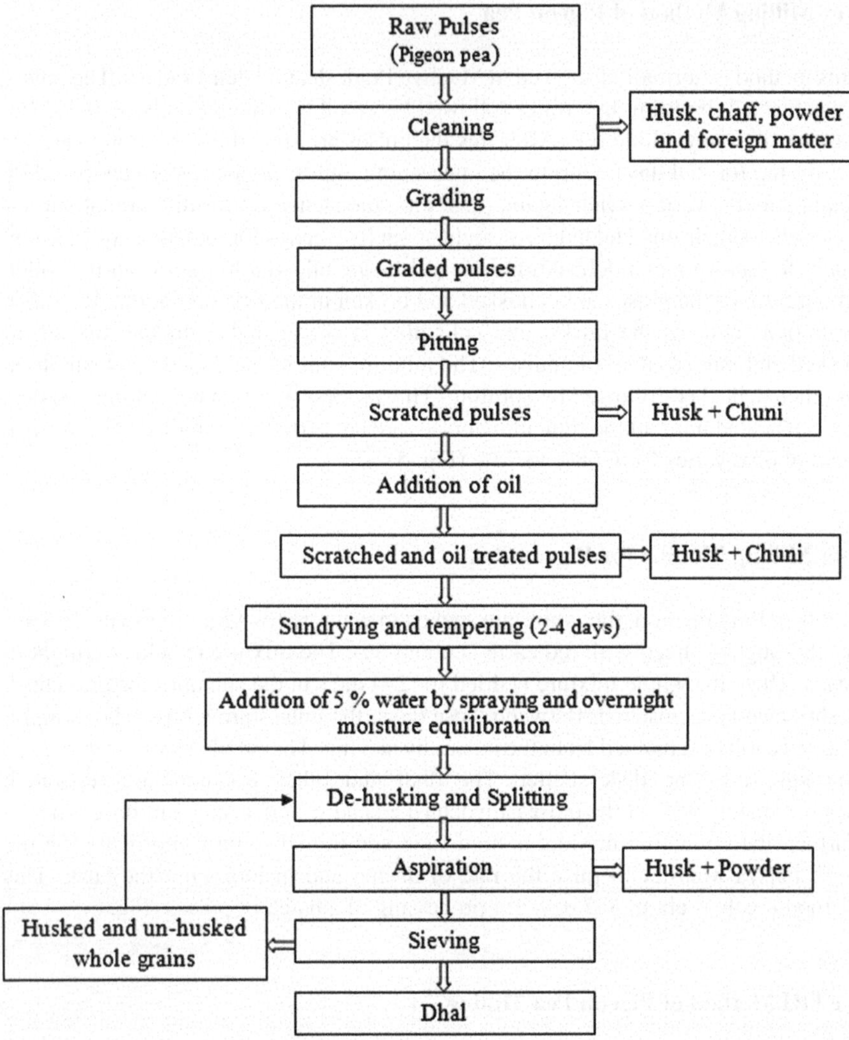

Fig. 5 Flow diagram of dry milling method of pigeon pea (Source: Chakraverthy 1995)

conditioned, and dried dehusked whole pulses are split in the emery roller. The mixture is classified into pulses of grade I, whole dehusked pulses, and small brokens. For subsequent splitting, the unsplit husked pulses are fed back to the conditioner. This method offers an average yield of 80%, in less time and with less cost of production compared to other methods (Fig. 7).

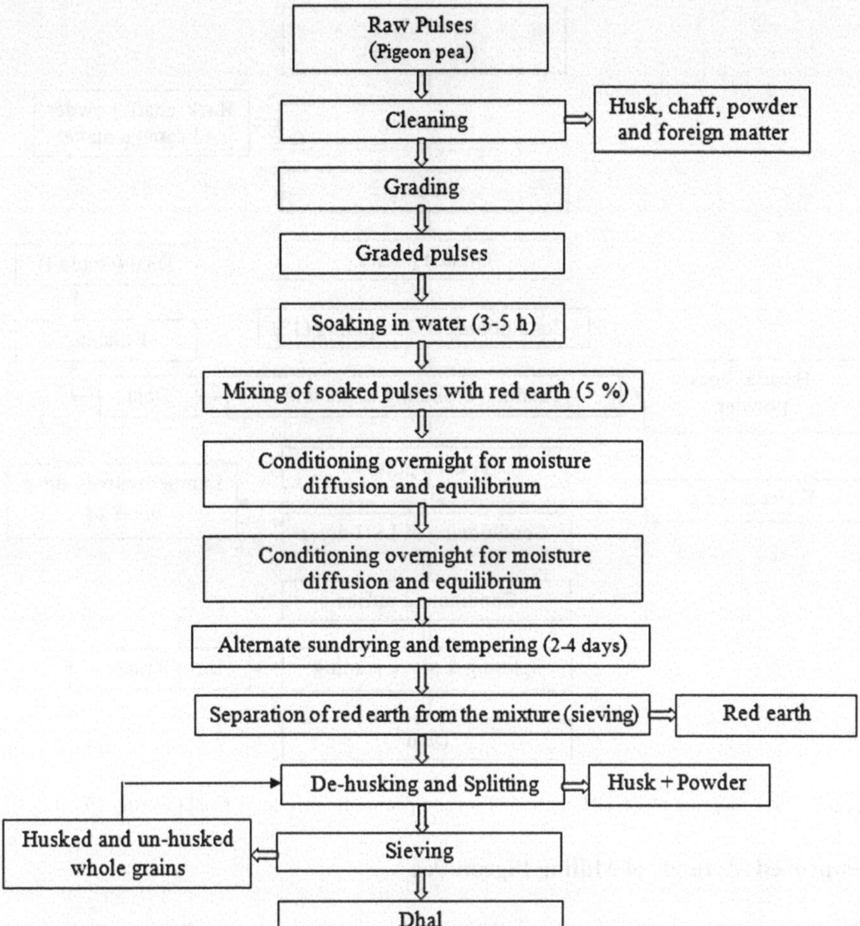

Fig. 6 Flow diagram of wet milling method of pigeon pea (Source: Chakraverthy 1995)

CIAE Method of Pigeon Pea Milling

This method avoids the use of edible oil in the process of pre-treatment. Pigeon pea, green gram, and black gram are first cleaned and fed for scratching to the roller mill developed at the Central Institute of Agricultural Engineering (CIAE). After cleaning, the scratched grains are soaked in tap water for 30 min at ambient temperature (for pigeon pea) or 1 h (for black gram and green). The liquid is drained off; the moisture content of the grains reaches 9–10% after drying. To produce dehusked split cotyledons, these conditioned grains are fed again to the roller mill.

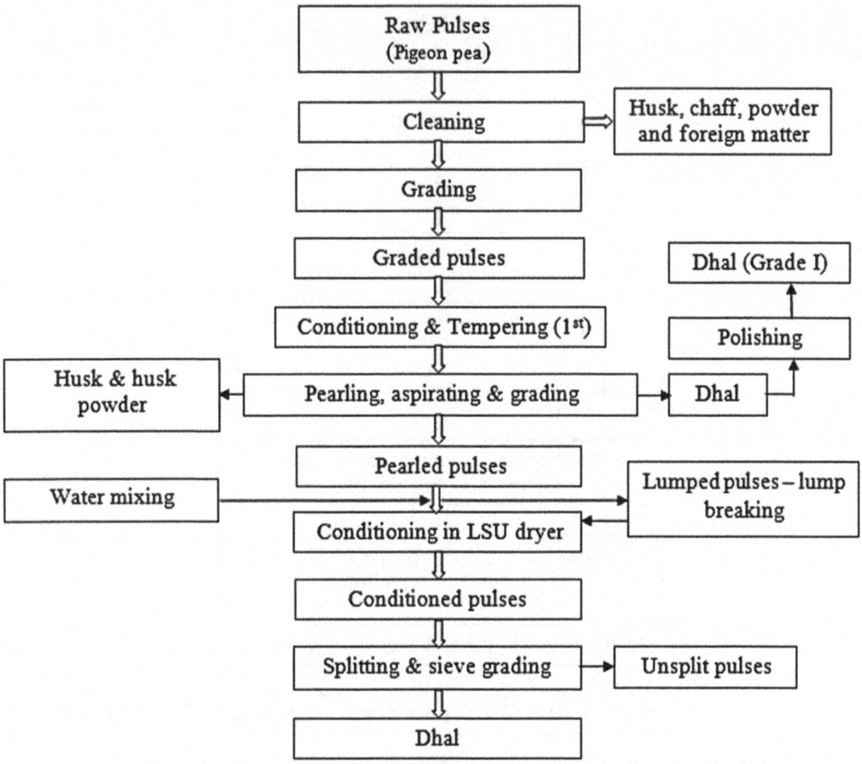

Fig. 7 Flow diagram of CFTRI method of pigeon pea milling (Source: Chakraverthy 1995)

Improved Methods of Milling Pigeon Pea

There are several drawbacks and difficulties in conventional technology and machines; thus, attempts have been made either to improve traditional methods and machines or to develop new ones. Although some machines built in other countries for the benefit of cereals and millets have been used to dehull grain legumes, dehulled grain recovery is not as good as with the traditional methods (CFTRI 1977).

In addition, the dehulling of grain legumes, particularly pigeon pea, can be effectively and economically accomplished in the two general steps of pre-milling to loosen husks and dehulling followed by splitting. At the Central Food Technological Research Institute, Mysore, India, effective technology and machinery have been established for dehulling pigeon pea and other grain legumes. The pigeon pea husk can be loosened and rendered brittle and pulverizable when exposed to a critical level of moisture under acceptable temperature conditions.

Temperature conditions and critical moisture level differ from cultivar to cultivar, but if pre-treated under proper conditions, adequate milling characteristics can be imparted to those most difficult to mill. The grain can be hardened and the degree of splitting can be significantly reduced by reducing the amount of moisture and

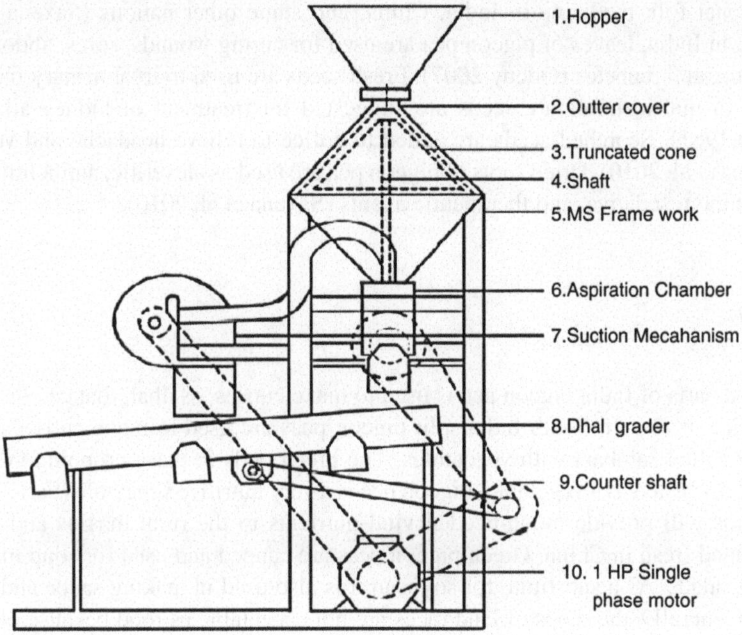

Fig. 8 CFTRI mini dhal mill for pigeon pea milling (Source: Goyal et al. 2005)

thus scouring losses. These results form a baseline for the improved technology developed at CFTRI (Fig. 8).

Product Development

Pigeon pea seeds are an important source of energy, proteins, and other nutrients in the diets of large population groups around the world, providing an excellent source of lysine, methionine, and tryptophan and other water-soluble vitamins (riboflavin, niacin, folacin) and minerals (phosphorus, iron, magnesium) (Ramcharran and Walker 1985). Potential sources of protein and fiber are found in processed grains, brokens, hulls, and powder, the by-products released during the dehulling process (Jayadeep et al. 2009). Food items that can be prepared from pigeon pea include fresh sprouts, tempe, ketchup, noodles, snacks, and various extruded food products (Saxena et al. 2002). In addition, the by-products of split and shrunken seeds are also used as a cheap livestock feed as a substitute to high-cost animal feed sources such as bone meal and fish meal (Chisowa 2002). Similarly, feed for cows, poultry, and pigs can be obtained through the by-products of pigeon pea such as seed coats, brokens, and powder obtained from the dhal mill (Samanta et al. 2013).

Besides its nutritional value, pigeon pea also possesses various medicinal properties from a number of polyphenols and flavonoids. It is an integral part of traditional folk medicine in India, China, and some other nations (Saxena et al. 2010). In India, leaves of pigeon pea are used for curing wounds, sores, abdominal tumours, and diabetes (Odeny 2007). Fresh seeds are used to treat urinary incontinence in males; immature seeds are suggested for treatment of kidney ailments (Duke 1981). Scorched seeds are added to coffee to relieve headache and vertigo (Saxena et al. 2010). Dried roots of pigeon pea are used as alexeritic, antihelminthic, expectorant, sedative, and therapeutic agents (Saxena et al. 2010).

Food

In most parts of India pigeon pea is used to make curries, as dhal, that can be eaten with rice or roti. In South India split pigeon peas are used to make curries (commonly called sambar) with vegetables. The green pods in these cropping systems can be harvested as a vegetable. Pigeon pea is a very nutritive vegetable (Faris 1987) and thus will provide much-needed vital nutrients to the rural masses and when consumed fresh in China. Green pigeon peas are canned and used for soup making and in salads. As a substitute for soybean it is also used in making sauce and bean curd. Generally, the seeds of land races are not acceptable as food because of their small size, high amounts of trypsin inhibitor, long cooking time, and puckery and odd taste in green (immature) and dry seeds, respectively. In the pigeon pea development program in Yunnan Province, the new varieties at the International Crops Research Institute for the Semi-Arid Tropics (ICRISAT) are performing well and their seed quality is acceptable. Therefore, to make their production sustainable, new uses of its consumption need to be developed (ICRISAT 1981). Several new processing technologies of products such as spicy-crisp pigeon pea, sweet bean paste, and pigeon pea starch were developed in China (Chaohong et al. 2001). The spicy-crisp pigeon pea was made through procedures of steeping, selection, and frying. The product is crisp and has a nice taste with a special flavour that met the National Standards. The sweet bean paste, golden yellow in colour, tastes good and feels smooth in the mouth. The starch products made from pigeon pea were found better in sensory index when compared with broad bean (*Vicia faba*).

Snacks

In the current scenario, a variety of snacks made from cereals, legumes, and fruits are available in the world market. Because pigeon pea seed contains 22% protein and eight important amino acids, necessary for the human body, its snacks and other processed products will be able to compete well in this enormous market. Some of the products with good potential include spicy-crisp grains, pigeon pea sweet paste, and noodles. The processing technology for spicy-crisp pigeon pea is established and it could be utilized immediately.

Bakery and By-Products of Pigeon Pea

Several researchers recommended the use of legume flour as a source of protein in the preparation of bakery products (Eneche 1999). To prevent protein-calorie malnutrition in the developing world, supplementation of cereals with protein-rich legumes is considered as one of the best solutions (Chitra et al. 1996). Pigeon pea flour has been tested and found to be suitable for consumption as bread, cookies, and chapattis because of its high levels of protein, iron (Fe), and phosphorus (P) content (Harinder et al. 1999). So, it has been recommended in school feeding programs and to the vulnerable sections of the population in developing nations. Protein-rich pigeon pea seeds have also been incorporated into cassava flour to produce acceptable extruded products (Rampersadet al. 2003). The underutilized by-product of pigeon pea (20% protein content) with great potential as a valuable protein can also be used as drug/nutraceutical carriers, which would be safe for human consumption compared to synthetic materials (Tapal and Tiku 2013).

A study was conducted by Lee et al. (2015) to produce nattokinase, a serine fibrinolytic enzyme, in pigeon pea by fermenting with *Bacillus subtilis*. Various strains of *B. subtilis* (14,714, 14,715, 14,716, 14,718) were tested for production of nattokinase, and *B. subtilis* 14,715 fermentation had the highest nattokinase activity in pigeon pea for 32 h. In addition, the levels of antioxidants (phenolics and flavonoids) and angiotensin-converting enzyme inhibitory activity were increased in *B. subtilis* 14,715-fermented pigeon pea, compared with those in nonfermented pigeon pea. In an animal model, both water extracts of pigeon pea and *B. subtilis*-fermented pigeon pea significantly improved systolic blood pressure (21 mmHg) and diastolic blood pressure (30 mmHg) during in vivo testing, which showed *B. subtilis*-fermented pigeon pea can be used as a dietary supplement to improve cardiovascular health and prevent hypertension (Lee et al. 2015).

The nutritional value of toasted pigeon pea was investigated by increasing its inclusion (100, 200, 300, and 400 g kg^{-1}) in iso-nitrogenous (35% crude protein) and iso-energetic (17.7 kJ g^{-1}) diets. In toasting of *Cajanus cajan* it may be concluded that it not only improved feed quality but also resulted in the best use at the 400 g kg^{-1} inclusion level and significantly reduced production costs. Future research can concentrate on other processing methods aimed at further improving the nutritional use of feed in the diet of different cultured fishes. Moreover, because conclusions were based on fingerlings performance within the 2 months of the study, longer-term studies are needed under real conditions of aquaculture in grow-out ponds. Higher levels of inclusions can also be researched to further offset the cost of feeding, which is estimated to be more than 70% of the total cost of production (Solomon et al. 2017).

Pigeon pea starch can be used as a food additive. Olagunju et al. (2018) tested the functionality changes of pigeon pea starch by the modification of natural starches during food additive manufacturing. Extracted starch from pigeon pea was modified using acetic anhydride at various degrees of substitution (DS) (0.05, 0.09, 0.14 DS) by esterification and tested for physiochemical changes using a Fourier-transform infrared spectroscopy (FTIR) spectrophotometer. Heat transition temperatures and enthalpy of gelatinization decreased and the 0.05 DS starch had the lowest retrogradation tendency (30.54%) among all the samples. The FTIR spectra showed characteristic absorption peaks at 1245, 1370, and 1732 cm^{-1} for C–C, C–O, and C–O stretch vibrations, respectively. Scanning electron microscopy (SEM) images collected from the samples showed modification using acetylation did not cause any changes in starch granule shape and size, but a significant rupture of starch granules was observed at 0.14 DS (Olagunju et al. 2019).

Future Trends in Processing and Product Development

Pigeon pea is a rich source of food proteins and occupies an important place among the pulses. Based on its amino acid profile and biological value, it is considered as the best natural source of essential nutrients and is recommended as a balanced diet. Pigeon pea can fulfil the nutritional gap of proteins when it is mixed with cereals for the poorer sections of developing economies that cannot afford a nonvegetarian diet. At present, the protein availability in developing countries is about one third of the normal requirements as the population is ever growing. Thus, various nutritional development programs at present are facing a tough challenge to meet the protein demand. Hence, the use of pigeon pea-derived by-products can be an alternative source to combat protein deficiency by the cereal–legume mutual supplementation principle. These by-products also possess therapeutic properties, which can be branded as a food with nutraceutical properties. So, food products prepared using an economic and cheap source such as pigeon pea by-products offer an efficient alternative strategy to reduce malnutrition at the global level.

References

Akporhonor, E. E., Egwaikhide, P. A., & Eguavoen, I. O. (2006). Effect of spouting on in-vitro digestibility of some locally consumed leguminous seeds. *Journal of Applied Sciences and Environmental Management 10*(3), 55–58.

CFTRI (1977). Pulse Milling Technique- Technical Bulletin. Central Food Technological Research Institute, Mysore, India.

Chakraverthy, A. (1995). *Post-harvest technology of cereals, pulses and oilseeds*. New Delhi: Oxford & IBH.

Chaohong, Z., Zhenghong, L., Saxena, K. B., Jianqiu, Z., Yong, G., Shi Ying, Y., & XuXiao, Z. (2001). Traditional and alternative uses of pigeon pea in China. *International Chickpea and Pigeonpea Newsletter, 8*, 55–57.

Chavan, U. D. (2012). *Post-harvest management and processing technology: Cereals, pulses, oilseeds, fruits and vegetables*. Rome: Daya House.

Chisowa, D. M. (2002). *Comparative evaluation of performance of growing rabbits fed Leucaena leucocephala-cereal basal diet supplemented with legume grains*. MSc. Thesis. University of Malawi, Bunda College of Agriculture.

Chitra, U., Singh, U., & Venkateswara Rao, P. (1996). Phytic acid, in vitro protein digestibility, dietary fiber and minerals of pulses as influenced by processing methods. *Plant Foods for Human Nutrition, 49,* 307–316.

Duke, J. A. (1981). *Handbook of legumes of the world economic importance.* New York: Plenum Press, pp 33–37.

Eneche, E. H. (1999). Biscuit-making potential of millet/pigeon pea flour blends. *Plant Foods for Human Nutrition, 54*(1):21–27.

FAOSTAT (2014). FAOSTAT statistics database—Agriculture 2014. Accessed on Oct 21, 2019 from http://www.fao.org/faostat/en/?#data/QC

Faris, D. G. (1987). *Vegetable pigeonpea: A promising crop for India* (No. Folleto 1z554).

Feng, C., Jin, S., Luo, M., Wang, W., Xia, X. X., Zu, Y. G., Li, L. P., & Fu, Y. J. (2015). Optimization of production parameters for preparation of natto-pigeon pea with immobilized *Bacillus natto* and sensory evaluations of the product. *Innovative Food Science & Emerging Technologies, 31*, 160–169.

Goyal, R. K., Wanjari, O. D., Ilyas, S. M., Vishwakarma, R. K., Manikantan, M. R., & Mridula, D. (2005). *Pulse Milling Technologies.* Technical bulletin No. CIPHET/Pub/05.

Harinder, K., Kaur, B., & Sharma, S. (1999). Studies on the baking properties of wheat: Pigeon pea flour blends. *Plant Foods for Human Nutrition, 54,* 217–226.

Howe, R. W., & Currie, J. E. (1964). Some laboratory observations on the rates of development, mortality and oviposition of several species of Bruchidae breeding in stored pulses. *Bulletin of Entomological Research, 55,* 437–477.

ICRISAT. (1981). *Proceedings of the International Workshop on Pigeonpeas, International Crops Research Institute for the Semi-Arid Tropics*, 15–19 December 1980, Patancheru, A.P., India (Vol. 1).

Jayadeep, P. A., Sashikala, V. B., & Pratape, V. M. (2009). Nutrients and certain lipid soluble bioactive components in dehusked whole grains (gota) and dehusked splits (dhal) from pigeon pea (*Cajanus cajan*) and their cooking characteristics. *International Journal of Food Sciences and Nutrition, 60*(4), 273–284.

Kimani, P. M. (2001). Pigeonpea breeding: Objectives, experiences, and strategies for eastern Africa. In *Status and Potential of Pigeonpea in Eastern and Southern Africa: Proceedings of a Regional Workshop, 12–15 Sep 2000, Nairobi, Kenya* (Vol. 130, p. 21). *B-5030 Gembloux, Belgium: Gembloux Agricultural University; Patancheru 502 324, Andhra Pradesh, India: International Crops Research Institute for the Semi-Arid Tropics.*

Kurien, P. P. (1971). Traditional and improved method of milling of dhal. In *Proceedings of Hat Milling Conference*. Mysore: Central Food Technological Research Institute.

Kurien, P. P., Desikachar, H. S. R., & Parpia, H. A. B. (1986). Some developments in pigeonpea milling technology. In *Pigeonpea Scientists Meeting* (pp 24–26).

Lee, B. H., Lai, Y. S., & Wu, S. C. (2015). Antioxidation, angiotensin converting enzyme inhibition activity, nattokinase, and antihypertension of *Bacillus subtilis* (natto)-fermented pigeon pea. *Journal of Food and Drug Analysis, 23*(4), 750–757.

Narain, M., Singh, B. P. N., & Saxena, R. P. (1986). Some developments in pigeonpea milling technology. *Paper presented at Pigeon pea Scientists Meet, 24–26 Nov. 1986, ICRISAT, Patancheru, A.P. India: International Crops Research Institute for the Semi-Arid Tropics.*

Odeny, D. A. (2007). The potential of pigeonpea (*Cajanus cajan* (L.) Millsp.) in Africa. *Natural Resources Forum, 31*(4):297–305.

Olagunju, A. I., Omoba, O. S., Enujiugha, V. N., Alashi, A. M., & Aluko, R. E. (2018). Pigeon pea enzymatic protein hydrolysates and ultrafiltration peptide fractions as potential sources of antioxidant peptides: An in vitro study. *LWT, 97,* 269–278.

Olagunju, A. I., Omoba, O. S., Enujiugha, V. N., Wiens, R. A., Gough, K. M., & Aluko, R. E. (2019). Influence of acetylation on physicochemical and morphological characteristics of pigeon pea starch. *Food Hydrocolloids, 100*, 105424.

Pandey, P. H. (1998). *Principles and practices of post-harvest technology.* Ludhiana: Kalyani.

Ramcharran, C., & Walker, A. D. (1985), Changes in the cooking quality and nutritional value of starchy legumes due to adverse storage conditions. *Journal of Plants Foods, 6,* 73–88.

Reddy, V. P. R. (1981). *Dehusking of arhar grains (Cajanus cajan).* M. Tech. thesis, Govind Ballabh Pant University of Agriculture and Technology, Uttar Pradesh, India.

Sahay, K. M., & Singh, K. K. (1994). *Unit operations of agricultural processing*. New Delhi: Vikas Publishing House.

Samanta, A. K., Jayapal, N., Kolte, A. P., Senani, S., Sridhar, M., Mishra, S., Prasad, C. S., & Suresh, K. P. (2013). Application of pigeon pea (*Cajanus cajan*) stalks as raw material for xylo-oligosaccharides production. *Applied Biochemistry and Biotechnology, 169*(8), 2392–2404.

Saxena, R. P., Singh, B. P. N., Singh, A. K., & Singh, J. K. (1981). Effect of chemical treatment on husk removal of arhar (*Cajanus cajan*) grain. In *Annual Convention of the Indian Society of Agricultural Engineering (ISAE)*, New Delhi, India.

Saxena, K. B., Kumar, R. V. & Rao, P. V. (2002). Pigeon pea nutrition and its improvement in Quality Improvement. *Crops Journal of Crop Production* (Basara, ed.), 5, 227–260.

Saxena, K. B., Kumar, R. V., & Gowda, C. L. L. (2010). Vegetable pigeon pea- a review. *Journal of Food Legumes, 23*(2):91–98.

Singh, U., & Jambunathan, R. (1990). Pigeonpea: post-harvest technology. In: Nene, Y. L., Hall, S. D., & Sheila, V. K., (eds.). *The Pigeonpea* (pp. 435-455). CAB International: Wallingford, Oxon, UK.

Snapp, S. S., Rohrbach, D. D., Simtowe, F., & Freeman, H. A. (2002). Sustainable soil management options for Malawi: Can smallholder farmers grow more legumes? *Agriculture, Ecosystems and Environment, 91*(1–3), 159–174.

Solomon, S. G., Okomoda, V. T., & Oda, S. O. (2017). Nutritional value of toasted pigeon pea, *Cajanus cajan* seed and its utilization in the diet of *Clarias gariepinus* (Burchell, 1822) fingerlings. *Aquaculture Reports, 7*, 34–39.

Srivastava, V., Mishra, D. P., Chand, L., Gupta, R. K., & Singh, B. P. N. (1988). Influence of soaking on various biochemical changes and dehusking efficiency in pigeon pea (*Cajanus cajan* L.) seeds. *Journal of Food Science and Technology, 25*(5), 267–271.

Tapal, A., & Tiku, P. K. (2013). Utilization of pigeon pea milling waste By-product as a protein source. *International Journal of Agriculture and Food Science Technology, 4*(8), 741–742.

Ricebean

Dadasaheb D. Wadikar and Rejaul Hoque Bepary

Introduction

Ricebean [*Vigna umbellata* (Thumb.), previously *Phaseolus calcaratus*] is a non-conventional and underutilized bean and an important crop for the generation of livelihood for poor rural and tribal farmers of South and Southeast Asia. Ricebean is also known as climbing mountain bean, *mambi* bean, oriental bean, and *Beziamah* in the Assamese language (Saini and Chopra 2012). Ricebean has a rich genetic diversity and high agricultural and nutritional potential in terms of being able to grow well in comparatively poor soils in hot and humid climates and resistance to storage pests and serious diseases (Chandel et al. 1978; Singh et al. 1980). Ricebean is a native crop of South and Southeast Asia (Ohwi 1965). It is mainly grown as a crop in India, Philippines, China, Myanmar, Malaysia, Korea, Indonesia, Fiji, Sri Lanka, Mauritius, Sierra Leone, Ghana, Zaire, Tanganyika, Jamaica, Haiti, and Mexico and also to a limited extent in the West Indies, USA, Queensland (Australia), and East Africa. In India, it is cultivated mainly as a rainfed crop in the Northeastern Hills, West Bengal, Sikkim Hills, Western and Eastern Ghats Hills, Chhota-Nagpur region and parts of Odisha in the Eastern peninsular tract, Kumaon Hills (Uttarakhand), and Chamba region (Himachal Pradesh) in the Western Himalaya. About 16 different grade colours, mainly black, red, cream, violet, purple, maroon, brown, chocolate, or mottled grains with greenish, brownish, or ash grey background, are available in ricebean grains (Arora et al. 1980; Chandel et al. 1988), which indicates the grain is a rich source of bioactive compounds. In Northeast India it is grown mainly in the tribal region of Assam, Meghalaya, Manipur, Mizoram, Arunachal Pradesh, and Nagaland. It is an important crop of shifting cultivation (1.7 m ha area) or kitchen gardens (Sarma et al. 1995) for reasons of attributes such as quick maturity, freedom from major insect and disease problems, and

D. D. Wadikar (✉) · R. H. Bepary
DRDO-Defence Food Research Laboratory, Mysore, India

A. Manickavasagan, P. Thirunathan (eds.), *Pulses*,
https://doi.org/10.1007/978-3-030-41376-7_16

producing easily cooked, good-tasting seeds. Ricebean seeds have a smooth shiny surface, are slender to oblong in shape, 6–8 mm in length, 3–5 mm in width, 3–4 mm thick; rounded at both ends and with concave, straight, or protruding hilum (de Carvalho and Vieira 1996; Bepary et al. 2018a). The proportion of cotyledon, testa, and embryo in ricebean ranges from 88% to 90%, 7% to 9%, and 0.3% to 0.5%, respectively. The world production of ricebean alone is about 1407 metric tons from an area of 1804 hectares with productivity of 2.25 q/ha. Very limited information is available on marketing of ricebean in the international trade. In the international trade of ricebean, Japan is the major importer whereas Thailand, Myanmar, China, and Madagascar are major exporting countries. In 1998–2000 about 1100 tons were exported annually. In recent years, Thailand has exported ricebean to Japan where it used in lieu of azuki bean (*Vigna angularis* (Willd.) Ohwi & Ohashi) (Tomooka et al. 2006). The proximate composition summarised in Table 1 by Bepary et al. (2016) shows that the ricebean is a carbohydrate-rich grain with about 20% protein and low levels of fat content. It has high-quality protein with all essential amino acids in a balanced manner. The starch of ricebean has the lowest glycemic index (GI) compared to other beans such as mung bean, pea, pigeon pea, soybean, and cowpea. Ricebean has higher levels of potassium, calcium, iron, and zinc with better bioavailability for calcium and zinc. The oligosaccharides (flatulence-producing saccharides such as raffinose, stachyose, verbascose, and ajugose) content in ricebean are lower than other pulses (Bepary and Wadikar 2019). The antinutritional factors such as trypsin inhibitor, α-amylase inhibitors, polyphenols, saponin, and phytic acid are greater, although tannin content is low as compared to other commonly consumed pulses. However, processing reduces these antinutrient components drastically. The hydrogen cyanide (HCN) and L-DOPA content in ricebean are lowest among the seven species of *Vigna*. Ricebean grain is rich in water-soluble vitamins (mg/100 g) such as thiamine (0.261 ± 0.164), ascorbic acid (0.903 ± 0.404), niacin (1.141 ± 0.713), pyridoxine hydrochloride (1.242 ± 1.593), pyridoxal-5-phosphate (3.353 ± 0.650), pantothenic acid (3.450 ± 2.609), folic acid (0.139 ± 0.071), and riboflavin (0.083 ± 0.038) (Bepary et al. 2019a). Ricebean bioactivities include hepato-protection, antiinflammation, anticancer, antifungal, antidiabetic, antihypertension, immunity booster, HIV-1 inhibition, and inhibition of mutagenic activities (Bepary et al. 2016). Much variation has been observed in phytic acid (%) content in ricebeans based on the reports wherein the percent of phytic acid is found to be 7.32–8.17 (Katoch 2013), 3.36 ± 0.57 (Kalidass and Mohan 2012), 2.02 ± 5.9 (Saharan et al. 2002), and 1.88–2.05 (Kaur and Kapoor 1992). The total polyphenols among the several ricebean cultivars ranged from 0.58 to 1.82 (Bepary et al. 2016). The biological value of ricebean protein is found to be 65.63 ± 1.43% with net protein utilization (NPU) of 57.31 ± 1.19% and true digestibility of 87.23 ± 0.32% (Singh et al. 1980). The ricebean grain thus has good nutritional content and wide scope for commercialisation through different processing techniques and value addition. Although postharvest management is crucial for all grains, for ricebean being traditionally grown in several parts of Southeast Asia varied postharvest practices are followed.

Table 1 Proximate composition of ricebean (g/100 g)

Author	Moisture	Carbohydrate	Protein	Fat	Crude fibre	Ash
Du et al. (2014)	9.8 ± 0.0	–	25.68 ± 1.19	1.58 ± 0.04	–	4.25 ± 0.16
Kaur et al. (2013)	–	–	19.67 ± 0.56	0.61 ± 0.36	–	4.11 ± 0.46
Katoch (2013)	–	54.22 ± 1.00	24.52 ± 0.58	2.61 ± 0.45	4.73 ± 0.49	–
Ren et al. (2012)	10.65 ± 0.70	–	25.99 ± 1.26	1.69 ± 0.09	–	3.26 ± 0.12
Pawar et al. (2012)	10.3 ± 0.07	61.6 ± 0.08	17.5 ± 0.05	0.51 ± 0.005	5.59 ± 0.06	4.5 ± 0.007
Kalidass and Mohan (2012)	5.56 ± 0.02	54.23 ± 1.60	26.12 ± 0.56	4.18 ± 0.50	4.74 ± 0.01	4.04 ± 0.04
Awasthi et al. (2011)	9.5–10.9	58.0–61.2	17.9–19.4	0.48–1.15	4.6–6.7	3.9–5.7
Buergelt et al. (2009)	10.26 ± 1.16	65.08 ± 3.49	22.53 ± 4.14	0.39 ± 0.16	4.71 ± 0.76	4.03 ± 0.49
Saharan et al. (2002)	–	61.85	18.2 ± 0.2	0.83	–	–
Saikia et al. (1999)	10.45 ± 0.13	60.41 ± 0.95	17.42 ± 0.37	0.49 ± 0.02	6.9 ± 0.55	4.33 ± 0.08
Kaur and Mehta (1993)	10.8	59.4	19.8	1.4	4.7	3.9
Kaur and Kapoor (1992)	10.41 ± 0.13	59.74 ± 0.74	17.88 ± 0.43	0.49 ± 0.04	7.09 ± 0.47	4.18 ± 0.10
Rodriguez and Mendoza (1991)	8.99 ± 0.16	58.06 ± 2.86	18.5 ± 2.53	3.76 ± 0.29	5.33 ± 1.84	5.36 ± 1.13
Revilleza et al. (1990)	8.97	59.01	16.82	3.46	4.31	7.43
Malhotra et al. (1988)	7.6 ± 0.54	62.32 ± 1.41	20.19 ± 1.78	2.98 ± 0.54	2.86 ± 0.83	4.06 ± 0.39
Chandel et al. (1988)	5.43 ± 0.32	63.76 ± 4.31	20.82 ± 2.64	1.14 ± 0.23	4.83 ± 1.88	4.01 ± 0.17
Singh et al. (1980)	–	–	17.81 ± 25.18	1.00 ± 1.60	3.3 ± 4.8	3.81 ± 4.31
Value range	***5.1–12.81***	**52.23–68.5**	**16.82–26.12**	**0.39–4.03**	**1.70–8.5**	**3.00–7.43**
Average	**9.10 ± 3.12**	**59.96 ± 2.81**	**20.78 ± 2.65**	**1.74 ± 1.42**	**5.07 ± 1.174**	**4.44 ± 1.03**

Adopted from Bepary et al. (2016)

Postharvest Processing

Postharvest operation is the integration of postharvest practices that follow the harvesting of ricebean grains. Included are judging of the maturity index, harvesting, threshing, drying, storing, processing (primary and secondary), and marketing to finally be mobilized toward the end user. Each of these practices decides the quality and safety of ricebean grains and its products and also the level of postharvest losses. Ignoring one of these practices can lead to severe consequences on quality, safety, and postharvest loss of ricebean grains. Each of these practices is discussed following.

Assessment of Maturity Index

Maturity index is the stage at which the ricebean grain can be harvested; usually, in this stage the seed reaches physiological maturity. Correct judgment of the ricebean maturity index is crucial to obtain a qualitative as well as quantitative harvest. If ricebean grains are harvested at a late or early stage of physiological maturity, grains may acquire a hard-to-cook property during the subsequent storage period (Acharya 2008). The maturity indexes of the ricebean fodder harvest are not the same as the maturity indexes of the ricebean grain harvest. Maturity indexes for ricebean grain harvest are determined by following factors: development of acceptable seed coat colour and pod colour, thousand grain weights and moisture content, increasing protein and phytic acid content, and crop duration. As these factors are dependent on variety, the nutritional status of the plant, environmental conditions, and soil fertility, it is necessary to determine the maturity index on a localized basis. Ricebean pods are very sensitive to shatter if allowed to remain on the plant for long time. Hence, ricebean needs to be harvested immediately after reaching the optimum maturity index. The generalized maturity index is considered to occur when pods turn colour from green to yellow or brown and 60% of the vine dries up. The pod shattering characteristic is common in the *Vigna* genus including ricebean. Excessive lignin deposition happens in ‘fibre-cap cells’ along the ventral suture tissue and the inner sclerenchyma tissue of the pod valves, which facilitates twisting forces in the pod wall if relative humidity of field is low, and hence pods are shattered (Rau et al. 2019). Ricebean takes 90–120 days to reach harvesting maturity, which strictly depends on the variety and the geographic and environmental conditions (Acharya 2008).

Harvesting

Harvesting of ricebean is defined as taking away the economic parts such as pods of the plant from the field when it reaches harvest maturity. Harvesting can be done either manually or by mechanical means. Methods of harvesting usually depend on the area of ricebean cultivation. Small and marginal cultivators generally harvest the grains by manual means. A commercial cultivator may use a combine harvester to harvest the grain. To reduce shattering loss, ricebean is harvested in the morning times when greater relative humidity makes the pods leathery and prevents shattering. Sowing time should be adjusted in such as a way that rainfall does not interfere with ricebean harvesting so as to prevent the development of hard-to-cook grain. Harvesting of ricebean pods is carried out by pulling out mature plants with the roots or cutting near the base of the mature plant. This harvesting practice is less labour intensive than harvesting by picking individual pods from plants. Harvesting the whole plant by cutting near its base has several advantages: reduced labour requirements, easier subsequent drying and threshing, use of the non-grain parts of the plant as fodder, less contamination of the grains with dust and soil-borne microorganisms, and land fertilization by the residual root nodules.

Pre-drying

Pre-drying is a unit operation of postharvest practices wherein harvested ricebean plants are dried to reduce moisture content for effective threshing. This practice is necessary when harvested plants have plenty of green foliage, non-uniform pod maturity, and high moisture (Lal and Verma 2007). For the maximum shattering effect of pods, the harvested plants are dried for 1–2 weeks in the sun. During the pre-drying period the harvested plants need protection from rain, dew, nocturnal humidity, and direct ground contact. Drying reduces the moisture content of pods, which facilitates twisting forces in the pod wall so that the pod shatters. Farmers dry the plants either in the backyard or on a roof top. The drying place should be free from food safety hazards. Careless drying results in contamination with physical hazards such as stone, cattle droppings, hair, and feathers.

Threshing and Cleaning

Threshing is a unique process in postharvest practices during which sound ricebean kernels are detached and separated from harvested plants. Before threshing, it is necessary to bring the moisture content to the optimal level by pre-drying. Threshing of ricebean at high moisture requires more impact force for grain detachment, creates difficulty in the removal of chaffs during winnowing, and acts as a source of pest and microbial cross-contamination (Mohan et al. 2011). Small and marginal farms thresh ricebean by beating the dried pods with bamboo or wooden sticks as soon as the pods reach the brittle stage while pre-drying (Lal and Verma 2007). In some places, threshing of ricebean is carried out by trampling cattle. A mechanical thresher can be used if a large quantity of harvested ricebean material is available. Cross-contamination with food hazards and physical damage to the grain needs to be prevented while threshing quality grains. As soon as the threshing operation is completed, the grains are cleaned to remove chaffs, very light seeds, broken grains, glumes, and other foreign materials by winnowing or dropping the grain from a bucket against the natural wind.

Drying

The moisture content of threshed ricebean grains remains high and must be lowered toward the safe level (9–12%) for grains storage. Insufficiently dried grains (if moisture exceeds 18%) is vulnerable for microbial spoilage and pest attacks and has high enzymatic activity during storage, which shortens storage life and reduces grains quality and market price (Mohan et al. 2011). Drying is considered as an important postharvest practice wherein excess moisture of the ricebean grains is removed by applying heat by natural sun radiation or from artificial hot air. Sun drying is a common drying practice followed by ricebean farmers immediately after threshing. Ricebean requires 2–3 days of sun drying to remove moisture for safe storage. Conventionally, ricebean grains are thinly spread in the home yard during the peak sunny period and stirred frequently for proper moisture removal (Paudel 2008). During the peak sunny period of the day, the relative humidity (RH) of the air remains low (below 70%) and temperature remains high, which facilitates the quick evaporation of moisture from the grains. The rate of drying is not uniform in the sun drying process because of the frequent changes in solar radiation intensity during drying days. A fast rate of drying creates cracks on grains, increases integument permeability, reduces germination percentage, and changes the seed coat colour (Scariot et al. 2017). Sun drying in a home yard is not a safe practice because it increases risk of contamination by microbial and physical hazards as well as pest infestation. The efficiency of the solar drying practice of ricebean can be increased by introduction of a small handheld moisture meter to monitor moisture, a hygrometer to monitor air relative humidity (RH), and a tarpaulin to prevent contamination. Ricebean grains that are harvested in the monsoon season have poor grain quality and short storage life because of the low solar radiation and high grain moisture (Joshi et al. 2008). This problem can be overcome by using artificial dryers with temperature control devices. In an artificial dryer, blowing hot air continuously transfers the heat and moisture away from grain surface. The efficiency of artificial dryer depends on dryer type as well as drying purpose. Drying at high temperature of high moisture ricebean grains should avoided to prevent severe damage to the grain.

Storage

Food and nutritional security from food grains, including ricebean, is only possible if the grains are stored properly. Ricebean grains are stored at different levels of the food chain for different purposes. Most Southeast Asian farmers store ricebean grains either for home consumption or as seed for future cultivation. Other storage purposes are handled by marketing agencies and processors. Ricebean can be stored for 3–4 years under cool and dry conditions. As other pulses, ricebean is more difficult to store than cereal grains because slight variations in the storage conditions cause drastic changes in physicochemical properties, seed colour, insect infestation, and microbial spoilage, with development of hard-to-cook seed. These changes ultimately influence the final quality of the processed grains. Six factors that determine the qualitative and quantitative changes of ricebean grains in storage are temperature, relative humidity, oxygen, light, moisture content of grain, and ricebean variety. In combination, these factors have severe effects causing postharvest loss. Grain moisture content is one of the prime factors that affect grain quality. Cotyledon and testa cracking was observed in canned beans prepared from grains stored at low moisture. Grains storage at high initial moisture is vulnerable for off-flavour and off-colour development, with reduced hydration capacity, increased cooking time, and the encouragement of microbial growth and pest infestation (Uebersax and Muhammad 2013). A moisture content of 5–14% is considered as safe storage when the grains are stored at a temperature range between above freezing point and below 30 °C. Within in this moisture range, the storage life is doubled with 1% decreases in moisture content (Lal and Verma 2007). The surrounding relative humidity of the storage system can change the moisture level of the grains during storage as the grain has a tendency to reach equilibrium moisture content. Grains storage at high initial moisture (above 18%) and high relative humidity (above 70%) promote the growth of moulds, including aflatoxin-producing mould, and pest infestation by bruchids. Most of the cultivars of ricebean are resistant to bruchids (*Callosobruchus* spp.) pest infestation, with a few exceptions (Somta et al. 2006). The yellow seeded variety of ricebean is slightly susceptible to bruchid pests, so it is necessary to maintain the grain at low moisture during storage. At the farm level, ricebean is stored in wooden and earthen pods for 1 year (Paudel 2008). These traditional storage systems are being rendered damp resistant and airtight by using indigenous technical knowledge. Ricebean grains harvested from northern sloping lands is susceptible to storage pest attacks. Marketing agencies and processors pack the ricebean in gunny bags stored in a well-ventilated godown.

Soaking

Ricebean soaking is a vital pre-processing unit operation that is usually carried out before cooking, germination, dehulling, and fermentation. The rate of water absorption capacity is the deciding factor for soaking time that depends on the anatomic features of the seeds, soaking medium, soaking conditions, and grain moisture content. The quality of soaked ricebean can be judged by such quality parameters as length/breadth ratio (LB ratio), water uptake ratio (weight increased ratio), swelling ratio (volume increased ratio), and solid loss percentage. Soaking ricebean overnight can increase the LB ratio (1.6 ± 0.2), water uptake ratio (2 ± 0.1), and swelling ratio (1.1 ± 0) and reduce hardness ($39 \pm 10\%$ reduction in hardness as compared to native grain hardness) (Hollington et al. 2010; Bepary et al. 2017). With increased soaking water temperature, the rate of water absorption is also increased. Soaking ricebean at 40–45°C can reduce soaking time from 12–16 to 7–7.5 h (Bepary and Wadikar 2018). High-temperature soaking induces a brown colour to the seed from leaching out of the water-soluble pigments of the seed coat and causes hard-to-cook grain. As in other pulses, hard-to-soak grains are also found in ricebean (Hollington et al. 2010). Ricebean grains are conventionally soaked overnight in tap water to prepare various traditional products such as dhal, paste, *pakauda*, and *biraula*. Soaking hydrates the starch granules and protein bodies of the cotyledon, which reduces cooking time by facilitating better starch gelatinization and protein denaturation. Soaking significantly improves overall quality of the protein, protein digestibility, starch digestibility, and mineral bioavailability by decreasing antinutritional factors. Soaking of ricebean in water for 12–16 h can reduce protein (3.62–12%), fat (1.62–3.92%), ash (2.2–4.72%), calcium (29%), and magnesium (25%), while increasing the crude fibre (+2.92–6.77%) insignificantly and improving starch digestibility by 35–69% and protein digestibility by 5.57–17.68% as compared to native grain (Kaur and Kapoor 1990b; Chau and Cheung 1997; Saharan et al. 2002; Pawar et al. 2012). Phytate:zinc, phytate:calcium, and calcium:zinc ratios are indicators for zinc availability in the diet, which is slightly improved after soaking in tap water (Kaur and Kawatra 2002b). The effects of soaking on sodium, potassium, zinc, and water-soluble vitamins are not available for ricebean. However, soaking can reduce potassium by 35%, sodium by 20%, zinc by 0%, thiamine by 4.87%, riboflavin by 17.9%, niacin by 12.74%, and folate by 28.5% while increasing vitamin C by 24.5% as compared to the raw bean (Rockland et al. 1977; Barampama and Simard 1995; Mubarak 2005; Kakati et al. 2010). Soaking of ricebean grains in water for 12–16 h can reduce the antinutritional factors such as tannin by 30.65%, polyphenolic compounds by 10–52%, trypsin inhibitor by 9–17%, amylase inhibitor by 11%, phytic acid by 6–33%, saponin by 3–9%, raffinose by 9.5–14%, and stachyose by 5–16.4% (Kaur and Kapoor 1990a; Chau and Cheung 1997; Kaur and Kawatra 2000; Saharan et al. 2002; Pawar et al. 2012). Most of the antinutritional factors are reduced by leaching out into the soaking water and enzymatic activities during soaking; the loss of saponin is not caused by leaching but rather by hydrolysis of the glycosidic linkage between the sapogenin and its sugar during soaking. The flavonoids (52%) and antioxidant activity (145%) are increased whereas oxalate (55%), hydrogen cyanide (HCN, 54.9%), and verbascose (49.5%) are decreased in soaked bean as compared to native bean (Oboh et al. 2000; Boateng et al. 2008; Okudu and Ojinnaka 2017).

Cooking

Cooking is one of the important and most commonly used processing operations for the utilization of ricebean. In most ricebean-consuming areas, cooking is carried out by boiling the overnight-soaked grains with water until reaching the appropriate softness for preparing household recipes. The introduction of modern cookery appliances allows cooking ricebean in a pressure-cooker without prior soaking. Cooking is one of the simplest types of heat processing that transforms native ricebean grain into an edible form. A cooked ricebean grain is considered edible when the grain becomes dark brown in colour with transverse splits and soft texture, with a characteristic cooked flavour and palatability (Bepary and Wadikar 2018). During cooking, the middle lamellae of native ricebean grains are broken down, resulting in the disruption and softening of the cell containing starch bodies in the protein matrix; this is followed by starch gelatinization and protein denaturation. Upon heating, starch gelatinization starts once the gelatinization temperature is reached, which is characterized by swelling of starch granules and loss of the birefringent property. The swelling power, pasting temperature, and peak melting temperature of ricebean starch are 71–75 and 62–70 °C and 7.5–14.8 g/g, respectively (Singh et al. 2012). Gelatinization temperature, peak melting temperature, and swelling power of starch are dependent on the amylose/amylopectin ratio. A higher content of amylopectin in ricebean leads to higher swelling power and greater softness of texture (Sánchez-Arteaga et al. 2015). During heating, the protein bodies present in cells start denaturing when denaturation temperature is reached, which causes coagulation of protein, thus reducing its solubility. Protein denaturation temperature is dependent on the molecular weight of the protein and mineral concentration. Proteins of low molecular weight are denatured first, and then high molecular weight and heat-resistant proteins are denatured subsequently as heating proceeds (Sánchez-Arteaga et al. 2015). High mineral content increases denaturation temperature. In easy-to-cook beans the protein denaturation temperature is always greater than the starch gelatinization temperature, meaning that the point for ending starch gelatinization is the point for protein denaturation. Although cooking increases the volume of individual cells by the action of starch gelatinization and protein denaturation, the whole cell remains intact because of the plasticizing property of the cell walls. At the end stage of cooking, the swollen intact cells initiate pulling the cells apart from each other, which leads to development of the soft texture.

The physicochemical transformations of native grains during cooking can be judge by certain qualitative parameters such as cooking time, texture, swelling ratio, solid loss, appearance, and flavour. These quality parameters are called cooking quality, which is dependent on variety, physicochemical composition of grain, and storage conditions. Cooking time is important in the cooking process, the time from adding the grains to boiled water till achievement of desired softness. Cooking time depends on grain hydration behaviour, variety, hardness, seed coat permeability, compactness of endosperm internal structure, etc. (Kaur et al. 2013). Short cooking times are always convenient to reduce energy requirements during cooking and to reduce nutritional losses. Average cooking time for unsoaked ricebean varieties is 47.72 ± 9.85 min (Saikia et al. 1999; Hira et al. 1988; Kaur et al. 2013). Cooking time for ricebean can be reduced by soaking. Bepary et al. (2017) reported that the average cooking time of ricebean varieties soaked for 12 h was 35 ± 5 min. By modifying soaking conditions, cooking time can be further be reduced. Soaking ricebean at 42 °C for 7 h can reduce cooking time by 37% as compared to conventional soaking (Bepary and Wadikar 2018). Cooking time is increased fivefold if ricebean is stored in high moisture and at high temperature, but the cooking time of this high moisture bean can be reduced by soaking with phytic acid or EDTA. Reduction in phytic acid content during storage increases cooking time, and increase in percentage phytic acid to percentage calcium ratio reduces cooking time (Kon and Sanshuck 1981).

Ricebean stored up to 1 year has good cooking qualities, hence is in the easy-to-cook category. After 1 year, stored ricebean begins to lose cooking quality because of changes in physiological conditions that induce hardening defects in the stored grain. These grains absorb sufficient water during soaking but do not soften sufficiently even after cooking for a long time and are termed hard-to-cook (HTC) grain. In easy-to-cook grain, individual cells separate easily, leading to tenderness; however, individual cells of HTC bean grain are not separated because the middle lamella is intact. Multiple physiochemical factors and mechanisms are involved in the development of HTC. Of these, storage environmental conditions and genetics are considered as prime factors. Beans stored at high temperature and high relative humidity generally develop HTC. HTC beans require more cooking time to soften the cotyledon, which ultimately increases energy consumption, with solid loss and decreased protein quality, destroying the heat-sensitive micronutrients. The percentage of water absorption, solid loss, and hardness in the unsoaked ricebean cooking process was 107.56 ± 9.42%, 8.47 ± 2.91%, and 27.34 ± 7.48 N, respectively (Hira et al. 1988; Saikia et al. 1999; Kaur et al. 2013). The percentage of water absorption, swelling ratio, cooked hardness, and degree of gelatinization can be improved by soaking. Bepary et al. (2017) reported the mean value cooking parameters in 12-h-soaked ricebean varieties cooked for 29–40 min as follows: hardness, 25 ± 1.96 N; percentage of water absorption, 121 ± 8; solid loss, 27 ± 1.58%; swelling ratio, 1.11 ± 0.13; and degree of gelatinization, 51 ± 2%.

Open Pan Cooking

Open pan cooking of ricebean for 20 min with 12 h prior soaking can decrease protein (12%), ash (5.45%), and fat (0.57%) whereas crude fibre (15.27%) and carbohydrate (4.78%) are increased (Kaur 2015). However, open pan cooking of ricebean for 50 min without prior soaking can reduce protein (24 ± 0.82%), ash (10.75 ± 1.26%), fat (53.75 ± 3.52%), crude fibre (55.55 ± 0.58%), and starch (16.5 ± 2%) (Saikia et al. 1999). The conventional cooking of ricebean can reduce significant amount of minerals and water-soluble vitamins. Open pan cooking of ricebean can reduce calcium (30.25 ± 4.6%), iron (55 ± 1%), phosphorus (38 ± 3%), potassium (13 ± 2%), sodium (7.75 ± 4.6%) magnesium (28.1 ± 5.6%), and zinc (17.5 ± 6.0%) (Saikia et al. 1999; El Maki et al. 2007). The conventional cooking of ricebean can reduce thiamine (20.34–20.98%), riboflavin (23.21–31.68%), niacin (30.58–34.76%), vitamin B_6 (27.48–33.31%), folic acid (31.73–40.54), and pentothenic acid (45%) as compared to native bean (Augustin et al. 1981; Hoppner and Lampi 1993; El Maki et al. 2007). The conventional cooking of ricebean also reduces the antinutritional factors and improves the antioxidant activity, in vitro protein digestibility (IVPD), in vitro starch digestibility (IVSD), in vitro iron bioavailability (IVIB), and in vitro calcium bioavailability (IVCB). The cooking of soaked ricebean reduces antinutritional factors by a significant amount as compared to cooking of unsoaked beans, increasing antioxidant activity, IVPD, IVSD, IVIB, and IVCB (Kaur and Kapoor 1990a, b; Chau and Cheung 1997). Cooking of ricebean can decrease polyphenol by 12.4–23%, flavonoids by 47.7%, phytic acid by 21.1–41.64%, saponin by 14.8–19.6%, tannin by 63.5%, trypsin inhibition activity (TIA) by 31.65%, amylase inhibition activity (AIA) by 23.07%, raffinose by 35.13%, and stachyose by 31.31%, and increases antioxidant activity by 72% (Kaur and Kapoor 1990a, b; Chau and Cheung 1997; Kaur and Kawatra 2000; Rani and Khabiruddin 2017). Cooking of unsoaked ricebean for 30 min can increasein vitro protein digestibility (IVPD) by 17.68% and in vitro starch digestibility (IVSD) by 100% (Chau and Cheung 1997).

Microwave Cooking

Microwave heating has gained popularity in food processing for its ability to achieve high heating rates, significant reduction in cooking time, more uniform heating, safe handling, and ease of operation. Application of microwave energy for cooking ricebean is a new approach and very little work is being carried out on this bean. Bepary et al. (2018b) first tried the microwave cooking of ricebean and then the cooking process has been optimized. For microwave cooking of ricebean, the cleaned grains need to be soaked in 250 mL distilled water at 42 °C for 7 h (Bepary and Wadikar 2018), and then the soaked grains are allowed to cook in a cylindrical polypropylene bowl (20 cm diameter and 10 cm height) with 400 mL salt solution (0.7%) in a domestic microwave oven at 2450 MHz with 900 W for 15 min. During cooking, the position of the bowl needs to be marked on the turntable of the microwave oven to uniform and repeatable power distribution, as the temperature is very sensitive to the position in the oven. Also, the bowl should be placed toward the periphery of the turntable to facilitate more uniform power distribution as the turntable rotates. During pulsed microwave input, the magnetron power should be kept constant for a period of 30 s followed by a pause of 5 s. The process parameters of the microwave cooking process such as microwave power, concentration of salt solution, and cooking time are being optimized based on the cooking quality of the bean. The cooking quality of ricebean is determined in terms of water uptake (%), solid loss (%), swelling ratio (%), hardness (g), protein denaturation (%), degree of gelatinization (%), and overall acceptability (OAA). Cooking of soaked ricebean at 900 W by using 0.7% salt solution for 15–20 min can reduce the cooking time by 43.43% and solid loss by 31% as compared to native ricebean. Bepary et al. (2018b) reported that the optimal microwave cooking process can increase the protein, fat, ash, polyphenols, and in vitro antioxidant activity and also slightly improve in vitro iron and calcium bioavailability, whereas crude fibre, calcium, phytic acid, and iron are decreased as compared to a conventional cooking process. Microwave cooking can reduce the content of minerals such as calcium (7.14%), iron (17.52%), magnesium (14.04%), phosphorus (6.65%), potassium (22.65%), sodium (32.5%), and zinc (8.65%) and can reduce the content of vitamin C (11.3%), thiamine (20.2%), riboflavin (40.8%), niacin (86.06%), and vitamin B_6 (18.58%) (Chung et al. 1981; Alajaji and El-Adawy 2006). Microwave cooking of spinach causes no significant loss of ascorbic acid but significant loss of folic acid as compare to conventional cooking (Klein et al. 1981). This cooking process causes reduction in functional, nonnutritional components and improves IVPD and IVSD, with less impact on IVIB, IVCB, and antioxidant activity. It is established that microwave cooking can decrease tannin (62.12%), saponin (47.25%), phytic acid (26.03%), TIA (100%), AIA (83.3%), oxalate (85.9%), HCN (83.2%), raffinose (60.62%), stachyose (60.71%), and verbascose (59.55%) while increasing IVPD (12.78%) and IVSD (319%) (Mulimani and Supriya 1994; Quinteros et al. 2003; Feng et al. 2003; Khatoon and Prakash 2004; Mubarak 2005; Alajaji and El-Adawy 2006; Khattab and Arntfield 2009).

Dehulling

Dehulling is an important unit operation in processing and value addition for legumes including ricebean. It is defined as removal of the seed coat from cotyledons before use. The seed coat of ricebean is hard and hydrophobic, which acts as a hurdle during soaking and cooking and also imparts a bitter taste and dark colour to the cooked bean, ultimately reducing the acceptability and palatability of ricebean. However, dehulling of ricebean is not a common practice because the seed is difficult to remove from the cotyledons (Andersen et al. 2009). Hence, ricebean can be categorized as a hard-to-dehull group. In hard-to-dehull beans, mucilage and gum with high uronic acid content form a strong bond between the hulls and the cotyledons (Ramakrishnaiah and Kurien 1985). Mucilage and gum are chemically known as the cellulosic microfibril network that is attached to the matrix of protein and nonstarch polysaccharides (NSP) located between hull and cotyledon. In some ricebean growing locations, the bean is dehulled by peeling the overnight soaked grain followed by sun drying. These dehulled grains are used to prepare several traditional products such as *pakora (halwa)*. Parvathi and Kumar (2006) studied the dehusking characteristics of ricebean using a dhal stone mortar and power-operated mini dhal mill; the dhal recovery percentage by stone mortar and power-operated mini dhal mill was reported as 70% and 45%, respectively. Kaur and Kawatra (2002a,b) prepared dehulled spilt ricebean. Whole ricebeans were split into two halves with a grain pearler, the split grain halves were soaked for 12 h, and then the hull was removed from the spilt halves by hand rubbing between the palms and then sun dried for 24 h. Dehulling of bean can reduce fibre content, antinutritional factors (tannin, polyphenol), and improve nutritional qualities such as protein digestibility, mineral bioavailability, and protein quality. The cooking quality, appearance, and texture are also improved in dehulled beans (Joyner and Yadav 2015; Deshpande et al. 1982). Dehulling also facilitates the production of higher quality flours, without browning/speckling, and also increases leavening ability. The dehulled legume has a higher value of true protein digestibility and net protein utilization as compared to whole legumes (Singh 1993). Dehulling of ricebean can decrease proximate components and minerals. The manual dehulling of split ricebean soaked for 12 h can decrease crude protein (8.76–9.32%), fat (13%), ash (11.82%), crude fibre (75–86%), total soluble sugar (66.29%), and total sugar (50%), and reduce sugar (82%) and increase carbohydrates (10.72%) (Kaur and Kawatra 2002a, b; Parvathi and Kumar 2006; Kaur 2015). Dehulling can cause loss of calcium (10.88%), magnesium (2.33%), iron (34.65%), phosphorus (0.44%), zinc (57%), copper (50%), manganese (14.7%), potassium (19.88%), and sodium (15%) (Saharan et al. 2001; Mubarak 2005; Parvathi and Kumar 2006). Dehulling causes little loss in thiamine, riboflavin, or niacin but causes more than 30% loss in vitamin C, vitamin B_6, folate, and pantothenic acid (Kik 1956; Nnanna and Phillips 1989; Ibeanu et al. 2012). The dehulling of ricebean can reduce antinutritional factors and improve antioxidant activity, IVPD, IVSD, IVIB, and IVCB. The dehulling process can decrease polyphenols (46.67%), tannin (11.76–85%), phytic acid (14.55%), saponin (12.94), raffinose (8.78%), stachyose (22.49%), verbascose (76.4%), oxalate (100%), HCl extractability of calcium (15%), and iron (25.26%), but increases flavonoids (8.93%), antioxidant activity (40%), TIA (15%), AIA (83%), IVSD (29–42%), and IVPD (3.5%) in ricebean (Deshpande et al. 1982; Akinyele and Akinlosotu 1991; Bishnoi and Khetarpaul 1993; Duhan et al. 2002; Kaur and Kawatra 2002a; Parvathi and Kumar 2006; Jayalaxmi et al. 2016; Rani and Khabiruddin 2017).

Germination

Germination or sprouting is one of the important household processing practices commonly followed in the ricebean growing belt of India and Nepal to improve taste and flavour. *Kwanti*, a popular dish in Nepal, is prepared from sprouted ricebean (Andersen et al. 2009). For germination of ricebean, grains are soaked overnight (6–12 h), and then the grains are allowed to develop rootlets under humid conditions at room temperature for 24–48 h (Sritongtae et al. 2017; Kaur and Kawatra 2000). Germination of 12-h soaked ricebean for 48 h at 37 ± 1 °C followed by pressure-cooking for 5 min can reduce the carbohydrate by 1.98%, fat by 3.4%, and ash by 0.18% whereas protein is increased by 2.99% and crude fibre by 14% (Kaur 2015). Germination of ricebean for 24 h at 30 ± 1 °C (73 ± 5% relative humidity) and followed by steam cooking for 10 min increases protein by 11% and reduces 45% carbohydrate, 40% of fat, and 0.37% of ash as compared to soaked bean (Sritongtae et al. 2017). Germination of ricebean soaked for 12 h at 30 °C for 24 h can reduce calcium by 4.03%, iron by 3.03%, and phosphorus by 0.65%, whereas germination for 48 h at 37 °C followed by pressure-cooking for 4 min reduces zinc by 2.67%. Magnesium is increased by 1.9% and potassium by 9.11%, and sodium is decreased by 3.3%, in germinated mung bean (72 h) (Mubarak 2005).

Germination of ricebean can increase the content of water-soluble vitamins such as vitamin C, thiamine, riboflavin, niacin, pyridoxine, folate, and pantothenic acid (Banerjee et al. 1954; Nnanna and Phillips 1989; Sattar et al. 1989; Ochanda et al. 2010; Hefni and Witthöft 2014). Germination of ricebean for 24 h at 30 ± 1 °C (73 ± 5% relative humidity) and followed by steam-cooking for 10 min increased thiamine by 210%, riboflavin by 87.5%, and niacin by 200% (Sritongtae et al. 2017).

Germination of ricebean can reduce nonnutritional components and increase digestibility (starch and protein), bioavailability of minerals (calcium and iron), and antioxidant activity. Germination of ricebean for 40–48 h can decrease polyphenol by 35%, phytate by 26.98–55%, tannin by 63%, saponin by 18.68%, trypsin inhibition activity by 42%, amylase inhibition by 57%, raffinose by 15%, stachyose by 75%, in vitro starch digestibility by 212%, in vitro protein digestibility by 12%, HCl extractability of calcium by 8.2%, and HCl extractability of iron by13%, as compared to native grain (Chau and Cheung 1997; Saharan et al. 2001; Kaur and Kawatra 2000; Kaur 2015). Germination can diminish flavonoids by 5.34% and verbascose by 51.21% (Pal et al. 2015; Oboh et al. 2000). The citric acid (1%) pretreatment followed by germination for 24 h followed by steam-cooking for 10 min can decrease phytate by 19% and increase total phenol by 105%, and DPPH (assay) by 111% as compared to 6-h soaked grain (Sritongtae et al. 2017), with increased total polyphenol and B-vitamins.

Roasting

Roasting of ricebean is usually not a common practice in its growing belt because expansion is limited during roasting. Roasting causes starch gelatinization, protein denaturation, and inactivation of antinutritional factors and improves the nutritional value of the bean. Roasting of ricebean develops some unique texture, flavour, and sensory attributes. Sometimes roasting is also practiced to reduce cooking time. It is traditionally carried out in hot sand or salt or without sand/salt. In the roasting process, grains are tempered with water and then roasted in four volumes of preheated sand or salt followed by cooling. Roasting of ricebean in sand heated to 250 °C for 2 min has negligible effect on proximate composition but it can reduce total soluble solid (10.29%), raffinose (25.67%), and stachyose (28.26%) (Kaur 2015; Kaur and Kawatra 2000). Roasting of ricebean at 103 °C can reduce protein (6.3%), TSS (7.3%), total free amino acids (22.8%), and tannin (25%) (Mankotia 2011). Roasting of ricebean can slightly reduce calcium, magnesium, phosphorus, potassium, and sodium and increase iron and zinc (Kaur and Kawatra 2002a, b; Jayalaxmi et al. 2016). Roasting can reduce the heat-, photo-, and oxygen-sensitive vitamins such as vitamin C (40%), thiamine (30%), vitamin B_6 (12.51%), beta-carotene (24.5%), and vitamin E (21.72 %) (Lawal 1986; Jayalaxmi et al. 2016; Chukwuma et al. 2016). Roasting causes little loss of polyphenol (16–48%), tannin (16.54%), saponin (22.22%), phytic acid (26.04%), raffinose (24.71%), stachyose (24.27%), verbascose (24.29%), and oxalate (11.81%), and little increase of flavonoids (15%), whereas it causes more loss of TIA (82.04%), AIA (78.9%), and HCN (66.93%), hence improving IVPD (16.96%) and IVSD (16.47%) (Siddhuraju et al. 1996; Siddhuraju and Becker 2007; Khattab and Arntfield 2009; Sharma and Punia 2017).

Fermentation

Fermentation of ricebean is limited; only in a part of Northeast India is this practice followed during pickle preparation. Fermentation of ricebean eliminates the beany flavour and improves the taste and palatability of foods in which complex food components (carbohydrates, proteins, lipids) are broken down into simple and absorbable fractions such as sugar, amino acids, and fatty acids by the action of microorganism (yeasts, fungi, or bacteria). Kaur and Mehta (1993) fermented soaked ricebean paste for 4 h to prepare *voda*. Kaur and Kawatra (2000) studied the ricebean fermentation process wherein the dhal of ricebean soaked overnight was made into a coarse paste and allowed to ferment for 12 h at 37 ± 1 °C in an incubator by using the natural microorganisms present in the bean. Gan et al. (2016) fermented ricebean by dissolving flour in sterile water (1:5 w/v) and incubating at 37 °C for 48 h (continuous agitation at 250 rpm) either with microorganisms existing in the ricebean or inoculated with *Lactobacillus paracasei* ASCC 279 and *Lactobacillus plantarum* WCSF1. Fermentation of ricebean for 12 h can reduce TSS (3.56%), protein (5.49%), ash (12.9%), fat (14%), fibre (62%), raffinose (30.4%), and stachyose (38%) while increasing carbohydrates (8%) (Kaur 2015; Kaur and Kawatra 2000). Fermentation of ricebean can increase total phenolic content (81–85%), ferric-reducing antioxidant power (FRAP) (26.25–33.2% and 2,2′-azino-*bis*(3-ethylbenzothiazoline-6-sulphonic acid) diammonium salt (ABTS) (17.9–20.7%) (Gan et al. 2016). It is reported that the fermentation process can increase potassium (11.62%), zinc (1%), vitamin C (243%), thiamine (26%), riboflavin (248%), and niacin (68%) and decrease calcium (23.9%), sodium (11.4%), and iron and magnesium negligibly (Barampama and Simard 1995; Nwanekezi et al. 2017). There is a substantial decrease in antinutritional factors and increase in IVPD, IVSD, and antioxidant activity after fermentation of bean. Application of the fermentation process in ricebean can diminish polyphenol (48%), tannin (28%), saponin (61%), phytic acid (67%), TIA (65.5%), oxalate (64%), HCN (86%), raffinose (71%), stachyose (71.7%), and verbascose (70.6%), which facilitates improvement of IVPD (45.7%), IVSD (86.6%), and antioxidant activity (49%) (Yadav and Khetarpaul 1994; Khattab and Arntfield 2009; Gan et al. 2016; Nwanekezi et al. 2017).

Extrusion

Extrusion is a high-temperature thermo-mechanical process that can improve the nutritional status of existing cereal-based snacks and breakfast foods by inclusion of pulses or beans, including ricebean. Very limited work has been carried out concerning extrusion cooking of ricebean. Prabhavat et al. (1996) prepared different types of extruded products from either ricebean only or a combination of ricebean flour blended with mung bean, black gram, rice, wheat, cassava, corn, or soy flour by using a twin-screw extruder. Lamichhane et al. (2013) extruded 17% moistened flour from ricebean (15%) in combination with maize flour (75%) and sorghum flour (10%) at 135 °C barrel temperature and 153 rpm screw speed to develop extruded snacks. Bepary et al. (2019b) studied the sole ricebean extrusion process. For extrusion cooking, total moisture of the flour needed is adjusted to 15–18% and the mixture is equilibrated for 16 h; then, extrusion cooking is carried out in a co-rotating twin-screw extruder (EB-10 model; M/s. Basic Technologies, Kolkata, India) at 110–120 °C of barrel temperature and 350–355 rpm of screw speed. Extrudates are collected when a steady state (constant temperature and torque) is achieved. Extrudates are dried at 60 °C for 1 h to reduce moisture content. Bepary et al. (2019b) also optimized the independent parameters involved in extrusion process-based quality characteristics of extrudates. The optimum process condition that gave extrudates a high expansion ratio but minimum extrudate density and breaking strength was 15% moistened flour extruded at 110 °C barrel temperature and 350 rpm screw speed. The water absorption index (WAI, cold and hot), water solubility index (WSI, cold and hot), swelling power (SP, cold and hot), oil absorption index (OAI), bulk density (BD), true density (TD), and colour (L, a, b) values for optimized extruded flour are significantly different ($p < 0.05$) than the native flour whereas the differences in values of carbohydrates, protein, fat, crude fibre, and ash are insignificant ($p > 0.05$). The value of carbohydrate, protein, crude fibre, TD, porosity, WAI (30 °C), WSI (30 °C), OAI (30 °C), and 'b' (redness) increased in extruded flour compared to native ricebean flour. However, the value of fat, ash, WAI (90 °C), WSI (90 °C), SP (30 and 90 °C), BD, L (lightness), and 'a' (greenness) decreased compared to native ricebean flour (Bepary et al. 2019b). Extrusion cooking can decrease antinutritional factors such as enzyme inhibitors, tannin, phytic acid, saponin, and oligosaccharides, which overall improve the nutritional qualities of extrudates by increasing nutrient digestibility and bioavailability. Extrusion cooking can reduce the percentage content of protein (2.2%), fat (47%), ash (4.87%), crude fibre (7.89%), and sodium (14.14%) whereas increasing the content of carbohydrates (3.3%), calcium (1.6%), iron (32.89%), magnesium (2.8%), phosphorus (4.15%), potassium (0.4%), and zinc (6.67%) (Alonso et al. 2001; Anuonye et al. 2012). Extrusion cooking also affects the vitamin content of bean, and the level of loss depends on extrusion conditions and the chemical composition of feed materials. Extrusion cooking can cause loss of vitamin C (14–68%), thiamine (3–78%), riboflavin (10–65%), niacin (20%), vitamin B_6 (10%), folate (45–65%), and pantothenic acid (9%) (Killeit 1994). The extrusion process can lessen the functional components such as polyphenols (45.89%), flavonoids (39.37%), and antioxidant activities (5.71%) but improve IVPD (21.87%), IVSD (128%), IVIB (39.8%), and IVCB (17.27%) (Alonso et al. 2001; Korus et al. 2007; Patil et al. 2015). The extrusion process greatly decreases the content of saponin (84.5%), TIA (86.12%), AIA (100%), oxalate (94.44%), HCN (78.18%), and raffinose (65.62%), with less impact on phytic acid (20.75%), stachyose (18.8%) and verbascose (20.83%) (Alonso et al. 2000, 2001; Anuonye et al. 2012).

Flaking

Flaking is an important processing technology known for creation of a crispy, flavourful, and tasty product that improves not only the nutritional and functional characteristics but also thc palatability of the products. It has been used commercially to produce ready-to-eat, ready-to-cook breakfast cereals and snacks for many years. Recently, this technology has been used for the development of protein-rich breakfast flakes as well as quick-cooking dhal from pulses or beans, including ricebean. Although this technology has been used to develop flakes from pigeon pea (Sethi et al. 2014; Nayak and Samuel 2012), lentil (Ryland et al. 2010), mung bean, cowpea, soybean, horsegram (Perera et al. 2017), and pea (Yang et al. 2008), very little information is available on ricebean flakes. Wadikar et al. (2018) developed production of ricebean flakes by using hydrothermal and extrusion processes. In the hydrothermal flaking process, soaked ricebean grains are cooked (6 min), then surface dried and flattened in a roller flaking machine. These flattened ricebean grains were toasted using fine salt (material to salt ratio, 1:4), and finally the salts were separated from the toasted flakes with a sieve. However, in the extrusion process, the 25% moistened flour is extruded at a screw speed of 345 rpm and barrel temperature of 55 °C after 16 h conditioning, and the extrudate strands (0.9–1.3 mm length) coming from the die are collected and immediately flattened in the roller flaking machine. These flattened extrudates are then toasted using fine salt (material to salt, 1:4). The ricebean flakes from the extrusion flaking process are significantly higher in crude fibre, iron, flavonoids, antioxidant activity, IVSD, IVCB, and IVIB than those from the hydrothermal flaking process (Wadikar et al. 2018). The thick hydrothermal flaking process can reduce fat (68%), fibre (30%), and ash (8.05%) while increasing carbohydrate (1.8%), moisture (18.31%), and protein (0.8%) (Sahu 2017). Hydrothermal roller flaking can increase iron (12–50.47%), zinc (3.52–19%), calcium (5.8–70%), and phosphorus (4.5%) and decrease magnesium (3.5%) and copper (10%) (Gruner et al. 1996; Wu et al. 2018). The degree of flaking (i.e., from thick to thin flakes) decreases starch, protein, lipid, and fibre components (Mujoo 1998). Hydrothermal flaking can cause loss of polyphenols (27.85–75.47%), phytate (40–88%), and tannin (70–94%), and improve IVSD (by severalfold), IVPD (some), IVIB (122–923%), and IVCB (70%) (Itagi et al. 2012; Wu et al. 2018; Mujoo and Ali 1998; Sailaya 1992). The loss of total polyphenols results from the loss of the bran layer from grains and thermal dissociation of conjugated phenolic into moderate reduction of hydroxycinnamic acid derivatives (Randhir et al. 2008). Flaking can decrease free radical scavenging activity, and flaking followed by toasting can increase free radical scavenging activity (Itagi et al. 2012). Protein digestibility is also increased but as compared to starch less so as starch is complexed with protein, thereby partly decreasing its susceptibility to hydrolysis (Mujoo et al.1998; Sailaya 1992).

Product Development

The purpose of ricebean cultivation and processing is the development, utilization, and consumption of ricebean-based products. Ricebean-based products have evolved in the confined growing belts based on traditional knowledge, consumer preferences, and availability over periods of time. Most of the traditional ricebean-based products are prepared by using simple conventional processing technologies such as soaking, cooking, sprouting, fermentation, and frying that are applied in other grains also. The consumption and utilization patterns of ricebean greatly vary among growing areas and depends on the food habits of each tribe. The consumption of rice bean and its products is an ancient practice in these areas. In Nagaland, it is believed that consumption of soaked ricebean by wrestlers can enhance their energy and can heal injuries acquired during wrestling competitions. The inhabitants of Khonoma, a village near Kohima, survived and remained defiant for a month by eating raw ricebean when British troops laid siege to the village in 1879. In Nepal, ricebean is considered as a hot food that is preferred during the winter season and is not served to individuals who are sensitive to cold. People have believed ricebean to be a sacred *dal*; hence, it has a distinct value in the religious aspects and cultural festivals of Nepalese society. Some popular products of ricebean served in those festivals are *khichadi* (ricebean and rice cooked together), *kwati* (soup from nine whole grain legumes), and *batuk* (Andersen et al. 2009).

Ricebean Curry

Curry is a traditional recipe of Southeast Asia that is eaten with rice and roti. For preparation of curry, grains soaked overnight are cooked and seasoned with spices, salt, and condiments in gravy form. In some places the unsoaked grains are pressure-cooked to prepare curry.

Sundal

Parvathi and Kumar (2006) developed this recipe from ricebean. It is prepared by boiling the whole grain that has been soaked for 12 h with water and salt until a soft texture is reached. Cooked grains are seasoned with onion, chilly, curry leaves, and coconut scraps; the resultant products are used as snacks. Sensory attributes are reported as appearance, 3.7, colour, 4.0, taste, 3.7, flavour, and OAA, 3.7.

Pulikulambu

Parvathi and Kumar (2006) developed this product from ricebean at Tamil Nadu Agricultural University, Coimbatore. First, whole grain is soaked for 12 h, then boiled in water with salt until texture is soft and set aside. Second, onion and garlic are fried in oil, followed by cooking for 5 min after adding tamarind juice, then cooked for 5 min after adding coconut paste, tomato paste, turmeric powder, coriander powder, chilli powder, and salt, and finally cooked for 2 min after adding the cooked ricebean. Sensory attributes were reported as appearance, 3.8, colour, 3.6, taste, 3.6, flavour, 3.4, and OAA, 3.9.

Ricebean Ball Curry

Parvathi and Kumar (2006) developed this curry product from ricebean. For preparation of this recipe, first whole grain soaked for 12 h is coarsely ground into paste without adding more water. The paste is mixed with desired amounts of chilli curry leaves, chopped onion, and salt and then steam-cooked as 30-cm-diameter paste balls. These steamed balls are used for curry by the same process and recipe as *Pulikulambu*. Sensory attributes were reported to be appearance, 3.8, colour, 3.6, taste, 3.5, flavour, 3.7, and OAA, 3.9.

Biramla

This traditional snack of Nepal is prepared by cooking overnight-soaked beans, followed by frying with desired spices, salt, and condiments (Andersen et al. 2009).

Ricebean Pork Curry

This curry, which is popular in Nagaland, is prepared by boiling the soaked beans with pork until the desired softness is developed. It is served with rice.

Ricebean Carp Decoction

This decoction is one kind of soup traditionally prepared in the southern part of China from ricebean and crucian crap (or carp). Soaked ricebean and fried crucian carp are boiled in water to reach full softness followed by further cooking after adding *kudzu* (root of *Pueraria thomsonii)* and orange (pericarp of *Citrus reticulata*). The finished soup is used as a decoction believed to have the functions of invigorating the spleen and eliminating diuresis, detumescence, and dampness (Wei et al. 2015).

Ricebean Decoction

This decoction is also a traditional product from China that is drunk as a tea and can reduce weight and enhance cosmetology. It is prepared by boiling the overnight-soaked ricebean in water until the cooking water has a soupy appearance (Wei et al. 2015).

Ricebean Sweetened Paste

Ricebean sweetened paste is a product of China and Japan where it is either consumed directly or used as a filling for desserts. For preparation of the paste, ricebean grains, after soaking for 2 h, are boiled in water until full softness develops, and the water is then allowed to evaporate. The cooked grains are crushed and mixed with crystal sugar and salad oil. After proper mixing, the paste is ready for use.

Ricebean Coix Gruel

This product is also from China where it is used medicinally to remove swelling and dampness, cure constipation, calm the mind, energize the heart, and refresh the spleen. It is the best food for patients with obesity or edema and for postpartum women. It prepared by boiling the soaked bean and coix seed initially with a hot flame, after which slow cooking ensues until the rice beans look like blowing sand and the coix seed is exploding. The resultant clear soup is seasoned with spices and beaten for smoothness (Wei et al. 2015).

Eromba

Eromba is in the traditional cuisine of Manipur. It is made from brinjal, cabbage, potato, bean, and onion chilli and served with ricebean. To prepare *eromba*, brinjal, cabbage outer leaf, soaked ricebean, and potato are boiled to softness and then these boiled vegetables are mashed into a paste. The water extract of roasted dry red chilli is added into the paste and mixed uniformly after adding salt.

Kwati Soup

This soup is a traditional dish from Nepal specially prepared from nine beans (black gram, ricebean, field bean, chickpea, soybean, field pea, garden pea, cowpea) on the occasion of the Gun Punhi festival. It is believed that *kwati*soup can cure cold and cough and is one of the best foods for women during maternity leave. Sprouted grains are boiled, and fried lovage seed is then added for special seasoning. The final product is served with flat bread (Andersen et al. 2009).

Nuggets

Nuggets or *Bori* is a popular traditional dish of India and Nepal made from pulse, specially from blackgram. In Nepal the nuggets are also prepared from ricebean and called *masaura*. For *masaura*, ricebean soaked overnight is made into a paste and mixed with taro petioles or corms; then a small ball is formed from the paste and dried to lower the moisture. Nuggets are cooked along with other vegetables in lean periods (Andersen et al. 2009).

Ricebean Kheer

For preparation of kheer, dehulled ricebean soaked for 3–4 h is cooked until it reaches a soft texture. The required amounts of jaggey, coconut milk, cardamom powder, and salt are added after mashing the cooked dhal, cooked to a slurry-like texture, and served as dessert.

Khichadi

Khichadi is a national dish of India and is also a popular dish in India, Pakistan, Nepal, and Bangladesh. It is prepared mainly from rice and legumes. Selection of cereals and pulses depends on the availability of these grains in the respective areas. In ricebean growing areas, *khichadi* is made from ricebean and rice. Split dehulled ricebeans are cooked with rice, spices, salt, and condiments (Andersen et al. 2009).

Ricebean Dhal

For preparation of dhal, split dehulled grain are boiled with turmeric and salt, and then fried seasoning such as other spices and condiments is added. It is generally served with rice and chapattis.

Pakora

Pakora is a popular fried snack of the Indian subcontinent. Gram flour is generally used to make the batter. Kaur and Mehta (1993) used ricebean flour for *pakora* preparation and found more crude fibre and calcium than in Bengal gram *pakora*. Seed coats of ricebean grains soaked overnight are removed; the grain is allowed to dry for 3–4 days, and the dried dehulled grain is then milled into flour. Ricebean flour and spinach leaf are mixed with water and then beaten for uniformity, followed by a second mixing after adding chopped onion, garlic, ginger, coriander leaves, green chillies, and salt. The shaped paste (0.5 in. thick, 2 in. diameter) is deep fried. Kaur and Mehta (1993) reported the nutritional composition of *pakora* as moisture, 15.5%; crude protein, 13.1%; fat, 16.8%; ash, 2.3%; crude fibre, 2.1%; calcium, 138.5 mg/100 g; iron, 6.7 mg/100 g; zinc, 3.1 mg/100 g; copper, 0.6 mg/100 g; while amino acids (g/16g N) were methionine,1.0; cystine, 1.0; tryptophan, 0.9; and lysine, 4.0.

Sepu vadi

Sepu vadi is a traditional dish of Himachal Pradesh. It is commonly prepared from lentils and blackgram with leafy vegetables and other spices, but in some ricebean growing areas it is also prepared from dehulled ricebean. Sharma (2014) prepared ricebean-based *sepu vadi* and found that it had more fat, crude fibre, and zinc than blackgram *sepu vadi* but with a lower overall acceptability (6.7) compared to blackgram (7.3). Wet grinding of soaked split dhal produces the paste. The paste is mixed with the desired quantity of garam masala and salt and shaped into large-sized balls, then steam-cooking ensues until the core of the balls is cooked properly. Rectangular pieces are made from the cooked balls and deep fried.

Papad

Papad is another traditional food product of the Indian subcontinent popular because of its crispness and taste. It is generally prepared from dehulled blackgram flour, but Sharma (2014) prepared ricebean *papad* and reported it had more crude fibre and zinc but overall acceptability was lower (6.5) than for blackgram *papad* (7.8). To prepare ricebean *papad*, ricebean whole flour is sieved and blended with desired quantities of Ajwan, cumin seed, sodium bicarbonate, chilly powder, and salt. Water is added into blended flour to make a dough that can be held for 2 h after covering with wet muslin cloth. Small dough balls are formed from soft dough made by hammering with a heavy pestle. Dough balls are shaped into a circular dish shape and dried in hot air at 60 °C, then deep fried in oil.

Boondi

Boondi is a dessert or savoury type of food available in the Indian subcontinent in various forms. It generally made from chickpea flour. Sharma (2014) prepared *boondi* from ricebean flour and found more crude fibre, ash, calcium, and zinc than in Bengal gram flour. For preparation, ricebean flour, baking powder, salt, gram masala, and chilli powder are blended uniformly; the flour blend is transformed into a lump-free batter after adding the desired quantity of water. The ready batter passes through a slotted spoon into heated oil where it forms small pearl-like balls after frying. The frying is continued to a light brown colour. *Boondi* is ready for use after draining the excess oil.

Bhujia

Bhujia is a popular crispy Indian snack usually made from moth bean, besan, and spices. Sharma (2014) developed ricebean-based *bhujia* and reported more crude fibre, ash, calcium, and zinc than with Bengal gram flour. For development of ricebean *bhujia*, the desired quantity of ricebean flour is kneaded with other ingredients such as boiled potatoes, salt, turmeric powder, *garam masala*, chilly powder, and citric acid into a smooth dough that is then covered for 15–20 min. The rolled dough is allowed to pass through a vermicelli-making machine fitted with a fine net; the machine's piston pushes the raw *bhujia* into already heated oil. The raw *bhujia* is fried till it turns golden brown in colour. The *Bhujia* is ready for use after removing excess oil.

Muruku

Muruku or *chakli* is a popular snack product of the Indian subcontinent, especially southern India and Sri Lanka. This crunchy savoury product is also available in Malaysia, Singapur, and Fiji and is usually made from rice flour and urad dal. Parvathi and Kumar (2006) developed ricebean-based *muruku* in the ratio 50:50 rice and dehulled ricebean with a sensory acceptability of 3.7 of 4.

Ladoo

Ladoo is a popular sweet product of the Indian subcontinent usually served at festivals and religious ceremonies. This sphere-shaped sweet product is made from sugar, ghee/butter/oil, and flour. Sharma (2014) developed a ricebean-based *ladoo* that had more crude fibre, ash, calcium, and zinc than Bengal gram flour *ladoo*. For making of *ladoo*, ricebean flour is roasted in ghee till golden brown and the aroma is released. The cooled roasted flour is mixed with powdered sugar and shaped into small ball-shaped *ladoo*. Parvathi and Kumar (2006) also developed a dehulled ricebean-based *sweetened ball* with 3.6 sensory acceptability of 4.

Supplementary Food Beverage

Four ricebean-based supplementary food beverages were developed by Parvathi and Kumar (2006) from starchy grain [rice, wheat, sorghum, and kodomillet (3 parts)], ricebean (2 parts), ground nut (1 part), and jaggary (2 parts). Of the four supplementary food beverages, that containing ricebean scored the highest in sensory acceptability (3.6 in the 4-point Hedonic scale).

Cake

Cake is a baked product usually served as a sweetened dessert. It is commonly made from soft wheat flour, sugar, eggs, margarine/butter, milk, and baking powder. Sharma (2014) prepared a ricebean-based cake with greater quantities of protein, crude fibre, ash, calcium, iron, and zinc than a cake made from wheat flour with comparable sensory acceptability.

Cookies

Kii et al. (2013) developed cookies by blending ricebean flour from 0% to 30% with wheat flour and found a significantly increased texture and calcium content with an insignificant change in sensory quality.

Munthiri Kothu

This product from Kanyakumari, Tamil Nadu, is usually made during the festival season. Parvathi and Kumar (2006) developed ricebean-based *munthiri kothu* with highest sensory acceptability. Roasted ricebean flour, fried coconut scrapings, and gingelly (sesame) seeds are blended uniformly. The blended flour is transformed into dough with the desired quantity of jaggery syrup: 15-mm balls are formed from the dough and coated in batter made from rice flour. Coated balls are fried in oil till reaching the characteristic brown colour. Shelf life is 2–3 months.

Halwa

Halwa is a popular sweet confectionary widely consume in the Indian subcontinent, Central Asia, and Middle East that is generally made from grain flour, ghee, and sugar. Sharma (2014) developed a ricebean-based *halwa* that contained more ash, crude fibre, calcium, and zinc than Bengal gram flour *halwa*. Ricebean flour was roasted in ghee till a golden-brown colour was reached, releasing a roasted aroma. Water and sugar were then added slowly, cooking with continuous stirring. When the desirable consistency is reached, *halwa* is ready for consumption.

Extruded Snacks

Prabhavat et al. (1996) prepared 17 different types of extruded products from ricebean, ricebean flour blended with mung bean, black gram, rice, wheat, cassava, corn, and soy flour using a twin-screw extruder (ZE 25X 33D) and coated with BQ flavour. Of 17 formulae, formulas no. 9 and 14 were reported with maximum sensory score (OAA, 7.4) whereas sole ricebean extrudates had OAA of 7.27. Extruded product prepared from formula 9 used 57.5% ricebean flour, 20% rice flour, 2.5% wheat flour, 20% corn grit; extruded product prepared from formula 14 used 28.6% ricebean flour, 23.8% black gram, 21.4% rice flour, 21.4% corn grit, and 4.8 full-fat soy flour. Formula 9 was reported as moisture, 7.16%; fat, 9.10%; protein, 15.23%; ash, 3.32%; crude fibre, 1.93%; and carbohydrate, 70.42%. Formula 14 proximate composition was 6.85%; fat, 11.95%; protein, 16.25%; ash, 3.88%; crude fibre, 1.24%; and carbohydrate, 66.87%. For formula 14 reported maximum protein efficiency ratio was 2.00 ± 0.11, followed by formula 9 (1.92 ± 0.13), which was far better than extrudates prepared from ricebean flour alone (1.60 ± 0.19).

Lamichhane et al. (2013) developed extruded snacks from 75% maize flour, 10% sorghum flour, and 15% ricebean flour with particle size 1090 μm. For extrusion, these flours are adjusted to moisture content of 17% and extruded at 135 °C barrel temperature and 153 rpm screw speed. The extruded products have expansion ratio 4.0, bulk density 80 kh/m^3, water-soluble index 21.48 ±0.82, and water absorption index 9.77 ± 0.52, moisture, 0.667%; crude protein, 2.97%; fat, 1.87%; ash, 2.54%; crude fibre, 12.06%; carbohydrate, 79.86%; and beta-glucan, 6.52 ± 0.61. The sensory quality of extrudates as colour, 7.5; taste, 7.0; texture, 7.5; and OAA, 7.5.

Vada

Kaur and Mehta (1993) developed ricebean-based vada. For preparation, bean grains are soaked overnight, ground to a coarse paste, and fermented for 4 h at room temperature. The fermented paste is beaten for 7–10 min and mixed properly after adding chopped onion, garlic, ginger, coriander leaves, green chillies, and salt.

Doughnut shapes (0.5 in. thick, 2 in. diameter) of paste are fried at 190 °C for 1 min, and the fried product is called vada. Nutritional composition of vada is moisture, 17.5%; crude protein, 19%; fat, 18.7%; ash, 2.5%; crude fibre, 4.3%; calcium, 282.6 mg/100 g; iron, 7.8 mg/100 g; zinc, 3.3 mg/100 g; copper, 9 mg/100 g; methionine, 9; cystine, 1.0; tryptophan, 0.7; lysine, 4.0 g/16 g N. The content of ash, crude fibre, calcium, iron, and zinc was found to be greater in rice vada than blackgram vada prepared from the same ingredients.

Ricebean Pickle

Ricebean pickle is usually prepared in Nagaland. Bean grains are soaked overnight, ground to a coarse paste, fermented, and mixed with spices and herbs.

Complementary Food

Ijeomah (2017) developed a ricebean-based complementary food from fonio (*Digitaria exilis*), ricebean, carrot, and crayfish. For preparation of final products, fonio powder was prepared by soaking then boiling followed by drying and milling, carrot powder slicing, blanching, drying, and milling; crayfish were prepared by cleaning and drying. Ricebean grain was prepared by soaking, sprouting, dehulling, and milling. Normal fonibean (FNBN)was prepared by blending 70 g fonio, 30 g ricebean, 30 g carrot, 30 g crayfish, 5 g sugar, and 1 g salt with 5 mL vegetable oil. Fonibean plus (FNBP) was prepared by blending a similar recipe of FNBN with 20 powdered milk, fonibean minus (FNBM) was designed by mixing fonio and sprouted ricebean, and fonibean untreated (FNBU) was prepared with unsprouted ricebean only with a recipe similar to FNBN. The overall acceptability of these products was reported as FNBP 8.52, FNBU 7.0, FNBN 8.33, and FNBM 5.62. The nutritional qualities of FNBP are as follows: body weight gain (g), 89.88 ± 0.04 g; biological value, 91.67 ± 0.20%; net protein utilization, 89.43a ± 0.40%; and digestibility, 97.56 ± 0.06%. The nutritional qualities of FNBN are body weight gain (g), 70.45 ± 0.01 g; biological value, 88.76 ± 0.35%; net protein utilization, 86.81 ± 0.10%; and digestibility, 97.80 ± 0.06. Although the ricebean used in the sample of FNBU was not sprouted, the sample showed a better quality protein than the sample of FNBM that contained sprouted ricebean.

Convenient Food Multimixes

Convenient food multimixes (Baruah et al. 2018a) were developed from ricebean (3 parts), foxtail millet (4 parts), flax seed (1 part), and tomato powder (1part) with maximum energy density of 446 kcal. For preparation of food mixes, milling of sprouted ricebean, toasted flax seed, and toasted foxtail millet was followed by sieving through a 100 mesh sieve. The product was prepared by blending all ingredients and compared with a product prepared with unsprouted ricebean. The food mix prepared from sprouted ricebean was higher in protein (17.86%) and lower in fat (3.91%), phytate (2.09%), and saponin (0.23%) than the food mix prepared from unsprouted ricebean.

Probiotic Food Multimixes

Baruah et al. (2018b) developed probiotic food multimixes from ricebean (3 parts), foxtail millet (4 parts), flax seed (1 part), and tomato powder (1 part), inoculated with *Lactobacillus plantarum* and *Lactobacillus rhamnosus*, which had maximum energy density (446 kcal). For preparation of food mixes, milling of sprouted ricebean, toasted flax seed, and toasted foxtail millet was followed by sieving through a 100 mesh sieve. The product was prepared by blending all ingredients and compared with a product prepared without probiotic microbes of ricebean. The food mix prepared from probiotic inoculation was reported as higher in protein (17.86%) and lower in fat (3.91%) and crude fibre (12%), with 30 days probiotic microbial viability compared to the noninoculated products.

Ojojo

Ojojo is a crispy-crust traditional fried product of Nigeria that is usually prepared from water yam with salt and spices. Okoye et al. (2018) developed ricebean-based *ojojo* by replacing water yam flour by 10–50%. To prepare *ojojo*, ricebean grain is soaked, dehulled, dried and milled into flour. A thick batter is made from the desired quantity of water yam, ricebean flour, and water. Onions (chopped or ground), seasoning cubes, fresh pepper, salt, and garlic (ground) are added to the batter and mixed uniformly. Batter formed into ball shapes is deep fried in ground nut oil until golden brown in colour and served after draining off oil. *Ojojo* prepared by replacing 50% water yam flour with ricebean flour was reported to have maximum overall acceptability (7.2) and higher amounts of nutritional attributes such as protein, ash, crude fibre, calcium, magnesium, vitamin B_{12}, vitamin C, and phosphorus than the *ojojo* prepared from 100% water yam flour. The nutritional attributes of this ojojo were reported as moisture, 18.56 ± 0.02%; crude protein, 10.84 ± 0.92%; fat, 14.78 ± 0.04%; ash, 2.81 ± 0.14%; crude fibre, 2.92 ± 0.14%; carbohydrate, 43.25 ± 0%; phosphorus, 83.04 + 0.03 mg; calcium, 75.05 + 0.0 mg; magnesium, 91.01 + 0.02 mg; vitamin B_{12}, 35.51 + 0.01 mg; and vitamin C, 39.26 + 0.37 mg per 100 g.

Future Trends in Processing and Product Development

Although ricebean is processed by various technologies such as soaking, conventional cooking, dehulling, germination, roasting, fermentation, and extrusion, the process parameters are not being optimized. Moreover, very limited studies are available on processes such as microwave cooking, flaking, and frying, which are not studied in terms of process or impact on functional components, nonnutritional components, digestibility (starch and protein), and mineral bioavailability (calcium and iron). Studies of the influence of the soaking process in ricebean on nutritional and nonnutritional components such as potassium, sodium, zinc, vitamin C, thiamin, riboflavin, niacin, folate, patothenic acid, flavonoids, oxalate, HCN, verbascose, and antioxidant activity are not available in the literature. Even though the processing effects of conventional cooking on functional components, nonnutritional components, and digestibility (starch and protein) for ricebean are available in part, there are wide variations in cooking processes. Information on the status of magnesium, zinc, vitamin C, thiamin, riboflavin, niacin, vitamin B_6, folate, pantothenic acid, oxalate, HCN, and mineral bioavailability (calcium and iron) in conventional cooking is sparse. The impact of dehulling on magnesium, potassium, sodium, vitamin C, thiamin, riboflavin, niacin, vitamin B_6, folate, patothenic acid, TIA, AIA, oxalate, HCN, raffinose, stachyose, verbascose, IVPD, IVSD, IVIB, IVCB, and antioxidant activity is also not reported. The sprouting of ricebean and functional components, nonnutritional components, and digestibility (starch and protein) is available in part but wide variations in the germination process have been observed. The available literature on the effect of ricebean germination on magnesium, potassium, sodium, vitamin B_6, folate, pantothenic acid, flavonoids, oxalate, HCN, and verbascose is sparse. The consequences of roasting on minerals, water-soluble vitamins, flavonoids, TIA, AIA, oxalate, HCN, raffinose, stachyose, verbascose, IVPD, IVSD, IVIB, IVCB, and antioxidant activity are not found in scientific texts. The effect of fermentation on calcium, iron, magnesium, phosphorus, potassium, sodium, zinc, vitamin C, thiamine, riboflavin, niacin, vitamin B_6, folate, pantothenic acid, tannin, flavonoids, saponin, phytic acid, TIA, AIA, oxalate, HCN, IVPD, IVSD, IVIB, and IVCB is not available in scientific reports. Hence, there is a need to study microwave cooking, extrusion cooking, flaking, and frying in terms of their influence on the functional components, nonnutritional components, digestibility (starch and protein), and mineral bioavailability (calcium and iron) in ricebean.

Most of the traditional ricebean-based products are prepared by using simple household processing methods such as soaking, cooking, sprouting, fermentation, and frying that are applied in other grains also. Other traditional products available in ricebean-growing areas are usually prepared using other food grains and ingredients, but recently such products are also trying with ricebean to increase the taste variability among consumers and to develop new products. Most of these products should be consumed immediately after preparation and are not suited for marketing because of the lack of a proper packing system. The existing developed products of ricebean also lack proper nutritional profiling and storage study. Hence, there is a need to develop new ricebean-based products in ready-to-eat (RTE) or ready-to-cook (RTC) form in proper packing modules that will remain viable for a long time. Apart from this, there is also a need to study the nutritional profile (minerals, and amino acids) and storage stability of such new products.

References

Acharya, B. K. (2008) *Cultivation and use of ricebean: A case study of Dang District, Nepal*. M. Phil. thesis submitted to University of Bergen, Norway.

Akinyele, I. O., & Akinlosotu, A. (1991). Effect of soaking, dehulling and fermentation on the oligosaccharides and nutrient content of cowpeas (*Vigna unguiculata*). *Food Chemistry, 41*(1), 43–53. https://doi.org/10.1016/0308-8146(91)90130-g.

Alajaji, S. A., & El-Adawy, T. A. (2006). Nutritional composition of chickpea (*Cicer arietinum* L.) as affected by microwave cooking and other traditional cooking methods. *Journal of Food Composition and Analysis, 19*, 806–812.

Alonso, R., Aguirre, A., & Marzo, F. (2000). Effects of extrusion and traditional processing methods on antinutrients and in vitro digestibility of protein and starch in faba and kidney beans. *Food Chemistry, 68*, 159–165.

Alonso, R., Rubio, L. A., Muzquiz, M., & Marzo, F. (2001). The effect extrusion cooking on mineral bioavailability in pea and kidney bean meals. *Animal Feed Science and Technology, 94*, 1–13.

Andersen, P., Kumar, N., & Acharya, B. P. (2009). Food security through ricebean research in India and Nepal (FOSRIN). Report 5. In P. A. Hollington (Ed.), *Ricebean food preparation and diets*. Bergen: Department of Geography, Universitet Bergen and Bangor, Wales, UK, CAZS Natural Resources, College of Natural Sciences, Bangor University.

Anuonye, J. C., Jigam, A. A., & Ndaceko, G. (2012). Effects of extrusion-cooking on the nutrient and anti-nutrient composition of pigeon pea and unripe plantain blends. *Journal of Applied Pharmaceutical Science, 2*(5), 158–162.

Arora, R. K., Chandel, K. P. S., Joshi, B. S., & Pant, K. C. (1980). Rice bean: Tribal pulse of eastern-India. *Economic Botany, 34*(3), 260–263.

Augustin, J., Beck, C. B., Kalbfleish, G., Kagel, L. C., & Matthews, R. H. (1981). Variation in the vitamin and mineral content of raw and cooked commercial *Phaseolus vulgaris* classes. *Journal of Food Science, 46*, 1701–1706.

Awasthi, C. P., Thakur, M., Dua, R. P., & Dhaliwal, Y. S. (2011). Biochemical evaluation of some promising varieties/genotypes of rice bean [*Vigna umbellata* (Thunb.; Ohwi and Ohashi)]. *Indian Journal of Agricultural Biochemistry, 24*(1), 39–42.

Banerjee, S., Rohatgi, K., & Lahiri, S. (1954). Pantothenic acid, folic acid, biotin and niacin contents of germinated pulses. *Journal of Food Science, 19*(1–6), 134–137. https://doi.org/10.1111/j.1365-2621.1954.tb17431.x.

Barampama, Z., & Simard, R. E. (1995). Effects of soaking, cooking and fermentation on composition, in vitro starch digestibility and nutritive value of common beans. *Plant Foods for Human Nutrition, 48*, 349–365.

Baruah, D. K., Das, M., & Bhattacharyya, R. (2018a). Formulation and quality evaluation of ricebean (*Vigna umbellata*) based convenient food multi mixes. *International Journal of Home Science, 4*(2), 216–221.

Baruah, D. K., Das, M., & Sharma, R. K. (2018b). Nutritional and microbiological evaluation of ricebean (*Vigna umbellata*) based probiotic food multi mix using *Lactobacillus plantarum* and *Lactobacillus rhamnosus*. *Journal of Probiotics & Health, 6*, 200. https://doi.org/10.4172/2329-8901.1000200.

Bepary, R. H., & Wadikar, D. D. (2018). Optimization of ricebean cooking parameters for the production of instant/convenience foods using response surface methodology. *Journal of Food Processing Preservation, 42*(3), e13547.
Bepary, R. H., & Wadikar, D. D. (2019). HPLC profiling of flatulence and non-flatulence saccharides in eleven ricebean (*Vigna umbellata*) varieties from north-East India. *Journal of Food Science and Technology, 56*(3), 1655–1662.
Bepary, R. H., Wadikar, D. D., & Patki, P. E. (2016). Ricebean: Nutritional vibrant bean of Himalayan belt (north East India). *Nutrition and Food Science, 46*(3), 412–431.
Bepary, R. H., Wadikar, D. D., Neog, S. B., & Patki, P. E. (2017). Studies on physicochemical and cooking characteristics of ricebean varieties grown in NE region of India. *Journal of Food Science and Technology, 54*(4), 973–986.
Bepary, R. H., Wadikar, D. D., & Patki, P. E. (2018a). Engineering properties of ricebean varieties from north East India. *Journal of Agricultural Engineering, 55*(3), 34–44.
Bepary, R. H., Wadikar, D. D., Semwal, A. D., & Sharma, G. K. (2018b). *Optimization of microwave cooking condition of ricebean (Vigna umbellata) by using response surface methodology*. Presented in International Conference on 'Recent Advances in Food Processing Technology,' 17–19 August, 2018, at IIFPT, Thanjavur, Tamil Nadu.
Bepary, R. H., Wadikar, D. D., & Patki, P. E. (2019a). Analysis of eight water soluble vitamins in ricebean (*Vigna umbellata*) varieties from NE India by reverse phase-HPLC. *Journal of Food Measurement and Characterization, 13*(2), 1287–1298.
Bepary, R. H., Wadikar, D. D., Vasudish, C. R., Semwal, A. D., & Sharma, G. K. (2019b). Optimisation and evaluation of ricebean (*Vigna umbellata*) extrusion process for downstream food processability. *Defence Life Science Journal, 4*(2), 130–139.
Bishnoi, S., & Khetarpaul. (1993). Effect of domestic processing and cooking methods on in vitro starch digestibility of different pea cultivars (*Pisum sativum*). *Food Chemistry, 47*, 177–182.
Boateng, J., Verghese, M., Walker, L. T., & Ogutu, S. (2008). Effect of processing on antioxidant contents in selected dry beans (*Phaseolus* spp. L.). *LWT - Food Science and Technology, 41*, 1541–1547.
Buergelt, D., von Oppen, M., & Yadavendra, J. P. (2009). Quality parameters in relation to consumer's preferences in rice bean. In *International Conference on Grain Legumes: Quality Improvement, Value Addition and Trade*, Kanpur, India, 14–16 February 2009.
Chandel, K. P. S., Joshi, B. S., Arora, R. K., & Pant, K. C. (1978). Rice bean: A new pulse with high potential. *Indian Farming, 28*(9), 19–22.
Chandel, K. P. S., Arora, R. K., & Pant, K. C. (1988). *Rice bean: A potential grain legume* (NBPGR Scientific Monograph No. 12). NBPGR, New Delhi.
Chau, C.-F., & Cheung, P. C.-K. (1997). Effect of various processing methods on antinutrients and in vitro digestibility of protein and starch of two Chinese indigenous legume seeds. *Journal of Agricultural and Food Chemistry, 45*, 4773–4776.
Chukwuma, O. E., Taiwo, O. O., & Boniface, U. V. (2016). Effect of the traditional cooking methods (boiling and roasting) on the nutritional profile of quality protein maize. *Journal of Food and Nutrition Sciences, 4*(2), 34–40.
Chung, S. Y., Morr, C. V., & Jen, J. J. (1981). Effect of microwave and conventional cooking on the nutritive value of colossus peas (*Vigna unguiculata*). *Journal of Food Science, 46*, 272–273.
de Carvalho, N. M., & Vieria, R. D. (1996). Rice bean [*Vigna umbellata* (Thunb.) Ohwi & Ohashi]. In E. Nkowolo & J. Smartt (Eds.), *Legumes and oilseeds in nutrition* (pp. 222–228). London: Chapman & Hall.
Deshpande, S. S., Sathe, S. K., Salunkhe, D. K., & Cornforth, D. P. (1982). Effects of dehulling on phytic acid, polyphenols, and enzyme inhibitors of dry beans (*Phaseolus vulgaris* L.). *Journal of Food Science, 47*, 1846–1849.
Du, S., Jiang, H., Yu, X., & Jane, J. (2014). Physicochemical and functional properties of whole legume flour. *LWT - Food Science and Technology, 55*(1), 308–313.
Duhan, A., Khetarpaul, N., & Bishnoi, S. (2002). Content of phytic acid and HCl-extractability of calcium, phosphorus and iron as affected by various domestic processing and cooking methods. *Food Chemistry, 78*(1), 9–14. https://doi.org/10.1016/s0308-8146(01)00144-3.

El Maki, H. B., Samia, R. S. M. A., Idris, W. H., Hassan, A. B., Babiker, E. E., & El Tinay, A. H. (2007). Content of antinutritional factors and HCl-extractability of minerals from white bean (*Phaseolus vulgaris*) cultivars: Influence of soaking and/or cooking. *Food Chemistry, 100*, 362–368.

Feng, D., Shen, Y., & Chavez, E. R. (2003). Effectiveness of different processing methods in reducing hydrogen cyanide content of flaxseed. *Journal of Science Food and Agriculture, 83*, 836–841.

Gan, R.-Y., Shah, N. P., Wang, M.-F., Lui, W.-Y., & Corke, H. (2016). Fermentation alters antioxidant capacity and polyphenol distribution in selected edible legumes. *International Journal of Food Science and Technology, 51*, 875–884.

Gruner, M., Horvatic, M., Gacic, M., & Banovi, M. (1996). Molar ration of phytic acid and zinc during cereal flake production. *Journal of the Science of Food and Agriculture, 70*, 355–358.

Hefni, M., & Witthöft, C. M. (2014). Folate content in processed legume foods commonly consumed in Egypt. *LWT - Food Science and Technology, 57*(1), 337–343. https://doi.org/10.1016/j.lwt.2013.12.026.

Hira, C. K., Kawar, J. K., Gupta, N., & Kochhar, A. (1988). Cooking quality and nutritional evaluation of the rice bean (*Vigna umbellata*). *Journal of Food Science and Technology, 25*(3), 133–136.

Hollington, P. A., Joshi, K. D., Witcombe, J. R., Harris, D., Mueller, R. A. E., Buergelt, D., Andersen, P., Yadavendra, J. P., Kumar, N., Neog, S. B., Bajracharya, J., Shrestha, R., Khadka, K., Gautam, R., Acharya, B. K., & Paudel, I. H. (2010) *Food security through ricebean research in India and Nepal (FOSRIN)*. Final report, April 1 2006–March 31 2010.

Hoppner, K., & Lampi, B. (1993). Pantothenic acid and biotin retention in cooked legumes. *Journal of Food Science, 58*(5), 1084–1088.

Ibeanu, V. N., Onyechi, U. A., & Ugwuanyi, G. U. (2012). Nutrient and dietary fibre profile of dehulled and undehulled seeds of sweet princess watermelon (*Citrullus lanatus*) consumed in Nigeria. *International Journal of Basic & Applied Sciences, 12*(6), 249–252.

Ijeomah, O. C. (2017). Protein quality and sensory attributes of fonio (*Digitaria exilis*)/ricebean (*Vigna umbellata*) based complementary food incorporated with carrot and crayfish. *Innovare Journal of Food Science, 5*(1), 5–7.

Itagi, H. B. N., Sathyendra Rao, B. V. S., Jayadeep, P. A., & Singh, V. (2012). Functional and antioxidant properties of ready-to-eat flakes from various cereals including sorghum and millet. *Quality Assurance and Safety of Crops & Foods, 4*, 126–133.

Jayalaxmi, B., Vijayalakshmi, D., Usha, R., Revanna, M. L., Chandru, R., & Gowda, P. H. R. (2016). Effect of different processing methods on proximate, mineral and antinutrient content of lima bean (*Phaseolus lunatus*) seeds. *Legume Research, 39*(4), 543–549.

Joshi, K. D., Bhandari, B., Gautam, R., Bajracharya, J., & Hollington, P. A. (2008). Ricebean: A multipurpose underutilised legume, proceeding of 5th Interntional syposium on New crops and uses: Their Role in a Rapidly Changing World (Edited by J Smartt and N Haq) 3-4 september 2007, University of Southamto, Southampton, pp. 234–248.

Joyner, J. J., & Yadav, B. K. (2015). Microwave assisted dehulling of black gram (*Vigna mungo* L). *Journal of Food Science and Technology, 52*(4), 2003–2012. https://doi.org/10.1007/s13197-013-1182-9.

Kakati, P., Deka, S. C., Kotoki, D., & Saikia, S. (2010). Effect of traditional methods of processing on the nutrient contents and some antinutritional factors in newly developed cultivars of green gram [*Vigna radiata* (L.) Wilezek] and black gram [*Vigna mungo* (L.) Hepper] of Assam. *India International Food Research Journal, 17*, 377–384.

Kalidass, C., & Mohan, V. R. (2012). Nutritional composition and antinutritional factors of little-known species of *Vigna*. *Tropical and Subtropical Agroecosystems, 15*(3), 525–538.

Katoch, R. (2013). Nutritional potential of rice bean (*Vigna umbellata*): An underutilized legume. *Journal of Food Science, 78*(1), C8–C16.

Kaur, M. (2015). Chemical composition of ricebean (*Vigna umbellata*): Effect of domestic processing. *Indian Journal of Applied Research, 5*(4), 311–313.

Kaur, D., & Kapoor, A. C. (1990a). Some antinutritional factors in ricebean (*Vigna umbellata*): Effects of domestic processing and cooking methods. *Food Chemistry, 37*(4), 171–179.
Kaur, D., & Kapoor, A. C. (1990b). Starch and protein digestibility of rice bean (*Vigna umbellata*): Effects of domestic processing and cooking methods. *Food Chemistry, 38*(4), 263–272.
Kaur, D., & Kapoor, A. C. (1992). Nutrient composition and anti-nutritional factors of rice bean (*Vigna umbellata*). *Food Chemistry, 43*(2), 19–124.
Kaur, M., & Kawatra, B. L. (2000). Effect of domestic processing on flatus producing factors in ricebean (*Vigna umbellata*). *Nahrung, 44*(6), 447–450.
Kaur, M., & Kawatra, B. L. (2002a). Evaluation of iron bioavailability from ricebean (*Vigna umbellata*) by using anaemic rats. *Nutrition Research, 22*, 633–640.
Kaur, M., & Kawatra, B. L. (2002b). Domestic processing on zinc bioavailbility from ricebean (*Vigna umbellata*) diets. *Plant Food for Human Nutrition, 57*, 307–318.
Kaur, M., & Mehta, U. (1993). Effect of partial and full substitution of Bengal gram and black gram dhals with rice bean on essential amino acids and mineral contents of vada and pakora. *Journal of Food Science and Technology, 30*(6), 454–456.
Kaur, A., Kaur, P., Singh, N., Virdi, A. S., Singh, P., & Rana, J. C. (2013). Grains, starch and protein characteristics of rice bean (*Vigna umbellata*) grown in Indian Himalaya regions. *Food Research International, 54*, 102–110.
Khatoon, N., & Prakash, J. (2004). Nutritional quality of microwave-cooked and pressure-cooked legume. *International Journal of Food Sciences and Nutrition, 55*(6), 441–448.
Khattab, R. Y., & Arntfield, S. D. (2009). Nutritional quality of legume seeds as affected by some physical treatments. 2. Antinutritional factors. *LWT - Food Science and Technology, 42*, 1113–1118.
Kii, S. V. M., Wijanarka, A., & Puruhita, T. K. A. (2013). Effect of ricebean flour (*Vigna umbellata)* and wheat flour mixing variations on physical properties, organoleptic properties and calcium levels of cookies production. *Medika Respati, 8*(1).
Kik, M. C. (1956). Nutritive value of rice, nutrients in rice bran and rice polish and improvement of protein quality with amino acids. *Journal of Agricultural and Food Chemistry, 4*(2), 170–172. https://doi.org/10.1021/jf60060a011.
Killeit, U. (1994). Vitamin retention in extrusion cooking. *Food Chemistry, 49*, 149–155.
Klein, B. P., Kuo, C. H. Y., & Boyd, G. (1981). Folacin and ascorbic acid retention in fresh raw, microwave, and conventionally cooked spinach. *Journal of Food Science, 46*(2), 640–641. https://doi.org/10.1111/j.1365-2621.1981.tb04931.x.
Kon, S., & Sanshuck, D. W. (1981). Phytate content and its effect of cooking quality of beans. *Journal of Food Processing & Preservation, 5*, 169–178.
Korus, J., Gumul, D., & Czechowska, K. (2007). Effect of extrusion on the phenolic composition and antioxidant activity of dry beans of *Phaseolus vulgaris* L. *Food Technology and Biotechnology, 45*(2), 139–146.
Lal, R. R., & Verma, P. (2007). *Post harvest management of pulses*. Kanpur: Indian Institute of Pulses Research.
Lamichhane, B., Ojha, P., & Paudyal, B. (2013). Formulation and quality evaluation of extruded product from composite blend of maize, sorghum and ricebean. *Journal of Food Science and Technology Nepal, 8*, 40–45.
Lawal, R. O. (1986). Effect of roasting on the chemical composition of the seeds of *Treculia africana*. *Food Chemistry, 22*, 305–314.
Malhotra, S., Malik, D., & Dhindsa, K. S. (1988). Proximate composition and anti-nutritional factors in rice bean (*Vigna umbellata*). *Plant Foods for Human Nutrition, 38*(1), 75–81.
Mankotia. (2011). *Biochemical evaluation of ricebean (Vigna umbellate (Thunb.) Ohwi and Ohashi) genotypes of Himachal Pradesh*. M.Sc. thesis, CSKHP Krishi Vishvavidyalaya, Palampur.
Mohan, R. J., Sangeetha, A., Narasimha, H. V., & Tiwari, B. K. (2011). Post-harvest technology of pulses. In B. K. Tiwari, A. Gowen, & B. McKenna (Eds.), *Pulse foods: Processing, quality and nutraceutical applications*. London: Academic Press–Elsevier.

Mubarak, A. E. (2005). Nutritional composition and antinutritional factors of mung bean seeds (*Phaseolus aureus*) as affected by some home traditional processes. *Food Chemistry, 89*, 489–495.

Mujoo, R., & Ali, S. Z. (1998). Susceptibility of starch to *in vitro* enzyme hydrolysis in rice: Rice flakes and intermediary products. *Lebensmittel-Wissenschaft und Technologie, 31*, 114–121.

Mujoo, R. (1998). *Effect of processing conditions on the properties of rice flakes*. PhD (Food Science) thesis submitted to University of Mysore.

Mujoo, R., Chandrashekar, A., & Ali, S. Z. (1998). Rice protein aggregation during the flaking process. *Journal of Cereal Science, 28*, 187–195.

Mulimani, V. H., & Supriya, D. (1994). Effects of infrared radiation, solar cooking and microwave cooking on alpha-amylase inhibitor in sorghum (*Sorghum bicolor* L.). *Plant Foods for Human Nutrition, 46*(3), 231–235. https://doi.org/10.1007/bf01088995.

Nayak, L. K., & Samuel, D. V. K. (2012). Process optimization for development of instant pigeon pea (*Cajanus cajan* L.) dal using sodium carbonate pretreatment. *Journal of Dairying Foods & Home Sciences, 31*(3), 191–194.

Nnanna, I. A., & Phillips, R. D. (1989). Amino acid composition, protein quality and water-soluble vitamin content of germinated cowpeas (*Vigna unguiculata*). *Plant Foods for Human Nutrition, 39*, 187–200.

Nwanekezi, E. C., Ehirim, F. N., & Arukwe, D. C. (2017). Combined effects of different processing methods on vitamins and antinutrients contents of pigeon pea (*Cajanus cajan*) flour. *Journal of Environmental Science, Toxicology and Food Technology, 11*(4), 73–81.

Oboh, H. A., Muzquiz, M., Burbano, C., Cuadrado, C., Pedrosa, M. M., Ayet, G., & Osagie, A. U. (2000). Effect of soaking, cooking and germination on the oligosaccharide content of selected Nigerian legume seeds. *Plant Foods for Human Nutrition, 55*, 97–110.

Ochanda, S. O., Onyango, C. A., Mwasaru, A. M., Ochieng, J. K., & Mathooko, F. M. (2010). Effects of malting and fermentation treatments on group B-vitamins of red sorghum, white sorghum and pearl millets in Kenya. *Journal of Applied Biosciences, 34*, 2128–2134.

Ohwi, J. (1965). *Flora of Japan*. Washington, DC: Smithsonian Institution.

Okoye, E. C., Njoku, E. U., & Ugwuanyi, G. R. (2018). Proximate composition, micronutrient contents and acceptability of "Ojojo" from the blends of water yam and ricebean flours. *International Journal of Food Sciences and Nutrition, 3*(5), 5–10.

Okudu, H. O., & Ojinnaka, M. C. (2017). Effect of soaking time on the nutrient and antinutrient composition of bambara groundnut seeds (*Vigna subterranean*). *African Journal of Food Science and Technology, 8*(2), 25–29.

Pal, R. S., Bhartiya, A., Arun Kumar, R., Kant, L., Aditya, J. P., & Bisht, J. K. (2015). Impact of dehulling and germination on nutrients, antinutrients, and antioxidant properties in horsegram. *Journal of Food Science and Technology, 53*(1), 337–347. https://doi.org/10.1007/s13197-015-2037-3.

Parvathi, S., & Kumar, V. J. F. (2006). Value added products from rice bean (*Vigna umbellata*). *Journal of Food Science and Technology, 43*(2), 190–193.

Patil, K. B., Chimmad, B. V., & Itagi, S. (2015). Glycemic index and quality evaluation of little millet (*Panicum miliare*) flakes with enhanced shelf life. *Journal of Food Science and Technology, 52*(9), 6078–6082. https://doi.org/10.1007/s13197-014-1663-5.

Paudel, I. H. (2008). *Conservation and commercialization prospect of ricebean landraces in Ramechhap district of Nepal*. Kirtipur: Tribhuvan University.

Pawar, V. D., Akkena, M. K., Kotecha, P. M., Thorat, S. S., & Bansode, V. V. (2012). Effect of presoak treatment on cooking characteristics and nutritional functionality of rice bean. *Journal of Food Legumes, 25*(4), 321–325.

Perera, O. D. A. N., Eashwarage, I. S., & Herath, H. M. T. (2017). Development of dietary fiber rich multi legumes flake mix. *Journal of Pharmacognosy & Natural Products, 3*, 135. https://doi.org/10.4172/2472-0992.1000135.

Prabhavat, S., Boonyasirikool, P., & Charunuch, C. (1996). Study on the production of snacks from rice bean by twin screw extruder. *Kasetsart Journal (Natural Science), 30*, 200–210.

Quinteros, A., Farré, R., & Lagarda, M. J. (2003). Effect of cooking on oxalate content of pulses using an enzymatic procedure. *International Journal of Food Sciences and Nutrition, 54*(5), 373–377.

Ramakrishnaiah, N., & Kurien, P. P. (1985). Non-starchy polysaccharides of pigeon pea and their influence on dehulling characteristics. *Journal of Food Science and Technology, 22*, 429–430.

Randhir, R., Kwon, Y. I., & Shetty, K. (2008). Effect of thermal processing on phenolics, antioxidant activity and health-relevant functionality of select grain sprouts and seedlings. *Innovative Food Science and Emerging Technologies, 9*, 355–364.

Rani, S., & Khabiruddin, M. (2017). Antioxidant potential of processed *Vigna umbellata* (L.) seeds: An Indian underutilized legume. *International Journal of Chemical Studies, 5*(4), 1407–1412.

Rau, D., Murgia, M. L., Rodriguez, M., Bitocchi, E., Bellucci, E., Fois, D., Albani, D., Nanni, L., Gioia, T., Santo, D., Marcolungo, L., Delledonne, M., Attene, G., & Papa, R. (2019). Genomic dissection of pod shattering in common bean: Mutations at non-orthologous loci at the basis of convergent phenotypic evolution under domestication of leguminous species. *The Plant Journal, 97*(4), 693. https://doi.org/10.1111/tpj.14155.

Ren, S. C., Liu, Z. L., & Wang, P. (2012). Proximate composition and flavonoids content and *in vitro* antioxidant activity of 10 varieties of legume seeds grown in China. *Journal of Medicinal Plants Research, 6*(2), 301–308.

Revilleza, M. J. R., Mendoza, E. M. T., & Raymundo, L. C. (1990). Oligosaccharides in several Philippine indigenous food legumes: Determination, localization and removal. *Plant Foods for Human Nutrition, 40*(1), 83–93.

Rockland, L. B., Miller, C. F., & Hahn, D. M. (1977). Thiamine, pyridoxine, niacin and folacin in quick-cooking beans. *Journal of Food Science, 42*, 25–28.

Rodriguez, M. S., & Mendoza, E. M. T. (1991). Nutritional assessment of seed proteins in rice bean [*Vigna umbellata* (Thumb.) Ohwi and Ohashi]. *Plant Foods for Human Nutrition, 41*(1), 1–9.

Ryland, D., Vaisey-Genser, M., Arntfield, S. D., & Malcolmson, L. J. (2010). Development of a nutritious acceptable snack bar using micronized flaked lentils. *Food Research International, 43*, 642–649.

Saharan, K., Khetarpaul, N., & Bishnoi, S. (2001). HCl-extractability of minerals from ricebean and fababean: Influence of domestic processing methods. *Innovative Food Science and Emerging Technologies, 2*, 323–325.

Saharan, K., Khetarpaul, N., & Bishnoi, S. (2002). Variability in physicochemical properties and nutrient composition of newly released ricebean and fababean cultivars. *Journal of Food Composition and Analysis, 15*, 159–167.

Sahu, B. (2017) *Comparative analysis of selected varieties of paddy for suitability of flaking*. M.Tech thesis, Indira Gandhi Krishi Vishwavidyalaya, Raipur (Chhattisgarh).

Saikia, P., Sarkar, C. R., & Borua, I. (1999). Chemical composition, anti-nutritional factors and effect of cooking on nutritional quality of rice bean *Vigna umbellata* (Thunb.) Ohwi and Ohashi. *Food Chemistry, 67*(4), 347–352.

Sailaya, Y. S. (1992) *Investigated the popping and flaking quality of sorghum cultivars in relation to physicochemical characteristics and in vitro starch and protein digestibility*. Thesis submitted to the Andhra Pradesh Agricultural University, pp. 1–13.

Saini, R., & Chopra, A. R. (2012). In vitro plant regeneration via somatic embryogenesis in ricebean *Vigna umbellata* (Thunb.) Ohwi and Ohashi: An underutilized and recalcitrant grain legume. *Journal of Environmental Research and Development, 6*(3), 452–457.

Sánchez-Arteaga, H. M., Urías-Silvas, J. E., Espinosa-Andrews, H., & García-Márquez, E. (2015). Effect of chemical composition and thermal properties on the cooking quality of common beans (*Phaseolus vulgaris*). *CyTA Journal of Food, 13*(3), 385–391. https://doi.org/10.1080/19476337.2014.988182.

Sarma, B. K., Singh, M., Gupta, H. S., Singh, G., & Srivastava, L. S. (1995). *Studies in rice bean germplasm* (Research Bulletin #34). ICAR Research Complex for North East Hill Region, Barapani (Umiam), India.

Sattar, A., Durrani, S. K., Mahmood, F., Ahmad, A., & Khan, I. (1989). Effect of soaking and germination temperatures on selected nutrients and antinutrients of mungbean. *Food Chemistry, 34*, 111–120.

Scariot, M. A., Tiburski, G., Júnior, F. W. R., Radünz, L. L., & Meneguzzo, M. R. R. (2017). Moisture content at harvest and drying temperature on bean seed quality. *Pesquisa Agropecuária Tropical, 47*(1), 93–101. ISSN 1983-4063—www.agro.ufg.br/pat.

Sethi, S., Samuel, D. V. K., & Khan, I. (2014). Development and quality evaluation of quick cooking dhal—A convenience product. *Journal of Food Science and Technology, 51*(3), 595–600. https://doi.org/10.1007/s13197-011-0534-6.

Sharma, S. (2014). *Nutritional quality, functional properties and value addition of rice bean (Vigna umbellata)*. PhD thesis, Chaudhary Sarwan Kumar Himachal Pradesh Krishi, Vishvavidyalaya, Palampur, India.

Sharma, R., & Punia, D. (2017). Effect of traditional processing methods on protein digestibility and starch digestibility of field pea (*Pisum sativum*). *International Journal of Food and Nutritional Science, 6*(3). e-ISSN 2320-7876 www.ijfans.com.

Siddhuraju, P., & Becker, K. (2007). The antioxidant and free radical scavenging activities of processed cowpea (*Vigna unguiculata* (L.) Walp.) seed extracts. *Food Chemistry, 101*, 10–19.

Siddhuraju, P., Vijayakumari, K., & Janardhanan, K. (1996). Chemical composition and protein quality of the little-known legume, velvet bean (*Mucuna pruriens* (L.) DC.). *Journal of Agricultural and Food Chemistry, 44*, 2636–2641.

Singh, U. (1993). Protein quality of pigeon pea (*Cajanus cajan* L.) as influenced by seed polyphenols and cooking process. *Plant Foods for Human Nutrition, 43*, 171–179.

Singh, S. P., Misra, B. K., Chandel, K. P. S., & Pant, K. C. (1980). Major food constituents of ricebean (*Vigna umbellata*). *Journal of Food Science and Technology, 17*(5), 238–240.

Singh, N., Kaur, S., Isono, N., Ichihasshi, Y., Noda, T., Kaur, A., & Rana, J. C. (2012). Diversity in characteristics of starch amongst ricebean (*Vigna umbellata*) germplasm: Amylopectin structure, granule size distribution, thermal rheology. *Food Research International, 46*, 194–200.

Somta, P., Talekar, N., & Srinives, P. (2006). Characterization of *Callosobruchus chinensis* (L.) resistance in *Vigna umbellata* (Thunb.) Ohwi & Ohashi. *Journal of Stored Products Research, 42*, 313–327.

Sritongtae, B., Sangsukiam, T., Morgan, M. R. A., & Duangmal, K. (2017). Effect of acid pretreatment and the germination period on the composition and antioxidant activity of ricebean (*Vigna umbellata*). *Food Chemistry, 227*, 280–288.

Tomooka, N., Kaga, A., & Vaughan, D. A. (2006). The Asian vigna (*Vigna* subgenus *Ceratotropis*) biodiversity and evolution. In A. K. Sharma & A. Sharma (Eds.), *Plant genome: Biodiversity and evolution* (pp. 87–126). Enfield: Science Publishers.

Uebersax, M. A., & Muhammad, S. (2013). Postharvest storage quality, packaging and distribution of dry beans. In S. Siddiq Muhammad & M. A. Uebersax (Eds.), *Dry beans and pulses production, processing and nutrition* (p. 75). Ames: Wiley.

Wadikar, D. D., Bepary, R. H., Semwal, A. D., & Sharma, G. K. (2018). *Optimization of process parameters for production of ricebean (Vigna umbellata) flakes by using response surface methodology*. Presented at 8th International Food Convention (IFCON 2018), 12–15 Dec 2018, CSIR-CFTRI, Mysore.

Wei, Y., Yan, J., Long, F., & Lu, G. (2015). *Vigna umbellata* (Thunb.) Ohwi et Ohashi or *Vigna angularis* (Willd.) Ohwi et Ohashi (Chixiaodou, rice bean). In Y. Liu, Z. Wang, & J. Zhang (Eds.), *Dietary Chinese herbs: Chemistry, pharmacology and clinical evidence* (pp. 551–560). New York: Springer.

Wu, G., Ashton, J., Simic, A., Fang, Z., & Johnson, S. K. (2018). Mineral availability is modified by tannin and phytate content in sorghum flaked breakfast cereals. *Food Research International, 103*, 509–514.

Yadav, S., & Khetarpaul, N. (1994). Some antinutrients and in vitro digestibility of starch and protein. *Food Chemistry, 50*, 403–406.

Yang, S. C., Zandstra, T., & van der Poel, A. F. B. (2008). Starch gelatinization and physical quality of pea flakes in canine dinners as affected by soaking, steam treatment and infrared radiation. *Journal of Animal Physiology and Animal Nutrition, 92*, 310–315.

Index

A. Manickavasagan, P. Thirunathan (eds.), *Pulses*,
https://doi.org/10.1007/978-3-030-41376-7

D

The manufacturer's authorised representative in the EU is Springer
Nature Customer Service Centre GmbH, Europaplatz 3, 69115 Heidelberg,
Germany. If you have any concerns regarding our products, please
contact ProductSafety@springernature.com

Printed and bound by CPI Group (UK) Ltd, Croydon, CR0 4YY
05/10/2026
02228124-0012